POLYPEPTIDE AND PROTEIN DRUGS
Production, Characterization and Formulation

ELLIS HORWOOD SERIES IN PHARMACEUTICAL TECHNOLOGY

Editor: Professor M. H. RUBINSTEIN, School of Health Sciences, Liverpool Polytechnic

UNDERSTANDING EXPERIMENTAL DESIGN AND INTERPRETATION IN PHARMACEUTICS
N. A. Armstrong & K. C. James
MICROBIAL QUALITY ASSURANCE IN PHARMACEUTICALS, COSMETICS AND TOILETRIES
Edited by S. Bloomfield *et al.*
DRUG DISCOVERY TECHNOLOGIES
C. Clark & W. H. Moos
PHARMACEUTICAL PRODUCTION FACILITIES: Design and Applications
G. Cole
PHARMACEUTICAL TABLET AND PELLET COATING
G. Cole
THE PHARMACY AND PHARMACOTHERAPY OF ASTHMA
Edited by P. F. D'Arcy & J. C. McElnay
GUIDE TO MICROBIOLOGICAL CONTROL IN PHARMACEUTICALS
Edited by S. P. Denyer & R. M. Baird
PHARMACEUTICAL THERMAL ANALYSIS: Techniques and Applications
J. L. Ford and P. Timmins
PHYSICO-CHEMICAL PROPERTIES OF DRUGS: A Handbook for Pharmaceutical Scientists
P. Gould
DRUG DELIVERY TO THE GASTROINTESTINAL TRACT
Edited by J. G. Hardy, S. S. Davis and C. G. Wilson
POLYPEPTIDE AND PROTEIN DRUGS: Production, Characterization and Formulation
Edited by R. C. Hider and D. Barlow
HANDBOOK OF PHARMACOKINETICS: Toxicity Assessment of Chemicals
J. P. Labaune
TABLET MACHINE INSTRUMENTATION IN PHARMACEUTICS: Principles and Practice
P. Ridgway Watt
PHARMACEUTICAL CHEMISTRY, Volume 1 Drug Synthesis
H. J. Roth *et al.*
PHARMACEUTICAL CHEMISTRY, Volume 2 Drug Analysis
H. J. Roth *et al.*
PHARMACEUTICAL TECHNOLOGY: Controlled Drug Release, Volume 1
Edited by M. H. Rubinstein
PHARMACEUTICAL TECHNOLOGY: Controlled Drug Release, Volume 2*
Edited by M. H. Rubinstein
PHARMACEUTICAL TECHNOLOGY: Tableting Technology, Volume 1
Edited by M. H. Rubinstein
PHARMACEUTICAL TECHNOLOGY: Tableting Technology, Volume 2*
Edited by M. H. Rubinstein
PHARMACEUTICAL TECHNOLOGY: Drug Stability
Edited by M. H. Rubinstein
PHARMACEUTICAL TECHNOLOGY: Drug Targeting*
Edited by M. H. Rubinstein
UNDERSTANDING ANTIBACTERIAL ACTION AND RESISTANCE
A. D. Russell and I. Chopra
RADIOPHARMACEUTICALS USING RADIOACTIVE COMPOUNDS IN PHARMACEUTICS AND MEDICINE
Edited by A. Theobald
PHARMACEUTICAL PREFORMULATION: The Physicochemical Properties of Drug Substances
J. I. Wells
PHYSIOLOGICAL PHARMACEUTICS: Biological Barriers to Drug Absorption
C. G. Wilson & N. Washington
PHARMACOKINETIC MODELLING USING STELLA ON THE APPLE™ MACINTOSH™
C. Washington, N. Washington & C. Wilson

* *In preparation*

POLYPEPTIDE AND PROTEIN DRUGS

Production, Characterization and Formulation

Editors

R. C. HIDER B.Sc., Ph.D.

D. BARLOW B.Sc., M.Sc., Ph.D.

both Department of Pharmacy, King's College, London

ELLIS HORWOOD

NEW YORK LONDON TORONTO SYDNEY TOKYO SINGAPORE

First published in 1991 by
ELLIS HORWOOD LIMITED
Market Cross House, Cooper Street,
Chichester, West Sussex, PO19 1EB, England

A division of
Simon & Schuster International Group
A Paramount Communications Company

Printed and bound in Great Britain
by Bookcraft (Bath) Ltd., Midsomer Norton, Avon

British Library Cataloguing in Publication Data

Polypeptide and protein drugs.
1. Amino acids. Polypeptides and proteins
I. Hider, R. C. II. Barlow, D.
547.75
ISBN 0–13–677253–6

Library of Congress Cataloging-in-Publication Data

Polypeptide and protein drugs: production, characterization, and formulation / editors, R. C. Hider, D. Barlow
p. cm. — (Ellis Horwood series in pharmaceutical technology)
Includes bibliographical references and index.
ISBN 0–13–677253–6
1. Peptides. 2. Proteins. 3. Pharmaceutical chemistry. I. Hider, R. C. (Robert Charles), 1943– . II. Barlow, D. (David), 1954– . III. Series.
RS431.P38P65 1990
615′.19–dc20 90–22554
CIP

Table of contents

Preface

Peptide and protein pharmaceuticals are experiencing a period of rapid development. The importance of these specialized drugs and formulations was recognized in the 1990 Easter School which was jointly organized by the Royal Pharmaceutical Society of Great Britain and the Chelsea Department of Pharmacy, King's College London. The topics discussed at the Easter School ranged from theoretical aspects of protein structure and folding through peptide synthesis and genetic engineering, to current methods of formulation and product legislation.

This volume represents a distillation of the various presentations made at the School and reflects some current interests in polypeptide pharmaceuticals as seen by the contributors. The book also includes a number of appendices which describe analytical and synthetic techniques which were presented in practical sessions at the School.

Particular thanks are accorded to Dr Derek Callam of National Institute of Biological Standards and Control and to Dr Tony Theobald, both of whom helped with the original planning of the course. Thanks are also due to the contributors to this volume, all of whom produced their manuscripts and returned page proofs with fine promptness and precision, and to the publishers for their helpful advice and understanding at all stages.

D. Barlow and R. C. Hider
Pharmacy, King's College London

1

Peptide and protein structure

J. M. Thornton
Department of Biochemistry, University College London
D. J. Barlow
Departments of Pharmacy and Pharmacology, King's College London

1.1 INTRODUCTION

Peptides and proteins carry out a wide range of biological functions and are ubiquitous in living systems. In humans they range from small molecules like oxytocin and vasopressin, through to large molecules like thrombin and elastase. As a general class of biomolecule, they show a high degree of structural complexity and diversity, and are thereby made highly specific in their individual activities. This in turn means that they have great importance and potential as pharmaceuticals. In some cases the natural molecules may be employed directly as pharmaceuticals, whereas in other cases they simply serve as extremely good lead compounds for drug development. The former category includes peptides such as insulin and erythropoetin, whilst the latter is illustrated by the ICI product Zoladex (which is an analogue of human luteinizing hormone releasing hormone, LHRH, see Chapter 16).

Since the biological properties and pharmaceutical actions of peptides and proteins are critically dependent upon their three-dimensional (3D) structures, the following sections are provided by way of introduction to, and reference for, the more specific topics considered in subsequent chapters. The amount of detail included is necessarily limited, and for further information the reader should consult Schultz & Schirmer (1979).

1.2 PEPTIDE AND PROTEIN COVALENT/PRIMARY STRUCTURE

All eukaryotic (and most prokaryotic) peptides and proteins are formed from a basic set of 20 L-α-amino (carboxylic) acids (Fig. 1.1, Table 1.1). The polymerization of the monomer units proceeds by condensation reactions, so that a linear peptide chain

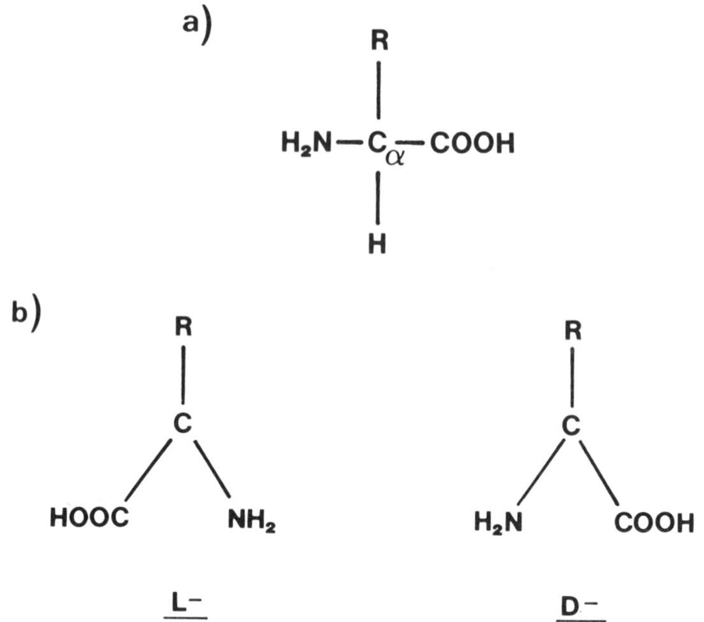

Fig. 1.1 — (a) General structural formula for an α-amino-acid, where R (referred to as the amino acid side-chain) represents a chemical group which distinguishes one amino acid from another (see Table 1.1). (b) L- and D-configurations of an α-amino-acid. Each molecule is seen in projection, viewed down and H-Cα bond. The configurations are determined with respect to the three-carbon sugar glyceraldehyde.

is produced in which each amino acid (residue) is linked to its neighbours by secondary amide/peptide bonds (see Fig. 1.2). The covalent structure of the resulting chains is commonly abbreviated as shown in Fig. 1.3, with an ordered list of the amino acids being presented using the standard one-letter or three-letter codes (see Table 1.1). These data, together with the details of any amino acid modifications (Fig. 1.4), and inter- or intra-chain cystine disulphide bonds (Fig. 1.5), are referred to as the primary structure of the peptide or protein (with the synonymous descriptions: amino-acid, peptide or protein sequence).

Those peptides that contain only a relatively small number of residues are usually described explicitly, e.g. dipeptide (two residues), tripeptide (three residues), etc., whereas those with 10–30 residues are termed oligopeptides, and those with more than 30 residues are termed polypeptides. The larger protein molecules (which are usually reckoned to contain a minimum of around 50 residues) comprise either a single polypeptide chain, or else several polypeptide chains (covalently) cross-linked by disulphide bonds.

1.3 PEPTIDE AND PROTEIN SECONDARY STRUCTURES

In most proteins (and some small peptides, see below) the peptide chain may become locally folded to form α-helices, reverse turns or β-sheets; these structures are

Table 1.1 — Structure and properties of 20 L-α-amino acids found in proteins

Amino acid	Three- and one-letter codes	Side chain/ structure	Side-chain pK^a	Hydro-pathy index[c]	Abundance[b] (%)	Hydro-philicity index[d]	Percentage exposed[e]	Comments
Alanine	ALA (A)	$-CH_3$	—	1.8	13	−0.5	15	
Arginine	ARG (R)	$-(CH_2)-NH-C-(NH)NH_2$	12.48	−3.5	9.9	3.0	67	Tryptic cleavage site residue
Asparagine[f]	ASN (N)	$-CH_2-CONH_2$	—	−4.5	5.3	0.2	49	*N*-glycosylation site residue; deamidated to Asp
Aspartic acid[f]	ASP (D)	$-CH_2-COOH$	3.86	−3.5	9.9	3.0	50	Prevalent at enzyme active sites
Cysteine	CYS (C)	$-CH_2-SH$	—	2.5	1.8	−1.0	5	Involved in formation of (cystine) disulphide bonds
Glutamic acid[f]	GLU (E)	$-(CH_2)_2-COOH$	4.25	−3.5	7.8	3.0	55	
Glutamine[f]	GLN (Q)	$-(CH_2)_2-CONH_2$	—	−3.5	10.8	0.2	56	Deamidated to Glu
Glycine	GLY (G)	$-H$	—	−0.4	7.8	0	10	Optically inactive; confers conformational flexibility to peptide main-chain; prevalent in reverse turns
Histidine	HIS (H)	$-CH_2-$ N N H	6.0	−3.2	0.7	−0.5	34	Prevalent at enzyme active sites
Isoleucine	ILE (I)	$-CH(CH_3)CH_2CH_3$	—	4.5	4.4	−1.8	13	
Leucine	LEU (L)	$-CH_2-CH(CH_3)_2$	—	3.8	7.8	−1.8	16	
Lysine	LYS (K)	$-(CH_2)_4-NH_2$	9.67	−3.9	7.0	3.0	85	Tryptic cleavage site residue
Methionine	MET (M)	$-(CH_2)_2-S-CH_3$	—	1.9	3.8	−1.3	20	Cyanogen bromide cleavage site residue
Phenylalanine	PHE (F)	$-CH_2-$	—	2.8	3.3	−2.5	10	Contributes to peptide/protein UV absorption; $\lambda_{max}=259$ nm; $E_{\lambda max}=1.1\times10^2$

Proline[g]	PRO (P)	N H COOH	—	−1.6	4.6	0	45	Disrupts secondary structures causing β-bulge in β-strands and kinks in α-helices; prevalent in reverse turns
Serine	SER (S)	$-CH_2-OH$	—	−0.8	6.0	0.3	32	O-glycosylation site residue
Threonine	THR (T)	$-CH(CH_3)OH$	—	−0.7	4.6	−0.4	32	O-glycosylation site residue
Tryptophan	TRP (W)	$-CH_2-$ N H	—	−0.9	1.0	−3.4	17	Major UV absorbing species in peptides/proteins; λ_{max}=279 nm; $E\lambda_{max}=5.2\times10^3$; fluorophore
Tyrosine	TYR(Y)	$-CH_2-$ OH	10.07	−1.3	2.2	−2.3	41	Contributes to peptide/protein UV absorption; λ_{max}=278 nm; $E\lambda_{max}=1.1\times10^3$
Valine	VAL (V)	$-CH(CH_3)_2$	—	4.2	6.0	−1.5	14	

[a]Data taken from Mahler & Cordes (1971).
[b]Percentage abundance in *Escherichia coli* proteins; data taken from Lehninger (1975).
[c]Hydropathy indices (H), taken from Kyte & Doolittle (1982), as used in the sequence-based prediction of transmembrane domains in integral membrane proteins; the amino acid sequence for the polypeptide is recorded numerically using the H values, and a 19-point moving average is then plotted against residue number; segments of the polypeptide which then give a mean H above 1.6 are considered as putative transmembrane regions.
[d]Hydrophilicity indices (h), taken from Hopp & Woods (1981) as used in the sequence-based prediction of protein antigenic determinants; the amino acid sequence for the polypeptide is recorded numerically using the h values, and a 6-point moving average is then plotted against residue number; the highest peaks in the plot are then used to identify potential antigenic determinants.
[e]Percentage of residues exposed to solvent on protein surfaces; exposed residues are identified as those having a solvent accessible surface area (calculated according to Lee & Richards (1971) greater than 60 Å^2.
[f]In the amino acid sequencing of peptides and proteins Asp and Asn residues are frequently not distinguished; the same also applies to Glu and Gln residues; in these instances the acids/amides are coded as Asx (or B) and Glx (or Z) respectively.
[g]Note that proline is actually an α-*imino*-acid.

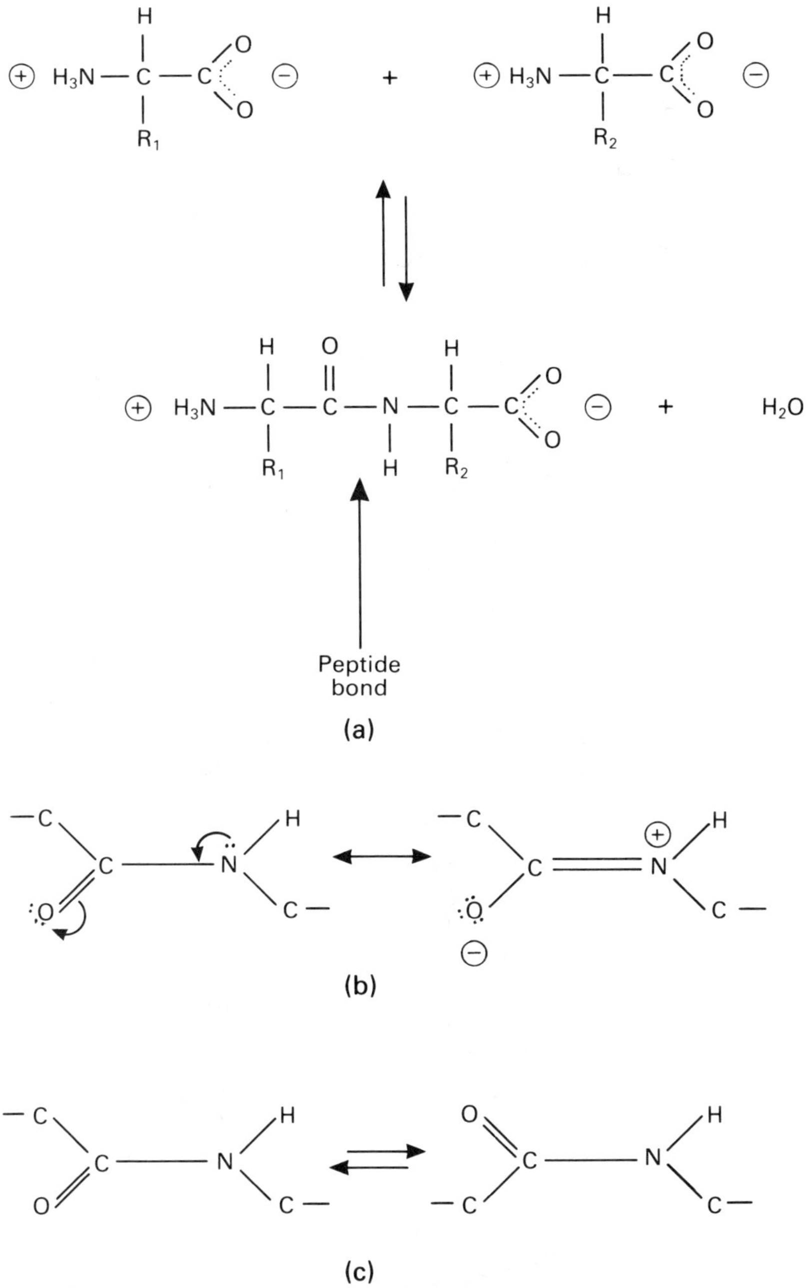

Fig. 1.2 — (a) Peptide bond formation. A condensation reaction involving the primary carboxyl group of one amino acid and the primary amino group of another produces (in this case) a dipeptide. The repeating (N, Cα, C′) unit constitutes the peptide backbone/main chain. (b) The two resonance forms of a peptide bond. The observed hybrid structure has roughly 60% single bond and 40% double bond character. (c) *cis*- and *trans*-forms of a peptide bond. The *trans*-form (shown on the left) is the predominant form found in proteins.

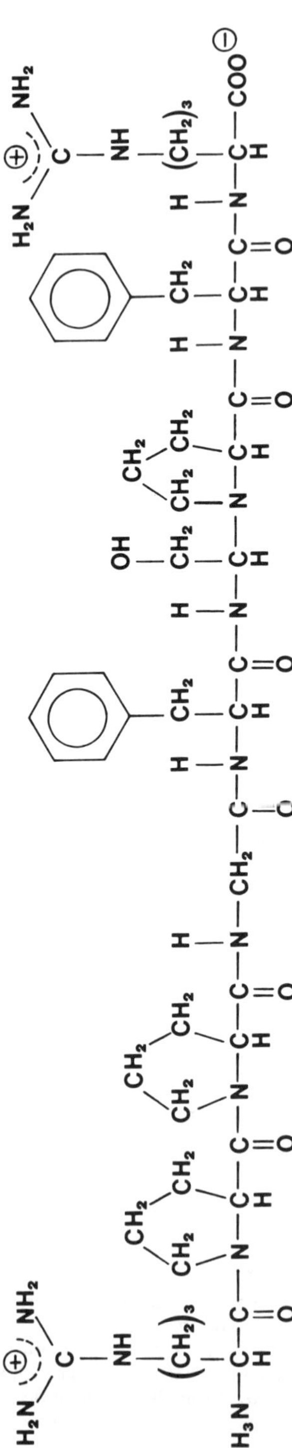

Fig. 1.3 — Amino acid sequence/primary structure of the nine-residue peptide bradykinin. An ordered list of the residues is presented (using the standard three-letter codes, see Table 1.1), beginning with the free amino(N) terminal amino acid, and ending with the free carboxyl(C)-terminal amino acid.

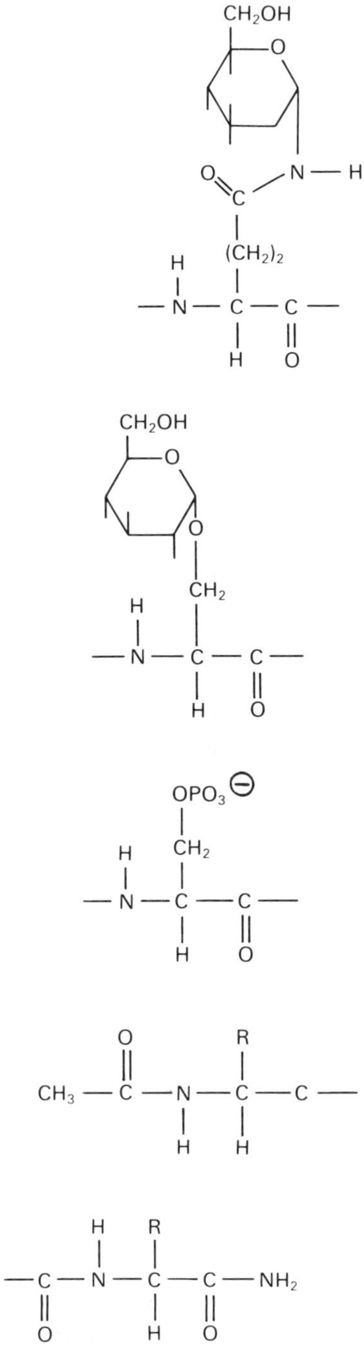

N-glycosylation (of Gln)

O-glycosylation (of Ser)

Phosphorylation (of Ser)

N-terminal acetylation

C-terminal amidation

Fig. 1.4 — Common modifications of the amino acid residues found in peptides and proteins.

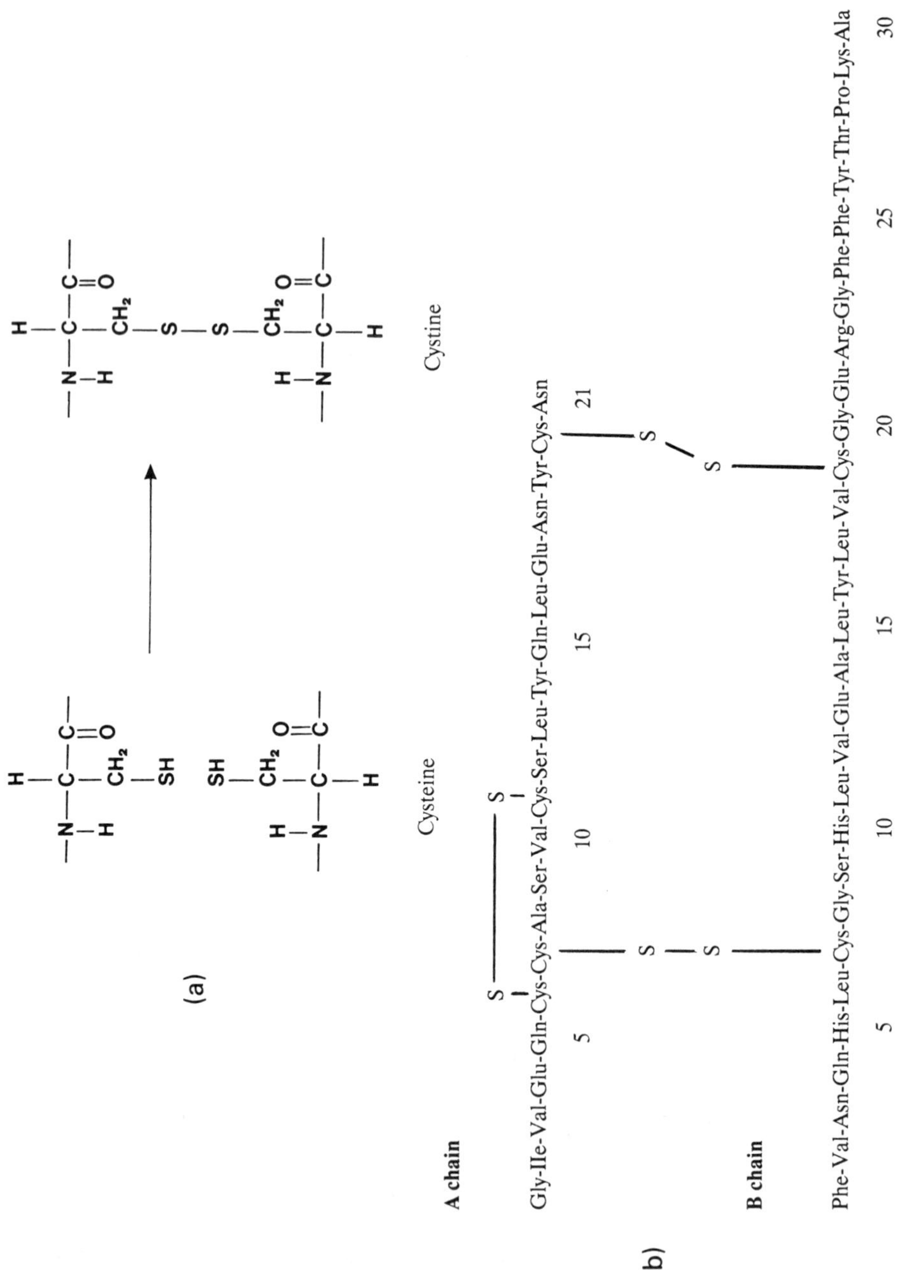

Fig. 1.5 — (a) Formation and structure of a cystine disulphide bond. (b) The arrangement of disulphide bonds/disulphide topology of bovine insulin; the protein consists of two polypeptide chains, referred to as the A and B chains; the two polypeptides are cross-linked by inter-chain disulphide bonds involving residues A7 and B7, and A20 and B19; polypeptide A also has an intra-chain disulphide involving residues A6 and A11.

termed secondary structures, and they are stabilized and characterized by regular patterns of hydrogen bonds involving the peptide CO and NH groups (see Fig. 1.6). Any given peptide will generally assume different secondary structures at different points along its length, and those regions which do not form secondary structures are (rather misleadingly) described as random coil regions.

1.4 GLOBULAR PROTEIN STRUCTURE

1.4.1 Formation of tertiary structure

After, or even during, the formation of secondary structures in the polypeptide chain(s) of a protein, there are non-covalent (Van der Waals, electrostatic, and hydrogen bond) interactions established between its amino acid side-chains. The net effect of these interactions is to cause the molecule to undergo further folding, to form a compact globular unit. The particular pattern of folding which is adopted is sometimes referred to as the chain topology, and the resulting conformation is described as the tertiary structure of the protein.

In all water-soluble proteins the tertiary structure is such that the residues with charged and dipolar side-chains are distributed mainly over the molecular surface, with the residues with apolar side-chains becoming buried and close-packed to form a 'hydrophobic core'.

1.4.2 Secondary structural motifs

As the tertiary structures of more and more proteins have been solved over the last 20 years (see Section 1.6) it has become apparent that there are recurrent structural motifs in which sequential elements of secondary structures are combined in preferred topologies (Fig. 1.7). These arrangements are known as supersecondary structures, and they are stabilized by close-packing and the formation of hydrogen bonds and/or a hydrophobic core. Fig. 1.8 shows how some samples of supersecondary structures involving sequential β-strands and Fig. 1.9 shows other examples based upon sequential α-helices. These structural motifs are extremely widespread, and occur in non-related proteins with many different functions.

In contrast, some structural motifs (with associated sequence motifs) are observed to have specific functions. One such motif is the E–F hand, which involves two perpendicular α-helices separated by a loop containing several negatively charged residues. The motif has the task of binding Ca^{2+} ions, and is found in a broad range of calcium-binding proteins that are thought to have diverged from a common ancestral motif, e.g. calmodulin and troponin C. However, it is also found in the apparently unrelated α-lactalbumin (which is structurally related to lysozyme) where it also binds calcium.

Another classic structural motif which has a specific function and characteristic sequence fingerprint is the α–loop–α structure which binds to DNA. The structure of this motif is well-conserved, with sequence changes which provide differing specificities in different proteins. (Note, however, that the same structural motif present in different proteins does not necessarily imply that the molecules share the same mode of binding to DNA; Brennan & Matthews 1989).

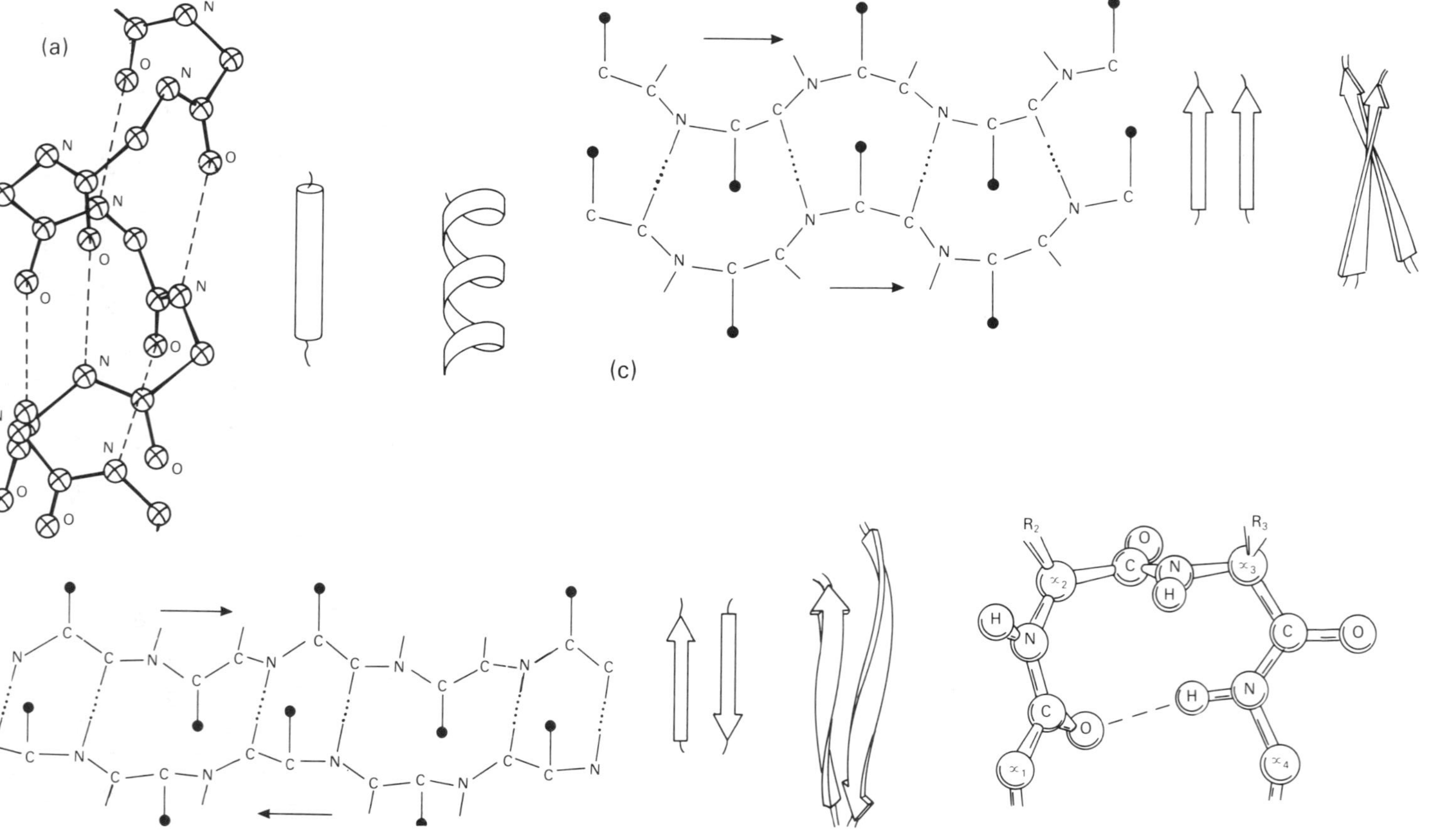

Fig. 1.6 — Examples of peptide/protein secondary structures: (a) α-helix. (b) anti-parallel β-pleated sheet; (c) parallel β-pleated sheet; (d) reverse turn. For (a)–(c) the right-hand figures show the simplified (cartoon) representations popularized by Richardson (1981). In the full structural representations the hydrogen bonds between the peptide CO and NH groups are indicated by dashed/dotted line. In the β-sheet diagrams the arrows indicate the (N- to C-terminal) chain directions, and amino acid side chains are abbreviated as filled circles.

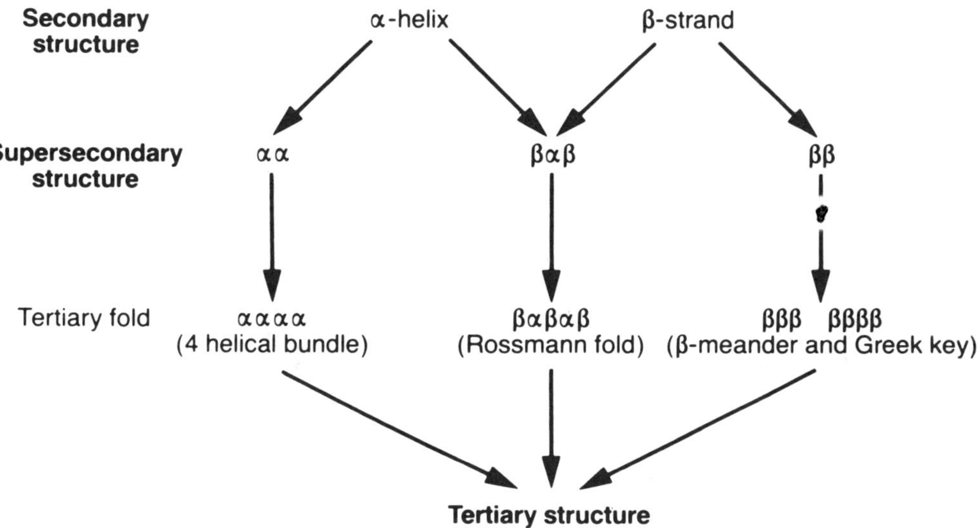

Fig. 1.7 — Hierarchy of protein structure.

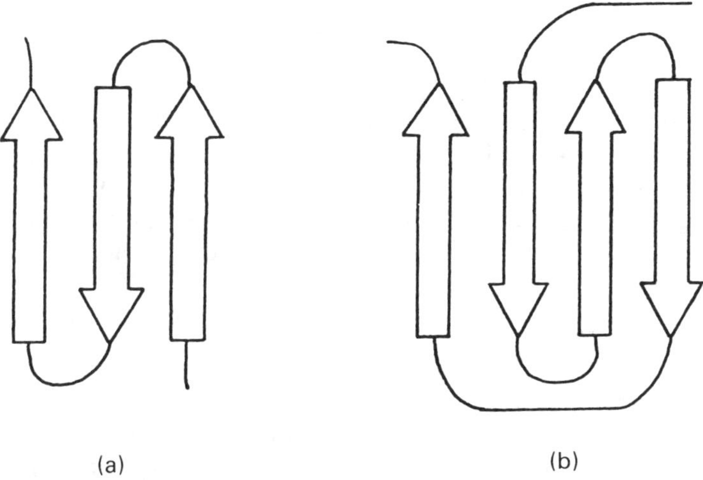

Fig. 1.8 — Schematic diagrams/cartoons showing the arangements of β-strands in the supersecondary structures: β-meander (a) and Greek key (b).

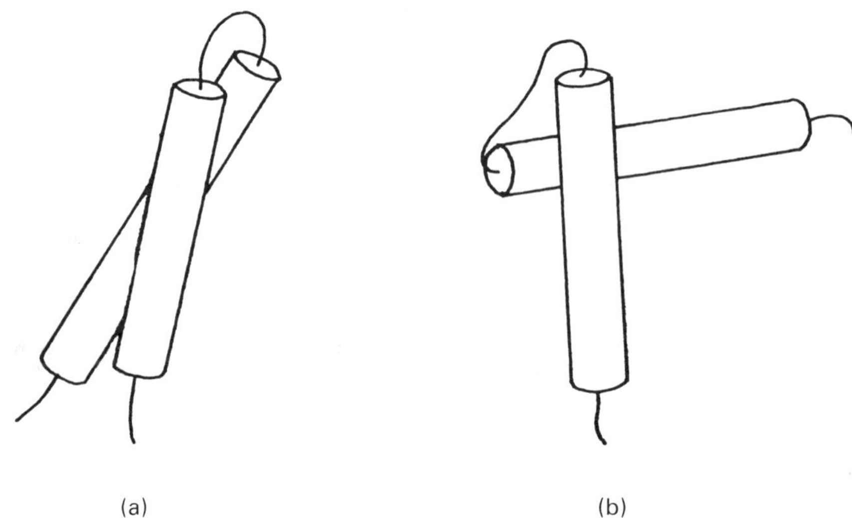

Fig. 1.9 — Schematic diagrams/cartoons showing the arrangements of α-helices in the supersecondary structures: α-hairpin (a) and αα-corner (b).

1.4.3 Protein domains and protein families

The relatively small supersecondary structures described above are unlikely to occur as independent entities. However, the majority of the larger proteins are observed to fold into 'domains' (Fig. 1.10) which have been defined as 'compact, local semi-independent units'. Such domains often fold autonomously, and can therefore be considered as stable topologies. They are classified into five major families, according to their secondary and supersecondary structural content (Richardson 1981, Fig. 1.11, Table 1.2). All recently discovered protein structures have been found to fall in to one of these five families, and are increasingly frequently observed as variations on a basic theme. For example, the structure of interleukin-1β has the same chain topology as soybean trypsin inhibitor (Priestle *et al.* 1988), and the chaperone protein, paperone, follows the same fold as that observed in immunoglobulin domains (Holmgren & Branden 1989). This recurrence of basic structural themes has led to the suggestion that there is a limited number of topologies to which all proteins will conform with only minor variations. If this is correct, it remains to be seen how many such topologies exist.

1.5 SMALL PEPTIDE STRUCTURES

1.5.1 3D structures

Most small peptides, and particularly those that are highly polar, do not adopt a well-defined conformation in aqueous solution; they have a limited capacity for establishing the intramolecular interactions which help stabilize folding intermediates, and so their structures (which although perhaps showing a strong conformational bias (Marqusee & Baldwin 1987)) vary dynamically over a range of conformations

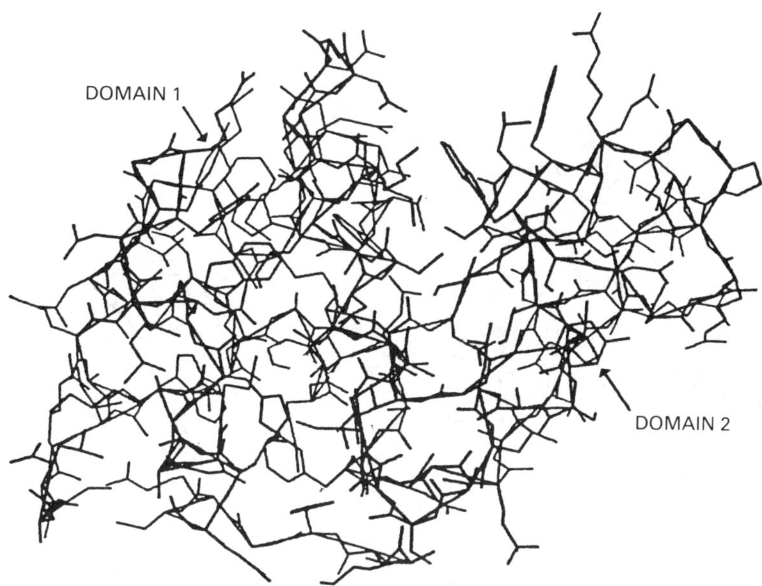

Fig. 1.10 — 3D structure of hen egg white lysozyme (data taken from Brookhaven Databank, Bernstein *et al.* 1977) showing the domain structure; the cleft between the two domains constitutes the active site of the enzyme.

(Blundell & Wood 1982). In order to carry out their biological functions, however, such molecules will adopt a particular conformation when they bind to their target receptor. With some peptides this 'active' conformation may be selected as the result of interaction with a specific macromolecule, and with others it may be selected as the result of self-association or because of the amphipathic nature of its folded structure at a lipid/water interface. The former category is thought to include peptides like atrial natriuretic factor (Theriault *et al.* 1987), and in the latter category are those such as glucagon and vasoactive intestinal polypeptide (Blundell & Wood 1982).

In the case of the small peptides which do adopt a well-defined structure in aqueous solution, there are usually conformational constraints that are imposed either by disulphide bonds, or by the cyclization of their peptide main-chains. The disulphide-linked molecules include α-helix-containing peptides like apamin (Pease & Wemmer 1988, Fig. 1.12), as well as β-sheet peptides like the transforming and epidermal growth factors (Brown *et al.* 1989, Cooke *et al.* 1987, Fig. 1.13). The peptides having a cyclic main-chain are illustrated by the antibiotic ionophore gramicidin S (Fig. 1.14).

The details of the 3D structures of such peptides are sometimes obtained by X-ray crystallography (as in the case of oxytocin, Wood *et al.* 1986), but more frequently are derived from 2D NMR studies (e.g. mast cell degranulating peptide, Kumar *et al.* 1988). In some instances circular dichroism studies are also instructive (Perkins *et al.* 1990), and since this technique is more straightforward and yet provides a highly sensitive means of detecting conformational changes, it is very often used in confirming the 3D structure of peptide pharmaceuticals (see Chapters 13 and 14).

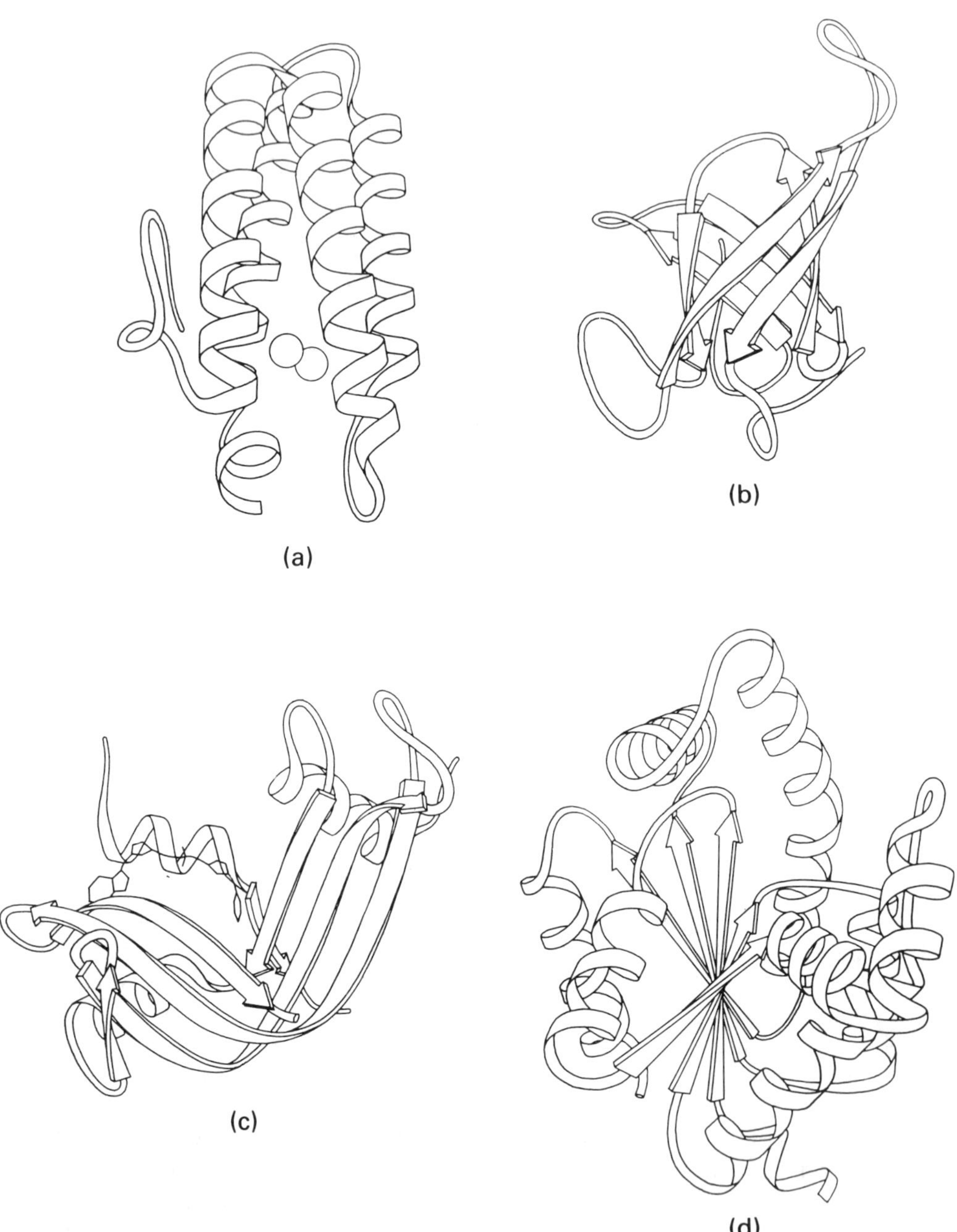

Fig. 1.11 — Cartoon diagrams illustrating four of the five types of protein domain: (a) all-α domain (myohemerythrin), (b) all-β domain (trypsin domain 1), (c) α/β domain (adenylate kinase), (d) $\alpha+\beta$ domain (ribonuclease-S). Diagrams are taken from Richardson (1981).

Table 1.2

Protein families

Classification of major structural patterns in domains:

Five major families:	All α
	All β
	Alternating α/β
	$\alpha+\beta$
	'Random' — disulphide or metal rich.

Each family can be further subdivided:

(1) All α	Four-helix bundles — hemerythrin
	Others — myoglobin, Ca^{2+}-binding proteins.
(2) All β	β-Sheets are antiparallel
	β-Sandwich structures:
	Up and down barrel, e.g. catalase domain 1
	Jelly roll, e.g. SBM virus coat protein
	Others, e.g. immunoglobulin
	Open-faced β-sandwich:
	e.g. phage T4 lysozyme domain 1
(3) Alternatively α/β	β-Strands form parallel β-sheets
	β-Barrel e.g. triose phosphate isomerase
	Standard $(\beta\alpha\beta\alpha\beta)_2$ structure
(4) $\alpha+\beta$	Include helices and strands, without alternation along sequence. β-Strands form antiparallel β-sheets.
	e.g. Lysozyme
	λ *cro* repressor protein
(5) Random	Low percentage of secondary structure.
	Fold is dominated by disulphide bridge connections (e.g. wheat germ agglutinin) or metal-binding sites (e.g. rubredoxin)

Structural repeats

There are many examples of structural repeats within proteins, where *gene duplication* has occurred

e.g.	Elastase	— whole domain is repeated
	γ-crystallin	— repeat within domain

1.5.2 Amino acid modifications

Among the natural small peptides, and even more commonly in the synthetic analogues designed specifically for use as pharmaceuticals, there are frequently 'non-protein' amino acids incorporated to improve their *in vivo* activity. Examples of these (shown in Fig. 1.15) include:

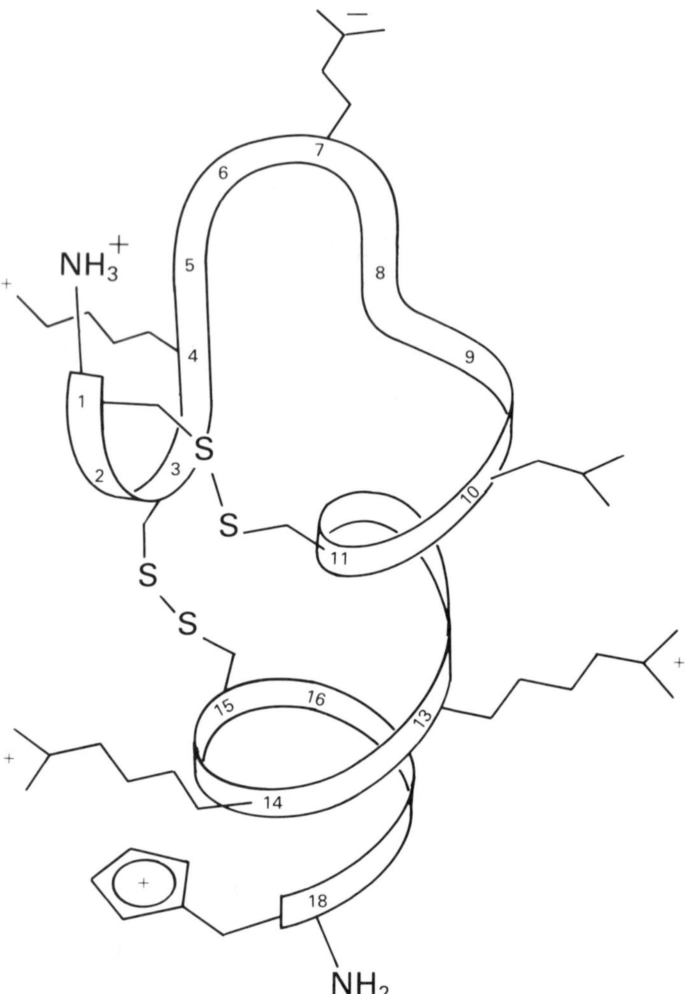

Fig. 1.12 — 3D structure of the bee venom peptide apamin (data taken from Freeman *et al.* 1986).

(1) D-amino acids, which are used to protect against proteolytic degradation, and to force particular conformations of reverse turn (as in somatostatin analogues, Veber *et al.* 1978);
(2) penicillamine residues, which are used to limit the conformational freedom in the immediate vicinity of disulphide bonds (as in antagonists of oxytocin, Wood *et al.* 1986);
(3) α-amino-isobutyric acid residues, which favour helical conformation (as seen in alamethicin, Fox & Richards 1982);

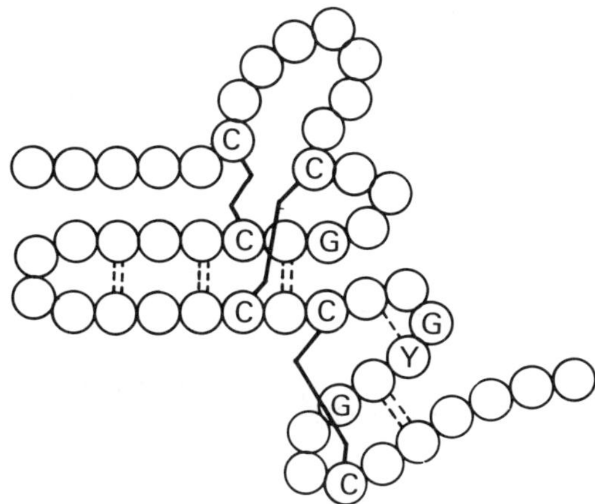

Fig. 1.13 — Secondary structure and disulphide topology of human epidermal growth factor (after Cooke *et al.* 1987).

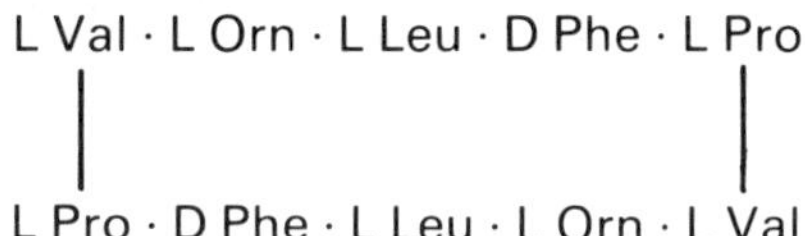

Fig. 1.14 — Structure of the antibiotic ionophore gramicidin S. Orn=Ornithine=Lys with three rather than four methylene groups.

(4) N-terminal pyroglutamate residues (as seen in LHRH, Chapter 16) which improve the half-life of peptides, again by limiting the rate of their proteolytic degradation.

1.6 POLYPEPTIDE/PROTEIN STRUCTURE DATA AND PREDICTION

1.6.1 Structure data

The atomic co-ordinates for about 400 polypeptide and protein structures (258 of which have non-identical sequences) are now available in the Brookhaven Protein Data Bank (Bernstein *et al.* 1977). The structures of these molecules have been solved mainly through X-ray crystallographic studies, but increasingly over recent years also through 2D-NMR. Although the NMR studies are still not as accurate as high resolution crystallography, they have the distinct advantage of not requiring crystalline material, and are particularly valuable for looking at the solution structures of smaller molecules (involving less than 150 residues).

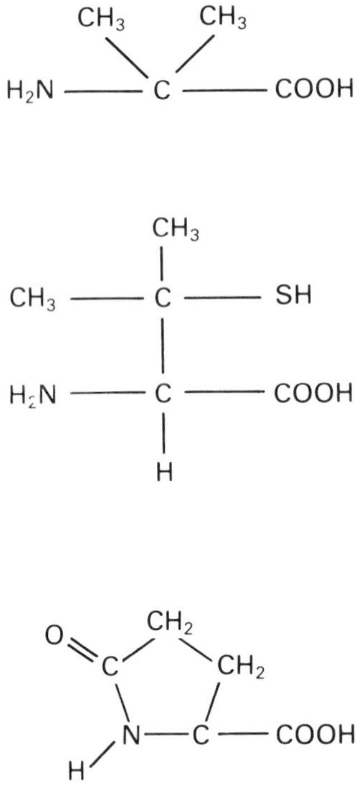

∝-Amino isobutyric acid

Penicillamine

Pyroglutamic acid

Fig. 1.15 — Non-standard amino acids often found in peptides.

Future structure determinations will undoubtedly be accelerated by the recent progress made in gene technology, which allows quantities of pure protein to be made on a scale of 10–50 mg. By comparison with sequencing, however, structure determinations are still likely to be slow and often difficult. For this reason, therefore, there has been, and will continue to be, considerable interest in the prediction of protein structures based on analyses of their amino acid sequences.

1.6.2 Structure prediction

There are currently over 20 000 protein sequences stored in various computer databanks. Many of these proteins will undoubtedly adopt a topology that has already been observed, and the critical problem in predicting their 3D structures, is thus to be able to recognize from their sequences which topology is most probable. Kinetic data show that under the correct conditions most small proteins fold spontaneously to their native state, and so their 3D structures would seem to be completely encoded by their amino acid sequences. In theory, therefore, it should be possible to analyse the sequence data and to correlate and then translate this into 3D structural information. Unfortunately, however, this task is still proving difficult, and the 'protein folding problem' is still not resolved. For example, the sequences of soybean trypsin inhibitor and interleukin-1β are so very different (only about 5% identity) that it is difficult to spot the analogy between them, even when it is known that they adopt the same 3D structure! As a further complication, it has now been demonstrated that some of the larger proteins may require assistance in folding, with help being provided by a 'chaperone' protein (Ellis 1989).

In spite of these difficulties protein structure predictions are still attempted, and in many cases are surprisingly successful.

Fig. 1.16 shows a basic step-wise approach to be followed when analysing a new

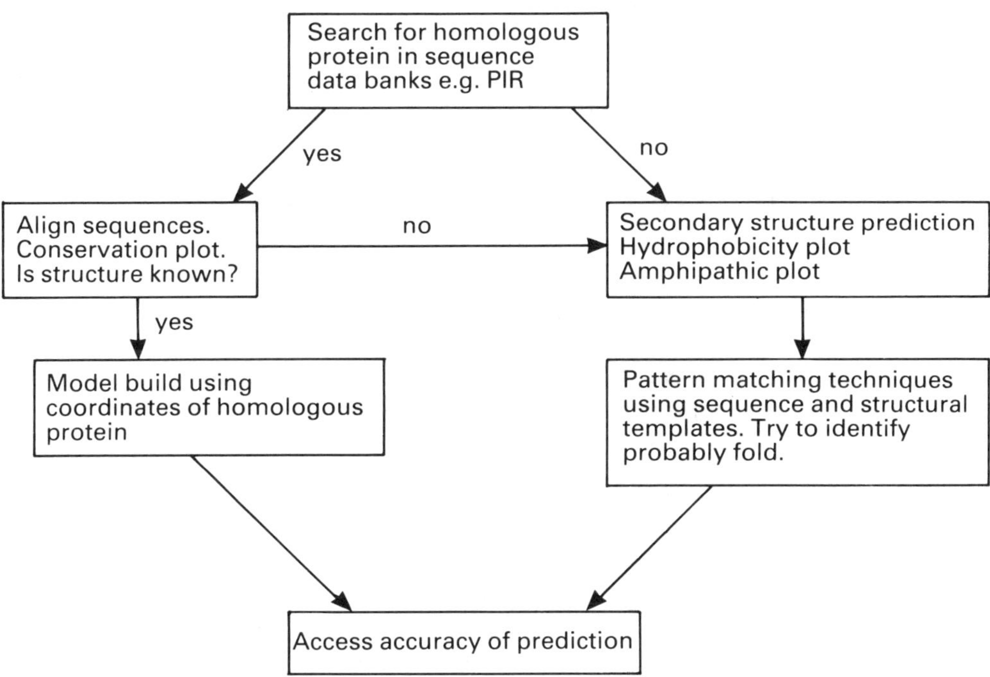

Fig. 1 16 — A stepwise approach to the analysis of amino acid sequences.

protein sequence (Thornton & Taylor 1989). The first step in the process is to search the sequence databanks to find if the new sequence is homologous to one that has already been determined. If matches are found for all or part of the new sequence, the homologous sequence found should be carefully aligned and conservation plots drawn to highlight important residues. If the coordinates of the homologous protein are known, this information can then be used directly to build a 3D model of the new protein using interactive computer graphics (see Barlow & Perkins 1990). This technique is currently the only technique which can predict protein tertiary structure with any accuracy.

If there are no homologous sequences matching the new protein sequence, or if there are no coordinates available for the homologous protein(s), it is necessary to resort to *ab initio* prediction methods, starting with secondary structure prediction. Among the various methods available, those most commonly employed are due to Chou & Fasman (1974) and Garnier *et al.* (1978). Hydrophilicity/hydrophobicity profiles are also widely used for predicting potential antigenic determinants and the locations of transmembrane regions in membrane bound proteins. (see Table 1.1).

The more recent developments in structure prediction have centred around the prediction of super secondary structural features, using database searches, template recognition and pattern matching (Thornton & Gardner 1989). For a single protein sequence, the sequence templates can very often be matched by eye, and so structural motifs like NAD-binding βαβ units, for example, can be spotted easily (Wierenga & Hol 1983). At this stage, if there is sufficient evidence in favour of one type of supersecondary or tertiary structure, it may be possible to model the new protein on the basis of a similar structure in the data bank, using computer graphics techniques (Thornton & Taylor 1989).

1.7 INFORMATION FROM PEPTIDE AND PROTEIN STRUCTURES

Knowledge of the 3D structures of peptides and proteins is now a prerequisite for many aspects of molecular biology. In addition to applications in the pharmaceutical development of peptide and protein drugs, such knowledge is also needed to:

(1) explore protein antigenicity and to design vaccines; e.g. antigenic determinants are correlated with protrusion and can be identified from the structure (Thornton *et al.* 1986);
(2) explain protein function; e.g. the detailed aspects of enzyme mechanisms are impossible to elucidate without a structure and not simple even when a structure is available;
(3) explore co-factor binding and molecular recognition;
(4) plan rational site-directed mutagenesis experiments, e.g. to explore the role of particular residues in protein function, to stabilize proteins by introducing disulphide bridges, or to change protein pH–activity profiles.

For these reasons there has been a resurgence of interest in peptide and protein structure and structure determination. Recent advances in X-ray (Eisenberg & Hill 1989) and NMR (Wright 1989) experimental techniques make structure determination faster and more reliable today than five years ago. Hopefully the next ten years will also see further progress in structure prediction so that we can fully exploit protein sequence information for rational drug and vaccine design.

REFERENCES

Barlow, D. J. & Perkins, T. D. J. (1990) Applications of interactive computer graphics in analyses of biomolecular structures. *Nat Prod. Reports* **7** 311–326.

Bernstein, F. C., Koetzle, T. F., Williams, G. J. B., Meyer, E. F., Brice, M. D., Rodgers, J., Kennard, O., Shimanuchi, T. & Tasumi, M. (1977) The protein data bank: a computer-based archival file for macromolecular structures. *J. Mol. Biol.* **112** 535–542.

Blundell, T. & Wood, S. (1982) The conformation, flexibility and dynamics of polypeptide hormones. *Ann. Rev. Biochem.* **51** 123–154.

Brennan, R. G. & Matthews, B. W. (1989) Structural basis of DNA–protein recognition. *TIBS* **14** 286–290.

Brown, S. C., Mueller, L. & Jeffs, P. W. (1989) ^{1}H NMR assignment and secondary structural elements of human transforming growth factor α. *Biochem.* **28** 593–599.

Chou, P. Y. & Fasman, G. D. (1974) Prediction of protein conformation. *Biochem.* **13** 222–244.

Cooke, R. M., Wilkinson, A. J., Baron, M., Pastore, A., Tappin, M. J., Campbell, I. D., Gregory, H. & Sheard, B. (1987) The solution of epidermal growth factor. *Nature* **327** 339–341.

Eisenberg, D. & Hill, C. P. (1989) Protein crystallography: more surprises ahead. *TIBS* **14**, 260–264.

Ellis, R. J. (1989) Molecular chaperones: proteins essential for the biogenesis of some macromolecular structures. *TIBS* **14** 339–392.

Fox, R. O. & Richards, F. M. (1982) A voltage-gated ion channel inferred from the crystal structure of alamethicin at 1.5 Å resolution. *Nature* **300** 325–330.

Freeman, C. M., Catlow, C. R. A., Hemmings, A. M. & Hider, R. C. (1986) The conformation of apamin. *FEBS Lett.* **197** 289–296.

Garnier, J., Osguthorpe, D. J. & Robson, B. (1978) Analysis of the accuracy and implications of simple methods for predicting the secondary structure of globular proteins. *J. Mol. Biol.* **120** 97–120.

Holmgren, A. & Branden, C. I. (1989) Crystal structure of chaperone protein PapD reveals an immunoglobulin fold. *Nature* **342** 248–251.

Hopp, T. P. & Woods, K. R. (1981) Prediction of protein antigenic determinants from amino acid sequence. *Proc. Natl. Acad. Sci. USA* **78** 3824–3828.

Kumar, V. N., Wemmer, D. E. & Kallenback, N. R. (1988) Structure of P401 (mast cell degranulating peptide) in solution. *Biophys. Chem.* **31** 113–119.

Kyte, J. & Doolittle, R. F. (1982) A simple method for displaying the hydropathy character of a protein. *J. Mol. Biol.* **157** 105–132.

Lee, B. & Richards, F. M. (1971) The interpretation of protein structures: Estimation of static accessibility. *J. Mol. Biol.* **55** 379–400.

Lehninger, A. L. (1975) *Biochemistry*, 2nd Ed., Worth, New York.

Mahler, H. R. & Cordes, E. H. (1971) *Biological Chemistry*, 2nd Ed., Harper & Row, New York.

Marqusee, S. & Baldwin, R. C. (1987) Helix stabilisation of Glu^-...Lys^+ salt bridges in short peptides of *de novo* design. *Proc. Natl. Acad. Sci. USA* **84** 8898–8902.

Pease, J. H. B. & Wemmer, D. E. (1988) Solution structure of apamin determined by nuclear magnetic resonance and distance geometry. *Biochem.* **27** 8491–8498.

Perkins, T. D. J., Hider, R. C. & Barlow, D. J. (1990) Proposed solution structure of endothelin. *Int. J. Peptide Protein Res.* **36** 128–133.

Priestle, J. B. (1988) Crystal structure of the cytokine interleuikin-1β. *EMBO J.* **7** 339–343.

Richardson, J. S. (1981) The anatomy and taxonomy of protein structure. *Adv. Prot. Chem.* **34** 167–339.

Schultz, G. E. & Schirmer, R. H. (1979) *Principles of Protein Structure*, Springer-Verlag, New York.

Theriault, Y., Boulanger, Y., Weber, P. L. & Reid, B. R. (1987) Two-dimensional NMR investigation of the water conformation of atrial natriuretic factor (ANF 101–126). *Biopolymers* **26** 1075–1086.

Thornton, G. M. & Gardner, S. P. (1989) Protein motifs and data-base searching. *TIBS* **14** 300–304.

Thornton, J. M. & Taylor, W. R. (1989) in *Protein Sequencing*. Findlay, J. B. C. & Geison, M. J. (Ed.), IRL Press, pp. 147–190.

Thornton, J. M., Edwards, M., Taylor, W. R. & Barlow, D. J. (1986) Location of 'continuous' antigenic determinants in the protruding regions of proteins. *EMBO J.* **5** 409–413.

Veber, D. F., Holly, F. W., Palereda, W. J., Nutt, R. F., Bergstrand, S. J., Torchiana, M., Glitzer, M. S., Saperstein, R. S. & Hirschmann, R. (1978) Conformationally restricted analogs of somatostatin. *Proc. Natl. Acad. Sci. USA* **75** 2636–2640.

Wieranga, R. K. & Hol, W. G. J. (1983) Predicted nucleotide-binding properties of P21 protein and its cancer associated variant. *Nature* **302** 842–884.

Wood, S. P., Tickle, I. J., Treharne, A. M., Pitts, J. E., Mascarenhas, Y., Li, J. Y., Husain, J., Cooper, S., Blundell, T. L., Hruby, V. J., Buku, A., Fischman, A. J. & Wyssbold, H. R. (1986) Crystal structure analysis of deamino-oxytocin: conformational flexibility and receptor binding. *Science* **232** 633–636.

Wright, P. E. (1989) What can two-dimensional NMR tell us about proteins? *TIBS* **14** 255–260.

2

Protein folding

Robert C. Hider
Department of Pharmacy, King's College, London University,
Manresa Road, London SW3 6LX, UK

2.1 INTRODUCTION

The folding of many proteins is a spontaneous process, the information for the native tertiary structure being totally located in the amino acid sequence. In the absence of disulphide bridges, proteins can fold to their native structure in a time ranging from a few seconds to tens of minutes (Baldwin 1975). This contrasts markedly with the long periods required to locate the native structure via a random search (Table 2.1).

Table 2.1 — Protein folding by random search

- Average number of amino acid (AA) conformations=10
  Thus a small protein (100 AAs) will have 10^{100} different conformations (There are only 10^{80} atoms in the Universe)
- Assume only 2 conformations per AA are critically important, then 10^{30} different conformations will have to be sampled
- Assume maximum rate of interconversion is 10^{-13} s
- Then average time to sample all conformations=10^{17} s (10^{9} years)

Table 2.2 — *In vivo* protein synthesis

- Polymerize amino acids in a sequence-specific manner
- Rate=20 s^{-1}
- Error rates <1 in 10^{4} amino acids (important for pharmaceuticals)
- Fold and secrete proteins with high efficiency
- *In vivo* rates of folding for proteins in the molecular weight range 6000–50000=1–10^{2} s

Indeed when the data in Table 2.1 are compared with the normal *in vivo* performance for protein synthesis (Table 2.2), it is clear that a considerable degree of directed folding must occur in order to produce native proteins with such high efficiency. This direction almost certainly originates via the formation of secondary structure nucleation centres (Fetrow *et al.* 1988, King 1989 and Matouschek *et al.* 1990). These centres may be induced when the protein is either in an expanded random coil or in a more condensed structure where interactions also occur between the hydrophobic side chains (Karplus and Weaver 1976, Baldwin 1989). Single nucleation centres are short lived (10^{-6} s) (McCammon *et al.* 1980) but when a favourable interaction with another centre occurs, the lifetime of the combined structural unit is much longer (Oas and Kim 1988, King 1989). Different amino acids have different propensities for the three major secondary structure forms namely, α-helix, reverse turn or β-strand (Chou and Fasman 1978). Precisely how the amino acid composition of a particular peptide segment determines the nucleation of a given secondary structure is not known. King (1989) describes a dispersed interactive model where, as folding proceeds, intra-segment interactions will be conditioned by the conformations adopted by adjacent sequences, thus permitting inter-segment interactions to occur in place of segment–solvent interactions. A particularly interesting example of how the secondary structure of a peptide segment is influenced by direct interaction with an adjacent peptide was recently provided by Oas and Kim (1988). Two fragments of a proteinase inhibitor that surround a disulphide bond were chemically synthesized. One fragment corresponds to an α-helical segment in the native structure and the other to a β-sheet (Fig. 2.1). Neither forms secondary structure when separate, but when joined by the cystine link, they fold to a structure similar to that found in the native protein (albeit only at low temperatures).

Typically helices and turns form prior to β-sheets (Miyazawa and Jernigan 1982). Indeed β-sheet formation can be associated with a large maximum in free energy, due to the associated loss in conformational entropy (Fig. 2.2(b)). This nucleation process permits the protein to fold in a cooperative manner, where there is efficient packing of both the polypeptide backbone and side-chains (Fig. 2.2(a)). As optimal packing of side-chains is achieved, water molecules are 'squeezed out' from the protein structure and hydrophobic interactions are maximized. The native structure, which contains little free space, is generally believed to be the most stable form of the protein. The packing of side-chains in most globular proteins is similar to that of an organic solid (26% free space, Table 2.3). Thus during folding the hydrodynamic volume of the protein decreases as the structure becomes more compact (Dill 1990). A general energetic scheme for the productive pathways of protein folding is given in Fig. 2.2(b). The reaction coordinate is expressed as compactness of the polypeptide chain (Goldenberg and Creighton 1985). The unfolded protein U is separated by relatively low energy barriers from the increasingly compact intermediate structures I, II and III (Fig. 2.2(a)). These structures are likely to be rapidly interconverting until a 'molten globular structure' (IV) is formed which is capable of undergoing further condensation to yield the native conformation. Direct evidence for the rapid dynamic exchange of such folded structures has been gained by ^{1}H-NMR spectroscopy (Wright *et al.* 1988, Bycroft *et al.* 1990). In many cases it could be that the formation of this critical structure (IV) is the rate-limiting step, as indicated in Fig.

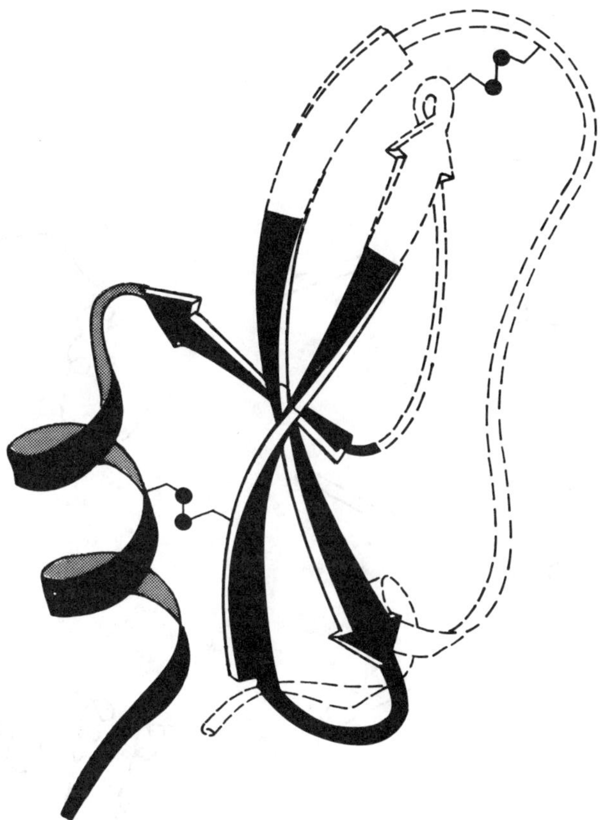

Fig. 2.1 — Secondary structure of polypeptide mimics. A model peptide containing two segments of the protein bovine pancreatic trypsin inhibitor (BPTI) was synthesized by Oas & Kim (1988). The two synthesized segments (═══) are joined by a cystine link (●—●). The rest of BPTI is indicated by (=====).

2.2(b). It is surprising that the free energy difference between the random and native structures is small (typically $-40\ \text{kJ mol}^{-1}$). Thus small shifts in solvent conditions and ligand–protein interactions can have a large influence on stability (Fischer and Schmid 1990). For this reason many extracellular proteins are cross-linked with disulphide links. The presence of such linkages strongly increases the stability of secondary structure (Oas and Kim, 1988).

It is clear that single domain proteins containing no or few disulphide links can be folded under *in vitro* conditions with high efficiency. Thus cytochrome *c* and erythropoietin can be folded in good yeild; staphylococcal nuclease refolds within 1 s and metmyoglobin within 10 s. However, the presence of three or more disulphide bonds, proline residues or the existence of a multidomain structure can present problems for protein folding under *in vitro* conditions (Blohm *et al.* 1988, West *et al.* 1990). Such examples are presented below.

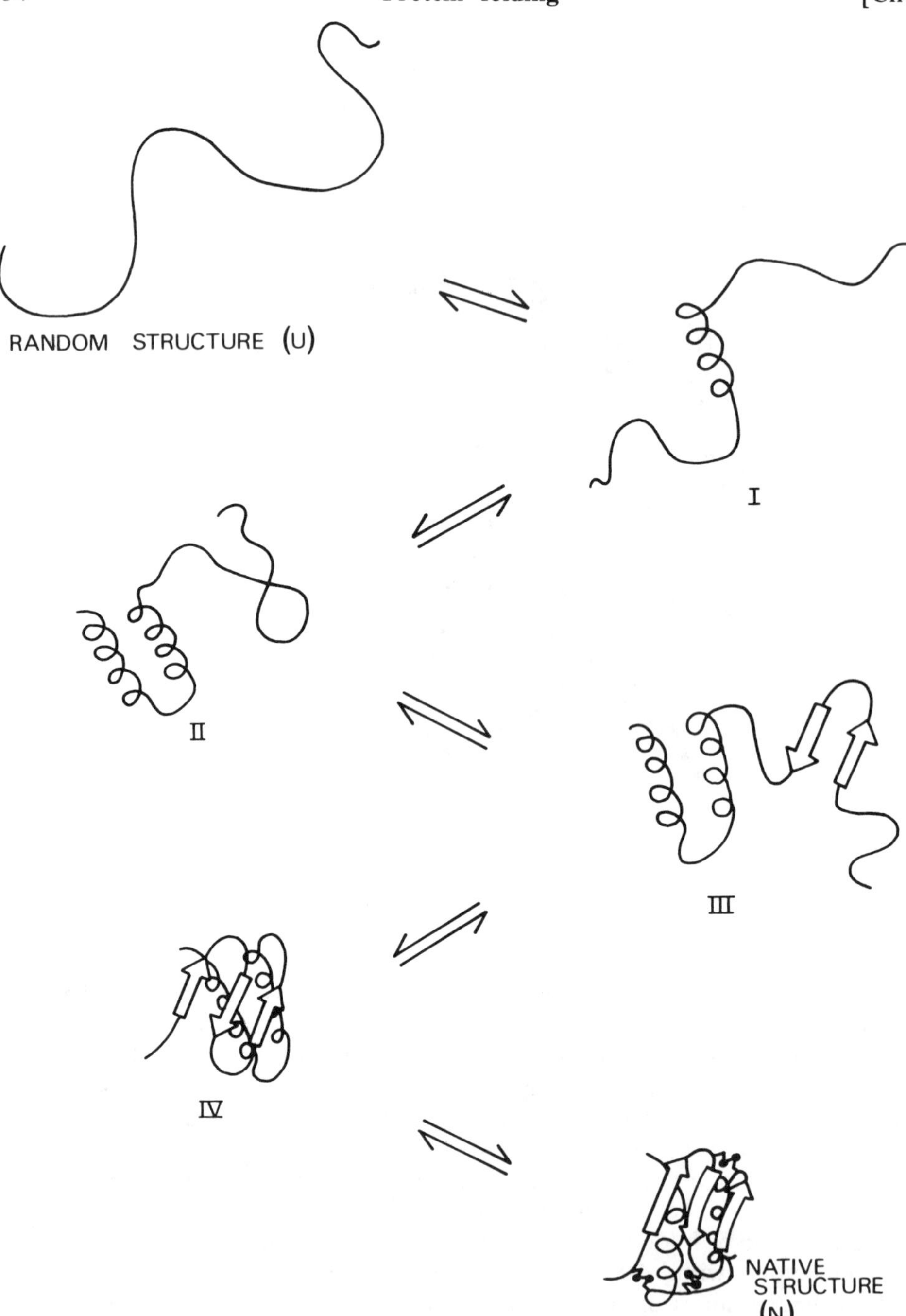

Fig. 2.2 — Schematic representation of protein folding. (a) Folding of a hypothetical protein starting from random structure (U), through intermediates I to IV which contain increasing proportions of secondary structure, to the native structure (N) which is cross-linked by three cystine links.

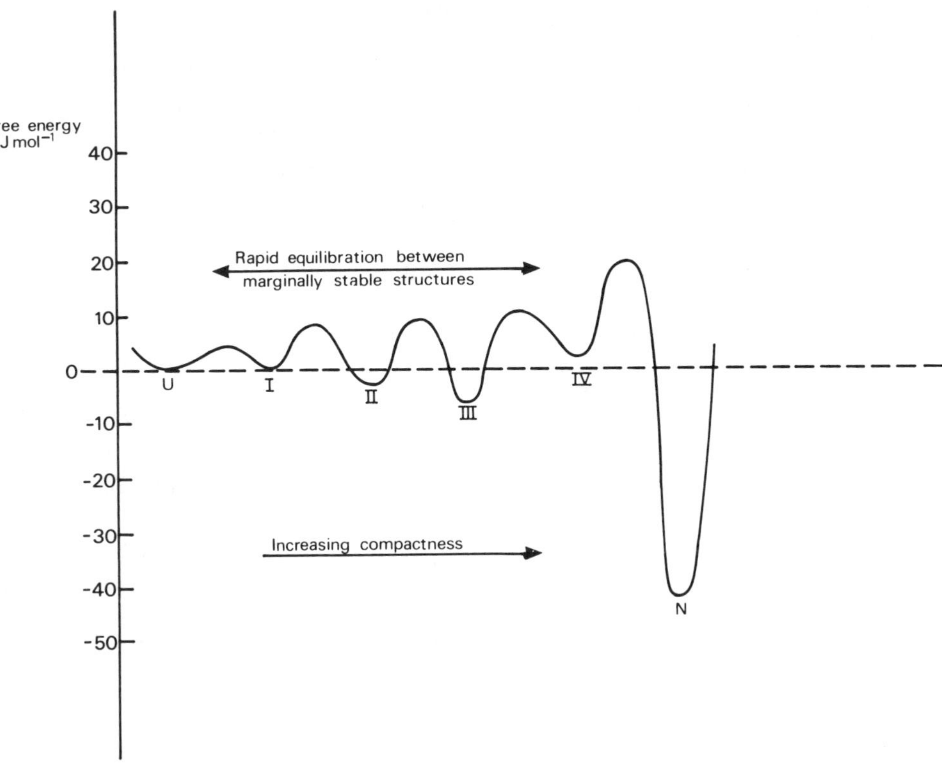

Fig. 2.2 — (b) Energetics corresponding to the folding processes outlined in (a). The native state is characterized by a deep narrow energy well. (Adopted from T. E. Creighton, *Proteins*, Freeman & Co. New York).

Table 2.3 — Packing of atoms

	Percentage of free space
Close packing of regular spheres	25%
Organic solids	22–30%
Proteins	18–32%
Water	42%
Cyclohexane	56%

2.2 DISULPHIDE-CONTAINING PROTEINS

As briefly indicated in the previous section, the free energy difference between the native and random structures of many proteins is rather small. As a consequence slight changes in solvent conditions, for example pH, ionic content or temperature, can lead to unfolding. In view of the extremely conserved nature of the intracellular environment this does not normally present a problem for intracellular proteins. The same assumption, however, cannot be made for proteins which are excreted into the extracellular environment, for instance plasma, the gastrointestinal tract or the extracellular space of another species (venom and salivary proteins). For this reason most secreted proteins, such as hormones, cytokines, enzymes, immunoglobulins, enzyme inhibitors and toxins are cross-linked with disulphide bonds (Table 2.4). Proteins of pharmaceutical interest generally fall into this class.

Table 2.4 — The number of ways in which $2n$ sulphydryl groups can combine to form n cystine links

Number of cystines	Number of combinations	Examples
1	1	Oxytocin
2	3	Growth hormone endothelin
3	15	Hirudin, proinsulin
4	105	Ribonuclease
5	945	α-Neurotoxins
6	10 395	
7	1.3×10^5	Phospholipase A_2
8	2×10^6	
10	6.5×10^8	
12	3×10^{11}	
14	2×10^{14}	
17	6×10^{18}	Albumin
23	2×10^{28}	IgG

Unfolded proteins with reduced disulphide bonds can spontaneously fold to the native structure, although the process requires a redox partner such as oxygen or glutathione. As the number of cysteine groups increases, the number of possible disulphide topologies increases exponentially (Table 2.4, Anfinsen and Scheraga 1975). This renders it more difficult to fold the native structure under *in vitro* conditions, although in principle this can be achieved by refolding in the presence of both oxidized and reduced forms of glutathione, thus allowing for the reshuffling of non-native disulphide bonds (Fig. 2.3). A further complication in the refolding of thiol-containing proteins is polymerization due to the formation of interchain cystine links. For this reason most refolding procedures are carried out at relatively low concentrations (not above 10^{-5} M). Using such conditions, reduced ribonuclease (four S–S bonds) can be folded to the native structure in 3 h (Anfinsen and Scheraga 1975) and bovine pancreatic trypsin inhibitor (three S–S bonds) within 30 min (Creighton, 1978). These are much longer than the times required for the refolding of cystine-free proteins of comparable size.

Proteins with more than four disulphide groups tend to be difficult to refold using such conditions. Indeed incorrect disulphide links are generally observed during the

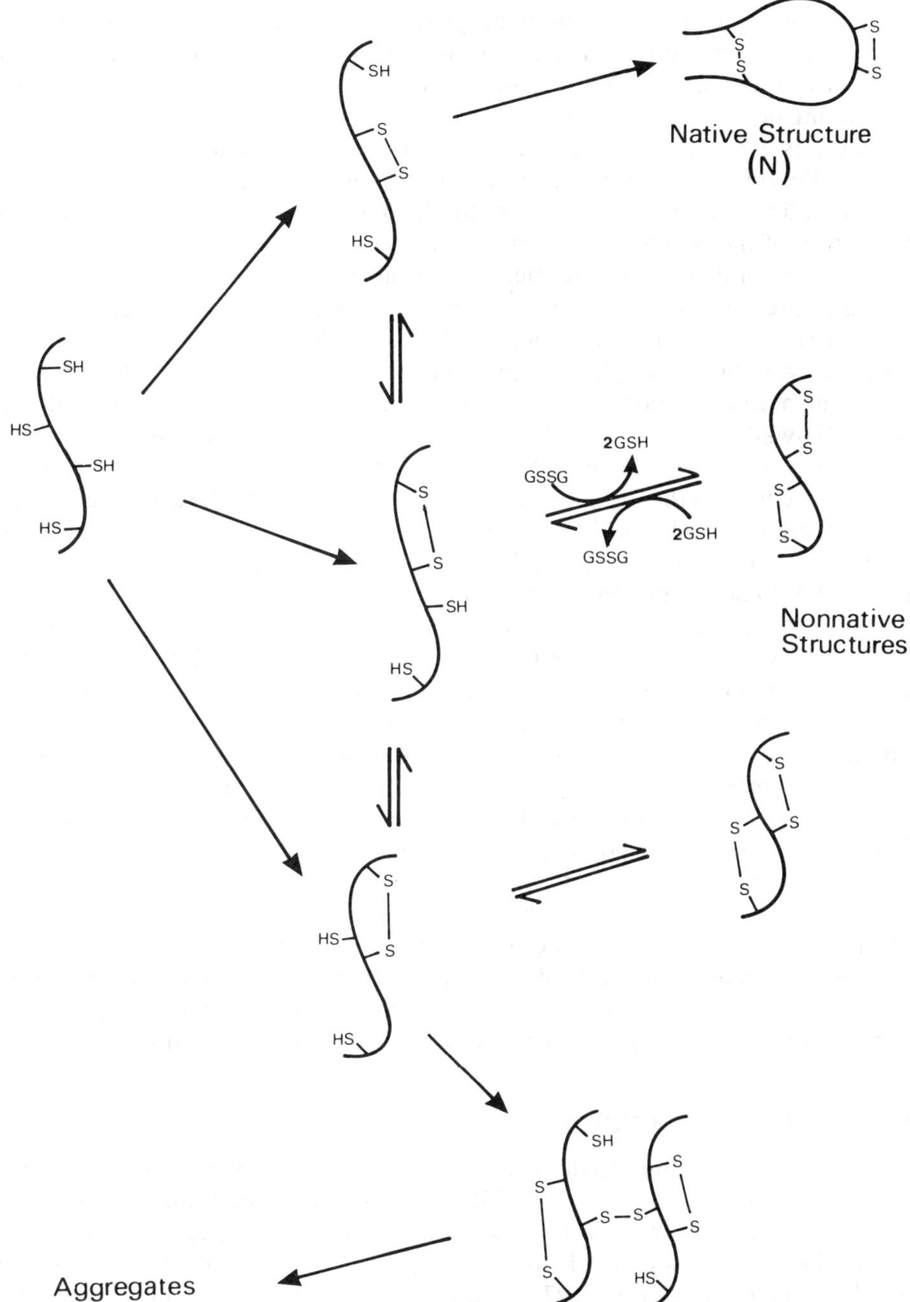

Fig. 2.3 — Schematic representation of a thiol-containing protein. A number of different monocystine-containing intermediates are possible, but only two lead to the native structure. Others have to undergo rearrangement before forming the native structure. Rearrangement can be facilitated by the simultaneous presence of reduced and oxidized forms of glutathione. Intermolecular cystine links lead to protein aggregation.

in vitro folding of cystine-containing proteins, the refolding of bovine pancreatic trypsin inhibitor being particularly well-documented (Creighton 1978, 1983, Creighton and Goldenberg 1984, Staley and Kim, 1990). The relevance of structures containing non-native disulphide links to *in vivo* folding, where the process is much quicker, is not clear. Eight intermediates have been characterized for bovine pancreatic trypsin inhibitor (Fig. 2.4) and some of these are located at 'cul-de-sacs' in the refolding pathway. The currently favoured *in vitro* pathway involves the formation of the non-native cystines, 5–14 or 5–38. Indeed it seems likely that the monocystine and dicystine intermediates are all in rapid equilibrium with each other, acting as precursors to either N^{-SH}_{-SH}, which is subsequently converted to the native form N (Fig. 2.4) (Creighton and Goldenberg 1984) or N* which is a native-like molecule with buried sulphydryl functions at positions 30 and 51. The N* species is stable in solution at room temperature for days and is only converted to the native form (N) when warmed at 37°C (States *et al.* 1980). The increase in temperature apparently enhances the mobility of side-chains, thereby permitting access to oxygen, facilitating the formation of the third cystine bond. The activation energy for this process is 220 kJ mol^{-1}. Such problems apparently do not occur under *in vivo* conditions. Clearly N* and possibly some of the suspected intermediates presented in Fig. 2.4 do not form under *in vivo* conditions.

Similar less detailed studies have been reported with snake neurotoxins (Bouet *et al.* 1982) which have either four or five disulphide links (Fig. 2.5, Dufton and Hider 1983). A greater number of intermediates are trapped with this class of molecule, as compared with the trypsin inhibitor, and a much longer time period is required to obtain a high yield of native protein (Fig. 2.6, Bouet *et al.* 1982). It seems likely that the more intermediates that accumulate in the refolding solution, the longer the *in vitro* refolding process takes to generate the native structure. Significantly, a homologous neurotoxin containing five disulphide bonds cannot be refolded under comparable bulk phase conditions (Smith and Hider, 1988).

Kinetic intermediates of folding are likely to be susceptible to aggregation by virtue of the increased amount of exposed hydrophobic surface. This transient low solubility of incompletely folded polypeptide chains is compounded in the presence of cysteine functions where covalent interpeptide links can form (Fig. 2.3). These critical problems are apparently minimized under *in vivo* conditions.

2.3 PROLINE ISOMERISM

The presence of proline residues can slow down the rate of protein refolding under *in vivo* conditions (Brandts *et al.* 1975). Unlike all the other protein amino acids, the energy difference between the *cis*- and *trans*-forms for the peptide bond preceding the proline is quite small, 8 $kJmol^{-1}$ as opposed to 80 $kJmol^{-1}$ for a non-proline amide (Shultz and Shirmer 1979). Consequently both forms will be present in an unfolded polypeptide chain. Clearly each residue on the N-terminal side of a proline must adopt either a *cis*- or a *trans*- conformation (Fig. 2.7) in the native structure and therefore isomerism must take place during the folding process. This frequently leads to a biphasic refolding rate, where those backbones containing the correct local conformation fold rapidly and those containing the incorrect conformation fold more

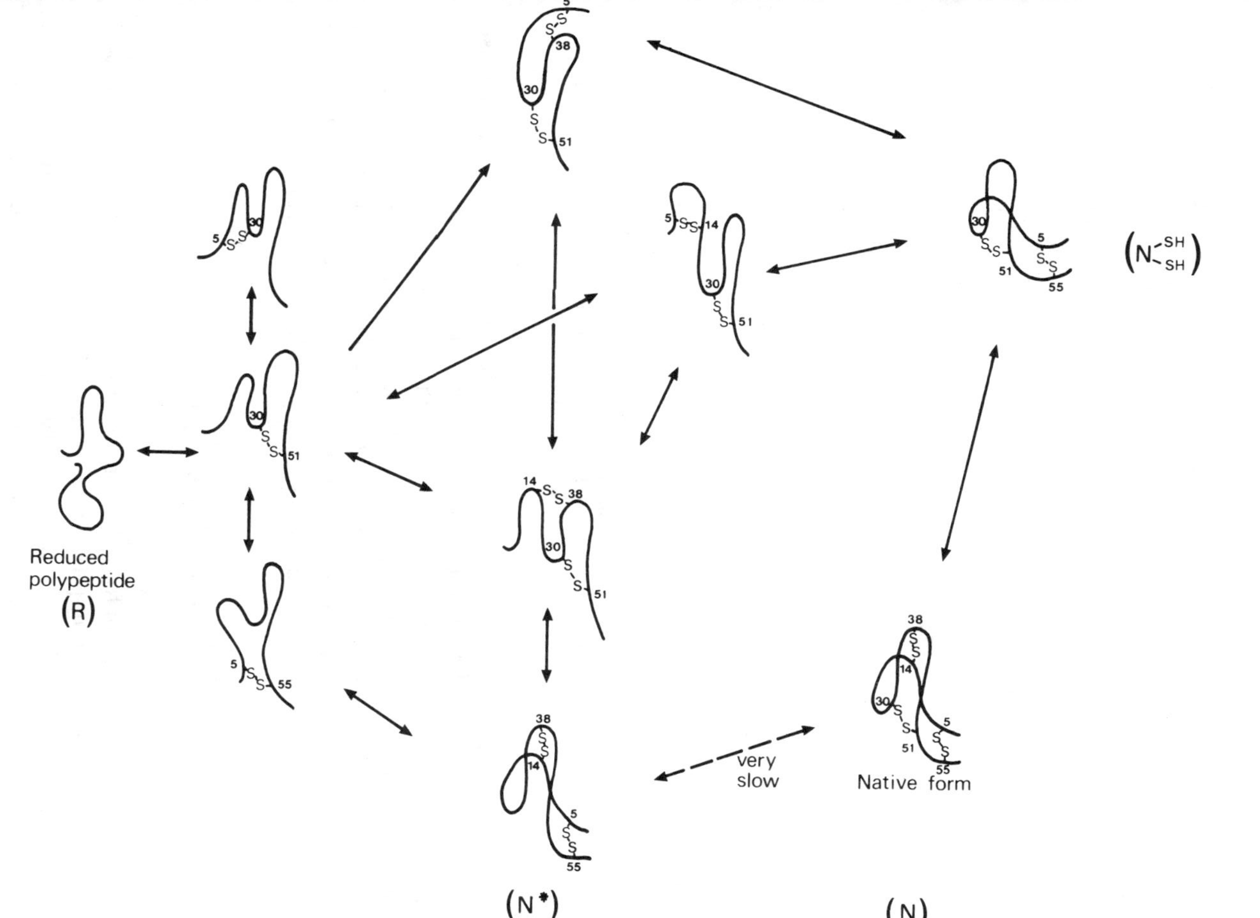

Fig. 2.4 — Folding intermediates of bovine pancreatic trypsin inhibitor. The folding process involves the formation of three disulphide bonds. There are 15 possible arrangements of these bonds. Under *in vitro* conditions the native structure (N) is formed *via* a range of partially oxidized intermediates which have been characterized by Creighton and co-workers. The numbers on the chain indicate the positions of the cystine links. (Adapted from Creighton & Goldenberg (1984)).

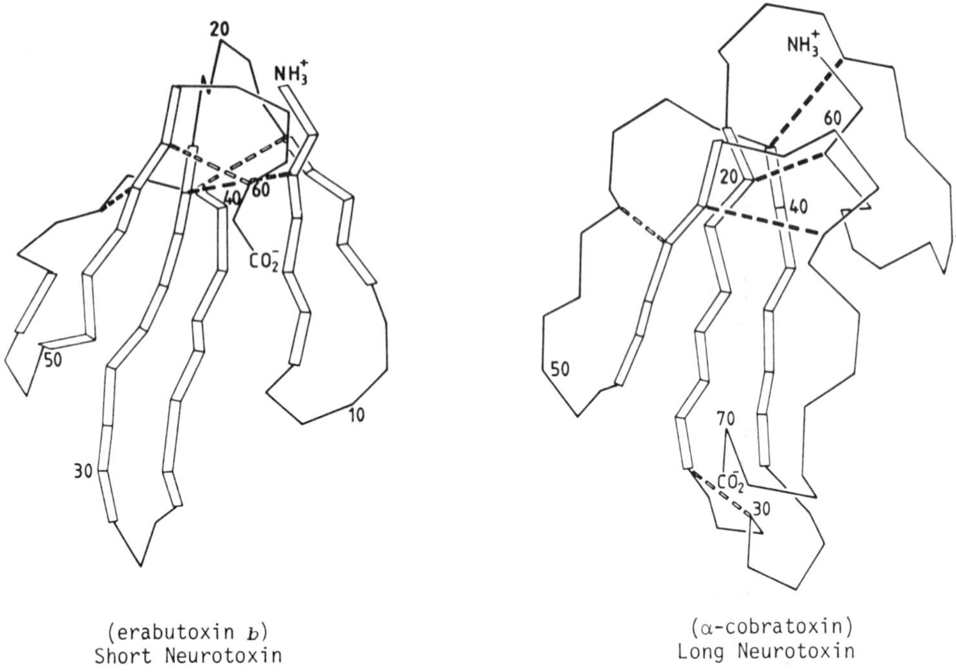

Fig. 2.5 — Structures of postsynaptically acting neurotoxins isolated from the venom of elapidae snakes. (a) Short neurotoxin (62 amino acids, four disulphide links). (b) Long neurotoxin (71 amino acids, five disulphide links). Both proteins are in rich antiparallel β-sheet. (Adapted from Dufton & Hider (1983).

slowly (Garel and Baldwin 1973, Lang *et al.* 1986). Recently two states of nuclease have been detected by ^{1}H-NMR, the major difference being related to the *cis*- and *trans*- conformation of the peptide bond before Pro-117. In contrast the mutant protein with Gly-117 only exists in one conformation as detected by ^{1}H-MNR (Evans *et al.* 1987). In similar fashion on substitution of Ser-54, Pro-55 by Gly-54, Asn-55 in ribonuclease T_1, one of the two slow phases in the folding of this protein is abolished and the kinetics of the entire process is dramatically simplified (Kiefhaher *et al.*, 1990).

2.4 PROTEIN FOLDING UNDER *IN VIVO* CONDITIONS

Cells use several membrane-bound enzymes and the membranes themselves to facilitate the efficiency of protein folding. For proteins which are normally exported from cells (disulphide-containing proteins) the biosynthesis and permeation of the protein through the cell membrane is tightly coordinated. A short peptide section on the N-terminus of the newly synthesized protein, the signal peptide, binds to the signal recognition particle (SRP) as it emerges from the ribosome (Fig. 2.8) (Randall and Hardy, 1989, Robinson *et al.* 1990). Polypeptide synthesis is arrested at this stage and is only reactivated when the ribosome binds to a membrane receptor,

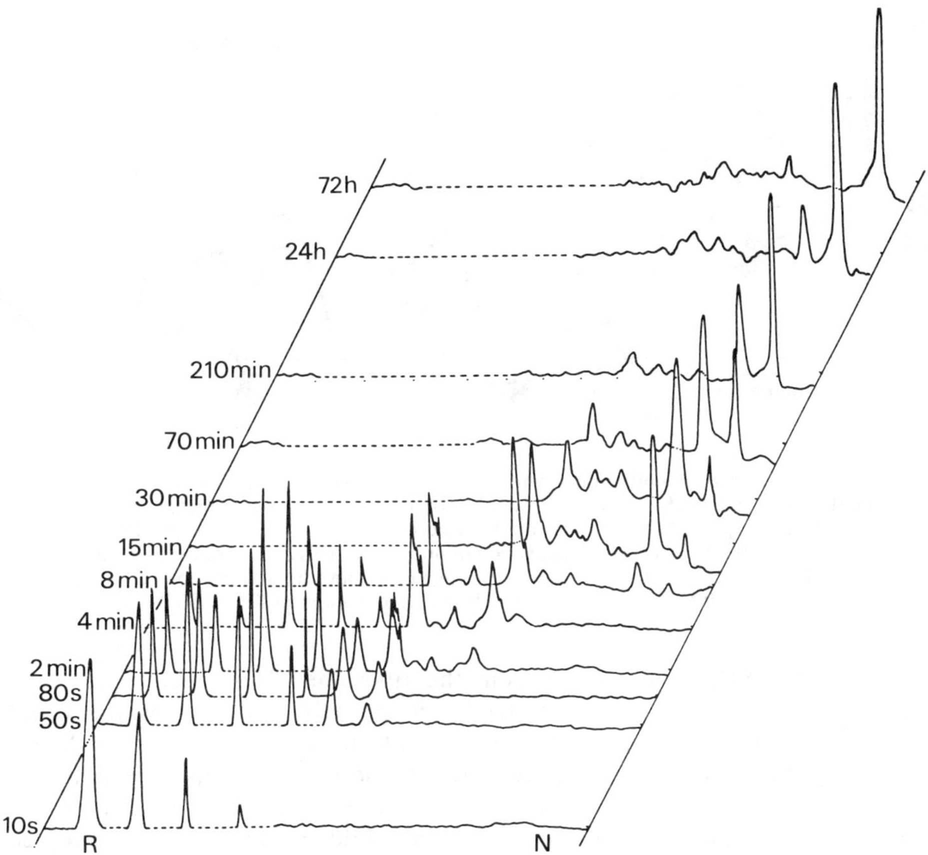

Fig. 2.6 — Refolding intermediates of a short neurotoxin, erabutoxin b. Densitometer tracings of the isoelectrofocusing patterns of trapped refolded toxin. Traces of the native protein appear after 8 min, but major contaminants are still present after 24 h. Twenty-six intermediates are identified. (Adapted from Bouet *et al.* (1982)).

Fig. 2.7 — *Cis*- and *trans*-proline.

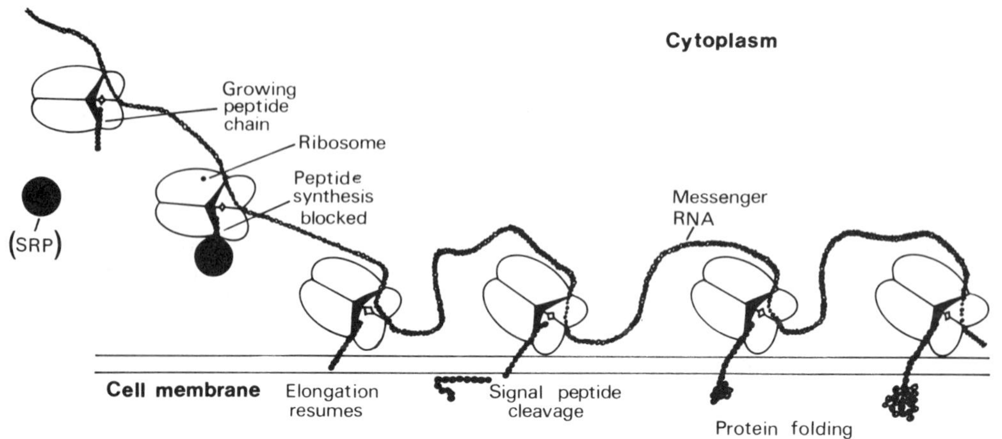

Fig. 2.8 — Membrane-bound protein synthesis. For secreted proteins the synthesis and folding is tightly coordinated. The N -terminal signal peptide binds to the signal recognition particle (SRP) which delays further synthesis until the ribosome binds to the membrane. Subsequent displacement of the SRP leads to reactivation of protein synthesis, the product being directed through the membrane. After cleavage of the signal protein, the nascent polypeptide chain folds as it emerges from the membrane. (Adapted from King (1989)).

whereupon the SRP is displaced from the ribosome. This mechanism ensures excretion of the protein through the membrane (probably via the ribosome receptor protein) (Fig. 2.8). The signal protein is enzymatically cleaved on the outside of the cell, leaving the growing nascent polypeptide to merge into the extracellular space. Indeed, the presence of the signal sequence retards protein folding until it is cleaved (Liu *et al.* 1989). The solubility of the partially folded protein chains in aqueous solvents is low and therefore aggregation is favoured. However, when a nascent polypeptide chain folds as it emerges from the surface of a membrane, the protein is effectively at infinite dilution and therefore aggregation is prevented.

The folding of proteins is fast and may well proceed at least 10 times more rapidly than the biosynthesis of the amino acid sequence (Schultz and Shirmer 1979, Table 2.2), therefore the protein will probably fold from the N-terminus as it emerges from the membrane (Fig. 2.8). The concept of protein folding being initiated from the N-terminus was first postulated by Philips (1967). In the case of larger proteins, domains fold independently, for instance pancreatic serine proteases (Higaki and Light 1986) and tryptophan synthase (Hurle and Matthews, 1987). A controlled folding process of this type readily accounts for the folding of the three domains present in albumin within 30 seconds (Peters and Davidson 1982) and for the folding of multidomain structures such as immunoglobulins (Bergman and Kuchl 1979). As soon as an entire domain is secreted, the single intradomain disulphide bond is formed.

The efficiency of the folding process is further enhanced by the activity of two enzymes, protein disulphide isomerase and prolyl isomerase. Both enzymes are located on the external surface of protein-secreting membranes.

2.4.1 Protein disulphide isomerase

Enzyme activity capable of accelerating the reoxidation of reduced proteins was first isolated from liver extracts in 1963 (Goldberger *et al.* 1963). The enzyme responsible for this activity, protein disulphide isomerase, which forms 2% of rat liver content (Freedman 1984, 1987), uses either oxygen or oxidized glutathione as an oxidant. Its function appears to be that of facilitating the formation of the correct set of disulphide bonds during the folding of secreted proteins. This is achieved by the rapid shuffling of incorrect disulphide bonds, thereby favouring the formation of the native cystines which are associated with the most stable structure. There are many reports of protein disulphide isomerase-catalysed formation of native disulphide-bonded proteins from the reduced form, including single domain proteins (Creighton *et al.* 1980), multidomain proteins (Freedman 1989) and hetero-oligomeric proteins such as immunoglobulins (Roth and Koshland 1981).

Many homologies have been reported between protein disulphide isomerase and other enzymes involved in co-translational and post-translational conversions (Freedman 1989). It seems likely that a family of closely related proteins carries out the multiple functions associated with the modification of secretory proteins.

2.4.2 Peptidyl-prolyl *cis-trans* isomerase

Conformational steps involving isomerism of proline amide bonds can be slow (Section 2.3) and in 1984 Fischer and coworkers isolated an enzyme which accelerates this isomerization. The protein, termed peptidyl-prolyl *cis-trans* isomerase (PPIase) is widely distributed. Catalysis of the isomerism is apparently achieved by distortion, whereby the transition state is characterized by a partial rotation of the C–N bond (Harrison and Stein 1990a). Thus the degree of catalysis of refolding depends on the sequence immediately before and after the proline residue (Fischer and Schmid 1990). Indeed, the existence of a range of PPIase molecules with differing specificities (Harrison and Stein 1990b) offers an explanation for the finding that not all proteins are influenced by the same enzyme; thus whereas the state of refolding of ribonuclease T_1 is markedly accelerated by liver PPIase, (Fischer *et al.* 1989), chymotrypsinogen is not (Lin *et al.* 1988). The presence of a family of PPIases with differing selectivity in the vicinity of ribosomes would be predicted to greatly facilitate the efficiency of protein folding.

Significantly, porcine kidney PPIase is identical to bovine cyclophilin, a protein which binds cyclosporin with high affinity (Takahashi *et al.* 1989). Cyclosporin abolishes the catalytic effect of PPIase on protein folding. FK506, another immunosuppressive drug, also binds to and inhibits a related prolyl isomerase (Tropschug *et al.* 1990). The physiological role of PPIases is not clearly established, but their importance is indicated by the finding that they appear to be the major cellular target of immunosuppressive drugs.

2.4.3 Control of translocation rates

As indicated above, translocation can be stopped by the binding of the signal recognition particle to the ribosome. However there are other possible mechanisms for temporal control and these are probably important for multidomain proteins. In principle there are two direct mechanisms whereby the nucleic acid sequence of mRNA can influence the rate of translation; one *via* the secondary structure of RNA (Baim *et al.* 1985) and the second by use of 'rare' codons (Morlon *et al.* 1983). In the former, it is proposed that protein synthesis is slowed while the secondary structure of mRNA is uncoiled and converted to a conformation which is acceptable by ribosomes; in the latter, 'rare' codons are believed to be associated with a low concentration of tRNA, thus again protein synthesis slows.

Direct evidence has been provided for the existence of clear translational pauses which result in the accumulation of nascent polypeptides in mammalian cells (Protzel and Morris 1974), in invertebrates (Candelas *et al.* 1983) and in micro-organisms (Varenne *et al.* 1982, Randall *et al.* 1980). Recently Purvis *et al.* (1987) have pointed out that a yeast pyruvate kinase gene contains an unusual arrangement of five consecutive rare codons located towards the end of the protein coding region. This region, which is located between domains A and C (Fig. 2.9), will almost certainly lead to a pause in pyruvate kinase synthesis. It has been suggested that such a pause

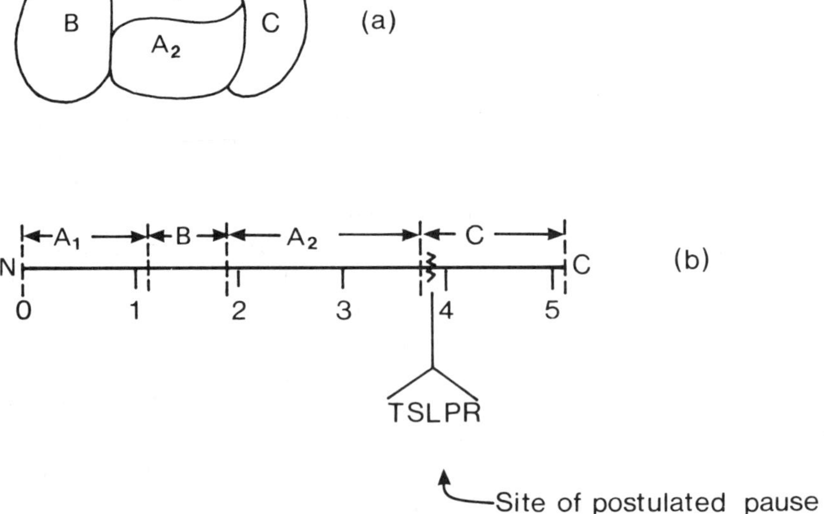

Fig. 2.9—Location of the proposed translational pause within pyruvate kinase. (a) Structure of pyruvate kinase, highlighting the three structural domains. (b) Linear presentation of pyruvate kinase indicating the various contributions to the three domains. Scale is in hundreds of amino acids. The pause, associated with the sequence TSLPR, is located between domain section A_2 and domain C. It has been suggested that domains A and B fold before domain C is synthesized.

might permit the two portions of domains A and B to interact to form a β-barrel before domain C is synthesized (Purvis *et al.* 1987).

2.5 RELEVANCE OF *IN VIVO* STUDIES TO *IN VITRO* FOLDING

It is clear that the subtle controls placed on protein folding by membranes, enzymes and temporal control are absent during *in vitro* procedures. Should these processes prove to be essential for correct folding, as seems likely for large multidomain proteins, then *in vitro* folding of such material would appear to have little practical potential. In contrast, single domain proteins, even those containing several disulphide links, can be successfully folded under *in vitro* conditions, although generally not at high concentrations, where aggregation competes with the natural folding pathway.

As indicated in section 2.2, the solubility of partially folded chains is low in aqueous solvents and this leads to aggregation problems. It has been suggested that 'chaperone' molecules are necessary for the correct folding of some proteins (Ellis & Hemmingsen 1989). These proteins (with molecular weights of at least 60 000) apparently bind transiently to partially folded or non-assembled peptide chains, thereby preventing uncontrolled aggregation (Hemmingsen *et al.* 1988). They have a clear role in the assembly of many oligomeric proteins, particularly when the genes for subunits are separated by a membrane, i.e. some subunits are coded either in the chloroplast or mitochondria and others in the nucleus (Ellis, 1987). They are reported to be capable of unfolding preformed proteins prior to intracellular membrane transfer (Wickner 1989) and of unfolding proteins which have been misfolded or aggregated (Ellis *et al.* 1989). Some are induced by heat shock and other forms of cell stress (Bardwell & Craig 1984).

The molecular mechanism by which chaperones facilitate folding is unknown. It seems likely that they provide extended surfaces suitable for the reversible binding of partially folded polypeptides. The three-dimensional structure of a large fragment of one of these molecules has recently been reported (Flaherty *et al.* 1990). In the author's view, it is unlikely that they have an important role in the folding of single domain or many single-chain multidomain proteins.

2.6 STRATEGIES FOR GENETICALLY ENGINEERED PRODUCTS

A wide range of proteins with pharmaceutical potential have been genetically engineered (Chapter 3, Blohm *et al.* 1988). Polypeptide production via recombinant DNA technology is described in Chapter 5 and the relative merits of using either intracellular or extracellular production is discussed. Clearly if post-translational modification is essential, for instance to achieve the pharmacological effect, optimal pharmacokinetics, protease resistance or solubility, then extracellular production will be necessary. Indeed mammalian cell lines may also be essential (Table 2.5, Yan *et al.* 1989; Wold, 1983). However if post-translational modification is not essential then intracellular production may offer advantages over and above those of extracellular production (Chapter 5).

2.6.1 Extracellular production

By the use of various *E. coli* signal peptides it is possible to engineer systems which permit secretion of the gene product (Chapter 5). The classic example of this strategy utilizes the signal peptide of the OmpA protein, a major outer protein of *E. coli*. This technique has been used to produce staphylococcal nuclease A (Takahara *et al.* 1985), subtilisin E (Ikemura *et al.* 1987, Takagi *et al.* 1988) and human growth hormone (Chapter 14, Becker & Hsiung, 1986; Hsiung *et al.* 1986). The proteins are

Table 2.5 — Post-translational modification

C-terminal CO_2H	Glycosylphosphatidyl inositol
Glutamate	γ-Carboxyglutamate
Proline	Hydroxylation
Lysine	Hydroxylation
Glutamate	Pyroglutamate
C-terminal CO_2H	Amidation
Tyrosine	*O*-Sulphate
Aspartate	β-Hydroxylation
Asparagine	β-Hydroxylation
Serine/threonine	Oligosaccharide

secreted into the periplasm of the bacteria in their native state in high yield (up to 10% of cell protein).

Similar techniques can be used with mammalian cell lines when glycosylation is required; as is the case with many large polypeptide hormones erythropoietin can be produced in this manner using hamster cells (Goto *et al.* 1988). Thus if protease attack can be minimized (Chapter 5) and separation from the host proteins be readily achieved, extracellular production is a useful method for the production of pharmaceutical proteins. Problems with folding are largely overcome by the utilization of the normal mechanisms employed with membrane-associated protein synthesis. However, if the use of rare codons for temporal control of protein folding is an important feature (section 2.4.3) then problems may result from the expression of a protein in a non-native species as the codon bias of the original organism is unlikely to be reflected in the tRNA pools of the host. Such problems, however, are unlikely to influence the folding of single-domain proteins.

2.6.2 Intracellular production

When engineered proteins are produced intracellularly, in high concentration, many tend to aggregate forming inclusion bodies (Chapter 5). Indeed the reducing environment leads to incomplete oxidation of cysteine functions which further facilitates the formation of intracellular aggregates (Fig. 5.1). An additional complication is the likelihood of a newly synthesized polypeptide chain forming a tight complex with a protein component of the host cell (Darby & Creighton, 1990). Such interaction is favoured for highly basic proteins. A number of pharmaceutically important proteins and peptides have been produced as inclusion bodies (Table 2.6).

Inclusion bodies are distinct granules which are visible with a phase contrast microscope. They are dense and therefore sediment readily with low speed centrifugation (Marston 1986, Schomer *et al.* 1985). Under these conditions, inclusion bodies sediment more rapidly than the bulk of all cell debris. These aggregates are formed from non-native states of the polypeptide chain and therefore conversion to the native state will, in general, require both unfolding and successful refolding reaction *in vitro*. As aqueous media are poor solvents for unfolded protein chains, solutions of urea, guanidinium hydrochloride or detergents are generally used to

Table 2.6 — Pharmaceutically important polypeptides produced as inclusion bodies

Polypeptide	Mol wt $\times 10^{-3}$	Mode of expression	Expression (%)	Number of cysteine residues
Somatostatin	1.5	Fusion	<0.1	2
Insulin (A-chain)	2.0	Fusion	20	4
Insulin (B-chain)	3.0	Fusion	20	2
β-Endorphin	3.9	Fusion	5	0
Urogastrone	7.5	Fusion	—	6
AIDS peptide 121	15	Direct	5	2
Growth hormone (Bovine)	22	Fusion	5	2
κ-Light chain (IgG)	24	Direct	0.5	5
β-Interferon	25	Direct	15	3
α-Heavy chain (IgG)	49	Direct	3	13

Taken from Marston (1986).
Definition of 'fusion' and 'direct' expression given in Chapter 6.

dissolve inclusion bodies. Indeed with some preparations, such treatment will selectively remove the endogenous proteins leaving the recombinant protein in up to 90% yield (Schomer *et al.* 1985). The product still requires solubilization and typical treatments are 5–8-M guanidinium chloride (insulin A and B chains, bovine growth hormone); 6–8-M urea (IgG heavy and light chains); detergents (β-interferon) and even organic solvents for some products. The effectiveness of a particular treatment (pH, temperature, contact time) is highly dependent on the polypeptide. With some proteins, successful refolding can be achieved on the surfaces of reversed micelles (Hagan *et al.* 1990). If conditions are selected such that there is only one (or fewer) protein molecules per micelle then the aggregation problem can be minimized.

If incorrect intra- or intermolecular disulphide bonds have been formed then disruption will be essential. Reversible *S*-sulphonation (Cabilly *et al.* 1984) and low molecular weight thiols (Shire *et al.* 1984, Marston, 1986) have been used to facilitate refolding in such circumstances. As discussed in section 2.2 the formation of correct disulphide bonds is critical for efficient conversion to the native state. In the bulk phase, such procedures will be limited to low concentrations (below 10^{-5} M). One method which circumvents this difficulty is centred on the use of an activated thiol-rich column. Fully reduced protein in guanidinium hydrochloride is applied to the column. This immediately forms cystine links with the column matrix. However, the high density of thiol functions on this column facilitates thiol–disulphide exchange. When eluted with a gradient of glutathione the native protein elutes from the column (Smith and Hider 1988). The method has been used successfully for neurotoxins containing up to 5 cystines. Such molecules cannot normally be refolded in the bulk phase even at a protein concentration of 10^{-6} M.

Correctly refolded proteins can be further purified by ion exchange, molecular sieve and reverse-phase chromatography (Shire *et al.* 1984).

2.7 ENGINEERED PROTEINS

Some effort has been directed at facilitating the efficiency of folding and in the design of more stable proteins using site-directed mutagenesis. Mutant proteins of β-interferon and interleukin-2 have been synthesized in *E. coli* in which one cysteine residue (of a total of three) was deleted or replaced by serine (Marston, 1986). It was hoped that these changes would minimize intermolecular cystine formation. However the products still produced insoluble aggregates which required renaturation and refolding.

In principle the introduction of an additional cystine link will enhance the stability of proteins (Pace, 1990); in practise, however, successful modifications of this type have been difficult to achieve. Seven different disulphide bonds have been individually added to subtilisin, but none of the resulting proteins were appreciably more stable than the wild type (Mitchinson & Wells, 1989). In contrast, a variant of T4 lysozyme with an additional disulphide bond has a dramatically enhanced thermal stability when compared with that of the wild type (Matsumura *et al.* 1989, Signor *et al.* 1990). In another example the enzyme thioredoxin has been modified by site-specific mutagenesis to form a protein with greatly enhanced stability towards denaturant-induced unfolding (Langsetmo *et al.* 1990). Aspartic acid-26 is replaced by alanine , thereby influencing electrostatic interactions within the protein.

It seems likely that it will be possible to design polypeptides with an enhanced stability and resistance to enzyme-catalysed cleavage. Such modifications could improve the pharmacokinetics of a protein.

Table 2.7 — Classification of pharmaceutically useful proteins

1 Single domain	Secretion not essential for successful folding
2 Single domain cross-linked by disulphide links	
3 Single domain+essential post-translational modifications	Secretion is likely to greatly facilitate efficient folding and modification
4 Multidomain	
5 Multidomain cross-linked by disulphide links	
6 Multidomain+essential post-translational modifications	

2.8 CONCLUSION

The major protein classes which can be genetically engineered are presented in Table 2.7. It is clear from the discussion in section 2.4 that efficient folding of most, if not all, multidomain proteins will require controlled folding starting from the

N-terminus during protein synthesis. Therefore secretion of the product from the producing organisms will be essential. The same conclusion holds for proteins which require post-translational modification for full activity and optimum pharmacokinetics. In contrast single-domain proteins, with and without disulphide cross-linking, can, in principle, be refolded with high efficiency. Under these circumstances production of the protein *via* insoluble 'inclusion bodies' may have advantages, as the proteins are readily isolated from the host cytoplasmic proteins in this form. Many potentially useful proteins are single-domain structures, cross-linked by multiple disulphide bonds. Currently, disulphide-rich proteins can only be efficiently refolded at high dilution; however, if thiol-containing columns are used then the proteins can be considered to fold at infinite dilution (Smith and Hider 1988). Such methods produce high yields of native products on the preparative scale.

REFERENCES

Anfinsen, C. B. & Scheraga, H. A. (1975) Experimental and theoretical aspects of protein folding. *Adv. Prot. Chem.* **29** 209–300.

Baim, S. B., Pietras, D. F., Eustice, D. C. & Sherman, F. (1985) A mutation allowing an mRNA secondary structure diminishes translation of *Saccharomyces ceverisiae iso*-I-cytochrome-c. *Mol. Cell. Biol.* **5** 1839–1846.

Baldwin, R. L. (1975) Intermediates in protein folding reactions and the mechanism of protein folding. *Ann. Rev. Biochem.* **44** 453–475.

Baldwin, R. L. (1989) How does protein folding get started? *Trends Biochem. Sci.* **14** 291–294.

Bardwell, J. C. & Craig, E. A. (1984) Major heat shock gene of *Drosophila* and the *Escherichia coli* heat-inducible *dna K* gene are homologous. *Proc. Natl. Acad. Sci. USA* **81** 848–852.

Becker, G. W. & Hsiung, H. M. (1986) Expression, secretion and folding of human growth hormone in *Escherichia coli. FEBS Letts.* **204** 145–150.

Bergman, L. W. & Kuchl, W. M. (1979) Formation of an intrachain disulphide bond on nascent immunoglobulin lightchains. *J. Biol. Chem.* **254** 8869–8876.

Blohm, D., Bollschweiler, C. & Hillen, H. (1988) Pharmaceutical proteins. *Angewandte Chemie* **27** 207–225.

Bouet, F., Ménez, A., Hider, R. C. & Fromageot, P. (1982) Separation of intermediates in the refolding of reduced erabutoxin b by analytical isoelectric focusing in layers of polyacrylamide gel. *Biochem. J.* **201** 495–499.

Brandts, J. F., Halvorson, H. R. & Brennan, M. (1975) Consideration of the possibility that the slow step in protein denaturation reactions is due to *cis–trans* isomerism of proline residues. *Biochemistry* **14** 4953–4963.

Bycroft, M., Matouschek, A., Kellis, J. T., Serrano, L. & Fersht, A. R. (1990) Detection and characterisation of a folding intermediate in barnase by NMR. *Nature* **346** 488–490.

Cabilly, S., Riggs, A. D., Pande, H., Shirely, J. E., Holmes, W. E., Rey, M., Perry, L. J., Wetzel, R. & Heyneker, H. L. (1984) Generation of antibody activity from immunoglobulin polypeptide chains produced in *Escherichia coli. Proc. Natl. Acad. Sci. USA* **81** 3273–3277.

Candelas, G., Candelas, T., Ortiz, A. & Rodriguez, O. (1983) Translational process during a spider fibroin synthesis. *Biochem. Biophys. Res. Comm.* **116** 1033–1038.

Chou, P. Y. & Fasman, G. D. (1978) Prediction of the secondary structure of proteins from their amino acid sequence. *Adv. Enzymol.* **47** 45–148.

Creighton, T. E. (1978) Experimental studies of protein folding and unfolding *Prog. Biophys. Mol. Biol.* **33** 231–297.

Creighton, T. E. (1983) An empirical approach to protein conformation stability and flexibility. *Biopolymers* **22** 49–58.

Creighton, T. E. & Goldenberg, D. P. (1984) Kinetic role of a meta-stable native-like two disulphide species in the folding transition of bovine pancreatic trypsin inhibitor. *J. Mol. Biol.* **179** 497–526.

Creighton, T. E., Hillson, D. A. & Freedman, R. B. (1980) Catalysis by protein-disulphide isomerase of the unfolding and refolding of proteins with disulphide bonds. *J. Mol. Biol.* **142** 43–62.

Darby, N. J. & Creighton, T. E. (1990) Folding proteins. *Nature* **344** 715–716.

Dill, K. A. (1990) Dominant forces in protein folding. *Biochemistry* **29** 7133–7155.

Dufton, M. J. & Hider, R. C. (1983) Conformational properties of the neurotoxins and cytotoxins isolated from elapid snake venoms. *Critical Rev. Biochem.* **14** 113–171.

Ellis, J. (1987) Proteins as molecular chaperons. *Nature* **328** 378–379.

Ellis, R. J. & Hemmingsen, S. M. (1989) Molecular chaperones: proteins essential for the biogenesis of some macromolecular structures. *Trends Biochem. Sci.* **14** 339–342.

Ellis, R. J., van der Vies, S. M. & Hemmingsen, S. M. (1989) The molecular chaperone concept. *Biochem. Soc. Symp.* **55** 145–155.

Evans, P. A., Dobson, C. M., Kautz, R. A., Hattull, G. & Fox, R. D. (1987) Proline isomerism in staphylococcal nuclease characterised by NMR and site-directed mutagenesis. *Nature* **329** 266–268.

Fetrow, J. S., Zehfus, M. H. & Rose, G. D. (1988) Protein Folding: New Twists. *Bio-Technology* **6** 167–171.

Fischer, G. & Schmid, F. X. (1990) The mechanism of protein folding. Implications of *in vitro* refolding models for *de novo* protein folding and translocation in the cell. *Biochem.* **29** 2205–2212.

Fischer, G., Bang, H. & Mech, C. (1984) Detection of enzyme catalysis for *cis-trans*-isomerisation of peptide bonds using proline containing peptides. *Biomed. Biochim. Acta* **43** 1101–1111.

Fischer, G., Wittmann-Liebold, B., Lang, K., Kiefhaber, T. & Schmid, F. X. (1989) Catalysis of proline isomerisation during protein-folding reactions. *Nature* **337** 268–270.

Flaherty, K. M., Deluca-Flaherty, C. & McKay, D. B. (1990) Three-dimensional structure of the ATPase fragment of a 70K heat-shock cognate protein. *Nature* **346** 623–628.

Freedman, R. B. (1984) Native disulphide bond formation in protein biosynthesis: evidence for the role of protein disulphide isomerase. *Trends Biochem. Sci.* **9** 438–441.

Freedman, R. B. (1987) Folding into the right shape. *Nature* **329** 196–197.

Freedman, R. B. (1989) Protein disulphide isomerase: multiple roles in the modification of nascent secretory proteins. *Cell* **57** 1069–1072.

Garel, J. R. & Baldwin, R. L. (1973) Both the fast and slow refolding reactions of the ribonuclease A yield native enzyme. *Proc. Natl. Acad. Sci. USA* **70** 3347–3351.

Goldberger, R. F., Epstein, C. J. & Anfinsen, C. B. (1963) Acceleration of reactivation of reduced bovine pancreatic ribonuclease by a microsomal system from rat liver. *J. Biol. Chem.* **238** 628–635.

Goldenberg, D. P. & Creighton, T. E. (1985) Energetics of protein structure and folding. *Biopolymers* **24** 167–182.

Goto, M., Akai, K., Murakami, A., Hashimoto, C., Tsuda, E., Ueda, M., Kawanish, G., Takahashi, N., Ishimoto, A., Chiba, H. & Sasaki, R. (1988) Production of recombinant erythropoietin in mammalian cells. *Bio-Technology* **6** 67–71.

Hagan, A. J., Hatton, T. A. & Wang, D. I. C. (1990) Protein folding in reversed micelles. *Biotechnol. & Bioeng.* **35** 955–965.

Harrison, R. K. & Stein, R. L. (1990a) Substrate specificities of the peptidyl prolyl *cis–trans* isomerase activities of cyclophilin and FK-506 binding protein: evidence for the existence of a family of distinct enzymes. *Biochem.* **29** 3813–3816.

Harrison, R. K. & Stein, R. L. (1990b) Mechanistic studies of peptidyl prolyl *cis–trans* isomerase: evidence for catalysis by distortion. *Biochem.* **29** 1684–1689.

Hemmingsen, S. M., Woolford, C., van der Vies, S. M., Tilly, K., Dennis, D. T., Georgopoulous, C. P., Hendrix, R. W. & Ellis, J. (1988) Homologous plant and bacterial proteins chaperone oligomeric protein assembly. *Nature* **333** 330–334.

Higaki, J. N. & Light, A. (1986) Independent refolding of domains in the pancreatic serine proteases. *J. Biol. Chem.* **261** 10606–10609.

Hsiung, H. M., Mayne, N. G. & Becker, G. W. (1986) High level expression, secretion and folding of human growth hormone in *E. coli. Bio-Technology* **4** 991–995.

Hurle, M. R. & Matthews, C. R. (1987) Proline isomerisation and the slow folding reactions of the α subunit of tryptophan synthase from *Escherichia coli. Biochim. Biophys. Acta* **913** 179–184.

Ikemura, H., Takagi, H. & Inouye, M. (1987) Requirement of pro-sequence for the production of active subtilisin E in *Escherichia coli. J. Biol. Chem.* **262** 7859–7864.

Karplus, M. & Weaver, D. L. (1976) Protein folding dynamics. *Nature* **260** 404–406.

Kiefhaher, T., Grunert, H. P., Hahn, U. & Schmid, F. X. (1990) Replacement of a *cis* proline simplifies the mechanism of ribonuclease T1 folding. *Biochem.* **29** 6475–6480.

King, J. (1989) Deciphering the rules of protein folding. *Chem. Eng. News.* 32–54.

Lang, K., Wrba, A., Krebs, H., Schmid, F. X. & Beintema, J. J. (1986) Folding kinetics of mammalian ribonucleases. *FEBS Letts.* **204** 135–139.

Langsetmo, K., Sung, Y. C., Fuchs, J. & Woodward, C. (1990) *E. coli* thioredoxin stability is greatly enhanced by substitution of aspatic acid 26 by alanine. Villafranca, J. J. (ed.) *Current Research in Protein Chemistry: Techniques, Structure and Function.* Academic Press, London, pp. 449–456.

Lin, L. N., Hasumi, H. & Brandts, J. F. (1988) Catalysis of proline isomerization during protein folding reactions. *Biochim. Biophys. Acta* **956** 256–266.

Liu, G., Topping, T. B. & Randall, L. L. (1989) Physiological role during export for the retardation of folding by the leader peptide of maltose-binding proteins. *Proc. Natl. Acad. Sci. USA* **86** 9213–9217.

McCammon, J. A., Northrup, S. H., Karplus, M. & Levy, R. M. (1980) Helix–coil transitions in a simple polypeptide model. *Biopolymers* **19** 2033–2045.

Marston, F. A. O. (1986) The purification of eukaryotic polypeptides synthesised in *Escherichia coli*. **240** 1–12.

Matouschek, A., Kellis, J. T., Serrano, L., Bycroft, M. & Fersht, A. R. (1990) Transient folding intermediates characterised by protein engineering. *Nature* **346** 440–445.

Matsumura, M., Signor, G. & Matthews, B. W. (1989) Substantial increase of protein stability by multiple disulphide bonds. *Nature* **342** 291–293.

Mitchinson, C. & Wells, J. A. (1989) Protein engineering of disulphide bonds in subtilisin BPN. *Biochem.* **28** 4807–4815.

Miyazawa, S. & Jernigan, R. L. (1982) Most probable intermediates in protein folding–unfolding with a noninteracting globule–coil model. *Biochem.* **21** 5203–5123.

Morlon, J., Lloubes, R., Varenne, S., Chartier, M. & Lazdunski, C. (1983) Complete nucleotide sequence of the structural gene for colicin A, a gene translocated at non-uniform rate. *J. Mol. Biol.* **170** 271–285.

Oas, T. G. & Kim, P. S. (1988) A peptide model of a protein folding intermediate. *Nature* **336** 42–48.

Pace, C. N. (1990) Measuring and increasing protein stability. *TIB TECH* **8** 93–98.

Peters, T. & Davidson, L. K. (1982) The biosynthesis of rat serum albumin. *J. Biol. Chem.* **257** 8847–8853.

Phillips, D. C. (1967) The hen egg-white lysozyme molecule. *Proc. Nat. Acad. Sci. (USA)* **57** 484–495.

Protzel, A. & Morris, A. J. (1974) Gel chromatographic analysis of nascent globin chains. *J. Biol. Chem.* **249** 4594–4600.

Purvis, I. J., Bettany, A. J. E., Santiago, T. C., Coggins, J. R., Duncan, K., Eason, R. & Brown, A. J. P. (1987) The efficiency of folding some proteins is increased by controlled rates of translation *in vivo*. *J. Mol. Biol.* **193** 413–417.

Randall, L. L. & Hardy, S. J. S. (1989) Unity in function in the absence of consensus in sequence: Role of leader peptides in export. *Science* **243** 1156–1159.

Randall, L. L., Josefsson, L. G. & Hardy, S. J. S. (1980) Novel intermediates in the synthesis of maltose-binding protein *Escherichia coli*. *Eur. J. Biochem.* **107** 375–379.

Robinson, A., Westwood, O. M. R. & Austen, B. M. (1990) Interactions of signal peptides with signal-recognition particle. *Biochem. J.* **266** 149–156.

Roth, R. A. & Koshland, M. E. (1981) Role of disulphide interchange enzyme in immunoglobin synthesis. *Biochem.* **20** 6594–6599.

Schomer, R. G., Ellis, L. F. & Schoner, B. E. (1985) Isolation and purification of protein growth *Escherichia coli* cells over producing bovine growth hormone. *Bio-Technology* **3** 151–154.

Shire, S. J., Bock, L., Ogez, J., Builder, S., Kleid, D. & Moore, D. M. (1984) Purification and immunogenicity of Fusion VPI protein of foot and mouth disease virus. *Biochem.* **23** 6474–6480.

Shultz, G. E. & Shirmer, R. H. (1979) *Principles of Protein Structure.* Springer Verlag, Berlin.

Signor, G., Matumura, J. A., Schellman, J. A. & Matthews, B. W. (1990) Engineering of multiple disulphide bonds dramatically stabilizes T4 lysozyme. In: Villafranca, J. J. (ed.) *Current Research in Protein Chemistry: Techniques, Structure and Function.* Academic Press, London, pp. 435–447.

Smith, D. C. & Hider, R. C. (1988) Thiol exchange catalysed refolding of small proteins utilizing solid-phase supports. *Biophys. Chem.* **31** 21–28.

Staley, B. P. & Kim, P. J. (1990) Role of subdomain in the folding of bovine pancreatic trypsin inhibitor. *Nature* **344** 685–688.

States, D. J., Dobson, C. M., Karplus, M. & Creighton, T. E. (1980) A conformational isomer of bovine pancreatic trypsin inhibitor protein produced by refolding. *Nature* **286** 630–632.

Takagi, H., Morinaga, Y., Tsuchiya, M., Ikemura, H. & Inouye, M. (1988) Control of folding of proteins secreted by a high expression secretion vector, p1N-111-*omp* A: 16-fold increase in production of active subtilisin E in *Escherichia coli. Bio-Technology* **6** 948–950.

Takahara, M., Hible, D. W., Barr, P. J., Gerlt, J. A. & Inouye, M. (1985) The *ompA* signal peptide directed secretion of staphylococcal nuclease A by *Escherichia coli. J. Biol. Chem.* **260** 2670–2674.

Takahashi, N., Hayano, T. & Suzuki, M. (1989) Peptidyl-prolyl. *cis–trans* isomerism is the cyclosporin A-binding protein cyclophilin. *Nature* **337** 473–475.

Tropschug, M., Wachter, E., Mayer, S., Schönbrunner, E. R. & Schmid, F. X. (1990) Isolation and sequence of an FK 506-binding protein from *N. crassa* which catalyses protein folding. *Nature* **346** 671–674.

Varenne, S., Knibrehler, M., Cavard, D., Marlon, J. & Luzdunski, C. (1982) Variable rate of polypeptide chain elongation for Colicins A, E2 and E3. *J. Mol. Biol.* **159** 57–70.

West, S. M., Kelly, S. M. & Price, N. C. (1990) The unfolding and attempted refolding of citrate synthase from pig heart. *Biochim. Biophys. Acta* **1037** 332–336.

Wickner, W. (1989) Secretion and membrane assembly. *Trends Biochem. Sci.* **14** 280–283.

Wold, F. (1983) In: *Post-Translational Covalent Modification of Proteins for Function* (Johnson, C. B. ed.) Academic Press, London pp. 1–17.

Wright, P. E., Dyson, H. J. & Lerner, R. A. (1988) Conformation of peptide fragments of proteins in aqueous solution: implications for initiation of protein folding. *Biochem.* **27** 7167–7175.

Yan, S. C. B., Grinnell, B. W. & Wold, F. (1989) Post-translational modifications of proteins. *Trends in Biochem. Sci.* **14** 264–268.

3

The current status of therapeutic peptides and proteins

A. F. Bristow
Division of Endocrinology, National Institute for Biological Standards and Control, Blanche Lane, South Mimms, Potters Bar, Herts, EN6 3QG

3.1 HISTORICAL DEVELOPMENT OF POLYPEPTIDE DRUGS

During the last hundred years or so, the parallel developments in medicine and biochemistry, and latterly in biotechnology, have resulted in a continuous evolution in therapeutic use of polypeptides, leading to increases in availability, quality and applications. The major steps in the development of pharmacologically active polypeptides are shown in Scheme 3.1. Many of the drugs still in widespread use today were originally identified as biologically active principles in tissue extracts. Insulin is the obvious example, indeed the first insulin preparations used to treat diabetes in the 1920s were crude extracts of dog pancreas. As the science of biochemistry developed through the 1950s and 1960s these biologically active principles were increasingly purified and defined as specific fractions, then as chromatographic bands or peaks and finally as primary structures (amino acid sequences) and in some cases crystal structures. Therapeutic preparations evolved along similar lines. Insulin for instance was initially purified by solvent extraction and zinc crystallization. During the 1970s process-scale chromatography was introduced leading to the so-called 'monocomponent' insulins with a purity of greater than 95%. In recent times two developments have had a major impact on therapeutic applications of pharmacologically active polypeptides. On the one hand peptides up to around 30 amino acids in length can now be manufactured by large-scale chemical synthesis (see Chapter 4). This has eliminated the need for processing biological tissue with all the inherent variability involved, and, more significantly, has allowed the design and manufacture of peptide analogues with super-potent or antagonistic properties. On the other hand, the development of recombinant DNA technology has resulted in the widespread availability of proteins and glycoproteins with virtually no limit on size or degree and nature of glycosylation (see Chapter 5). As a

PHARMACOLOGICALLY ACTIVE POLYPEPTIDES

Crude tissue extracts, e.g. insulin, oxytocin

Identification of active principle, purification of specific proteins from tissue

Chemical synthesis leading to availability of peptide drugs from non-biological sources	Recombinant DNA technology leading to widespread availability of proteins and glycoproteins
Structure–activity studies leading to peptide analogues with antagonistic or superpotency properties	Design of hybrid or mutated proteins
	Targeting
ANTIBODIES	**VACCINES**
Unfractionated sera (animal/human) anti-toxins anti-bacterial/viral anti-lymphocytic	Crude vaccines (related strains, heat-killed organisms)
	Attenuated strains
Purified immunoglobulins animal/human	Controlled tissue culture
Monoclonal antibodies murine	Acellular vaccines
Human monoclonals produced in heterohybridomas	Peptide vaccines (?)
Targeted drugs and diagnostics	rDNA vaccines
Humanized monoclonal antibodies produced in recombinant bacteria	
	Recombinant strains (?)

Scheme 3.1 — Development of peptide and protein pharmaceuticals.

result the number of proteins currently undergoing evaluation for potential clinical application is very large (Blohm *et al.* 1988). Many of these potential applications are discussed in this book.

Two other major classes of peptide and protein pharmaceuticals, namely antibodies (see Chapter 10) and vaccines (see Chapter 11), have undergone similar technology-related evolution. Antibodies were initially used in unfractionated sera, later becoming purified immunologlobin fractions (Scheme 3.1). Subsequently, monoclonal antibody technology gave rise to a large number of possibilities including targeting of drugs and toxins, the use of labelled antibodies for *in vivo* diagnostic purposes and using recombinant DNA technology, the production of 'humanized' antibodies in bacteria or other recombinant DNA host–vector systems. Developments in vaccine technology have seen the introduction of attenuated strains and the use of tissue culture, the production of acellular vaccines (e.g. pertussis vaccine) and production of immunogens by recombinant DNA technology (e.g. hepatitis B vaccine) (Scheme 3.1). There is currently much interest in precise definition of immunity-conferring peptide epitopes and in the engineering of recombinant strains for vaccination, for example vaccinia–HIV recombinants as AIDS vaccines.

In considering the current status of polypeptide pharmaceuticals, both the well-established drugs and also some examples which have resulted from the application of newer technologies will be discussed.

3.2 ESTABLISHED THERAPY USING POLYPEPTIDE PHARMACEUTICALS

The polypeptide drugs covered by the 1988 British Pharmacopoeia, together with examples of more recently licenced products such as erythropoietin, growth hormone, gonadorelin analogues and tissue plasminogen activator (tPA) are listed in Table 3.1. Polypeptides are administered in a variety of clinical situations. In replacement therepy they may be given long-term or for life and the patients' survival may depend on their administration. Thus insulin is given to regulate blood sugar levels in diabetics, clotting factors VIII and IX to control bleeding in haemophiliacs and erythropoietin to stimulate erythrocyte development in patients with renal failure, replacing the hormone which is normally produced in the kidney. Growth hormone (GH), a pituitary hormone, is administered between infancy and adolescence to children with GH insufficiency, to enable them to reach normal height. There are numerous examples of peptide and protein hormones which are given for

Table 3.1 — Therapeutic polypeptides: applications

Lifetime or long-term replacement therapy	Insulin, growth hormone, factor VIII, factor IX, erythropoietin
Acute or long-term stimulation of endocrine axes	Calcitonin, oxytocin, corticotrophin, desmopressin, vasopressin, glucagon, tetracosactide, gondadotrophins
Anti-tumour	Sandostatin, interferons, antibodies, gonadorelin analogues
Thrombolytic therapy	Tissue plasminogen activator (tPA), urokinase, streptokinase
Blood volume replacement	Albumin, plasma proteins
Active immunization	Vaccines
Passive immunization	Immunoglobulins, monoclonal antibodies, antitoxins
Miscellaneous	Aprotinin, trypsin, pepsin, hyaluronidase

acute or long-term stimulation of the endocrine axis. These include calcitonin to control osteoporosis, vasopressin and desmopressin as antidiuretics and the gonadotrophins for simulation of ovarian follicles either to induce ovulation for *in vivo* or *in vitro* fertilization programmes. The peptide hormone oxytocin is widely used to promote uterine contractions in labour. A number of polypeptides, including those listed in Table 3.1, have known or potential antitumour activity. Urokinase and streptokinase are established thrombolytic agents. Tissue plasminogen activator produced by recombinant DNA technology is now more or less an established therapeutic agent, although there is still considerable debate as to whether it offers a significant clinical advance over the other cheaper thrombolytic agents. The list of vaccines is of course very large. The 1988 British Pharmacopoeia contains monographs for 28 different bacterial and viral vaccines, and does not include the more recent recombinant-DNA derived hepatitis B vaccines. There are also six human immunoglobulin preparations and nine animal immunoglobulins in the Pharmacopoeia, together with new non-pharmacopoeial preparations such as antidigoxin. Immunological products therefore, for active and passive immunization, comprise a large part of the total use of polypeptides in therapy.

As well as exhibiting a wide range of clinical applications, therapeutic peptides and proteins exhibit a wide range of structure (Table 3.2). They range in size from a few amino acids (e.g. Sandostatin, 8 amino acids) to 330 000 kDa (Factor VIII). The list incorporates molecules with a range of known features of protein structure, including intra-chain disulphide bonds (e.g. oxytoxin) inter-chain disulphide bonds (e.g. insulin and antibodies), multi-subunit, non-covalently stabilized proteins (e.g. gonadotrophins), glycosylation, intra-chain cleavages (tissue plasminogen activator, urokinase) and non-naturally occurring structures such as D-amino acids and amino acid analogues (e.g. desmopressin).

Reflecting the wide range of protein structures used therapeutically is the diversity of sources of these materials (Table 3.3). Smaller peptides (up to 33 amino acids) are produced by chemical synthesis. Larger proteins may be purified from human blood or tissue, animal tissue, bacterial culture, tissue culture or culture of recombinant bacteria or cells. This situation, however, is rapidly evolving and we may reasonably expect to see a decline in the use of animal and human tissues as starting materials, and the manufacture of products such as albumin, Factor VIII, gonadotrophins and insulin by recombinant DNA technology. The various levels of purity required for these drugs tends to reflect the date of their licensing and introduction into therapy. Thus products of recombinant DNA technology utilize modern technology in manufacture and even large, complex molecules are expected to be extensively purified, usually to homogeniety and free from DNA, host-cell proteins and animal viruses. Older products such as animal immunoglobulins, urokinase and streptokinase have less stringent requirements.

3.3 RECENT DEVELOPMENTS IN PEPTIDE AND PROTEIN THERAPY

In order to illustrate potential applications of this new technology, the use of synthetic peptide analogues with modified or 'enhanced' activity, the use of cytokines in regulation of haematopoiesis, and the use of growth factors to promote wound healing will be discussed.

Table 3.2 — Therapeutic peptides: characteristics

Drug substance	Size	Other characteristics
Sandostatin	8 amino acids	
Oxytocin	9 amino acids	Cyclic nonapeptides
Vasopressin	9 amino acids	Cyclic nonapeptides
Desmopressin	9 amino acids	Cyclic nonapeptides
Gonadorelin	10 amino acids	
Tetracosactide	24 amino acids	
Glucagon	29 amino acids	
Calcitonin	33 amino acids	
Corticotrophin	39 amino acids	
Protamine	Various	Highly basic
Insulin	51 amino acids	Two-chain structure
Aprotinin	6 kDa	
Interferons	15–30 kDa	Various structures, some glycosylated
Growth hormone	22 kDa	
Trypsin	24 kDa	
Gonadotrophins	30 kDa	Two subunits, glycosylated
Erythropoietin	34 kDa	Heavily glycosylated
Pepsin	35 kDa	
Streptokinase	46 kDa	
Urokinase	54 kDa	A 33-kDa variant exists
Factor IX	60 kDa	
Albumin	67 kDa	
Tissue plasminogen activator	68 kDa	One or two chains, glycosylated
Immunoglobulins (Mabs, antitoxins)	150 kDa	Four-chain structure
Vaccines	Various (usually high mol. wt)	
Factor VIII	330 kDa	Glycosylated

3.3.1 Pharmaceutical peptide analogues

Examples of modification of the primary structure of peptide hormones are given in Table 3.4. Arginine vasopressin (AVP) is a cyclic nonapeptide with a rather short-lived antidiuretic effect in man. The analogue 1-desamido-8-D-arginine vasopressin (ddAVP, desmopressin) is 1200 times more potent than the parent compound in its antidiuretic activity and is used in the treatment of severe diabetes insipidus. Indeed the clinical dose of desmopressin (10–20 μg intranasally, twice daily) make it one of the most potent of all pharmacologically active agents in man. Somatostatin is a hypothalamic 14-amino-acid peptide which was originally identified as an inhibitor of pituitary growth hormone secretion, although is now known to inhibit a number of

Table 3.3 — Therapeutic polypeptides: source and purity

Drug	Source	Comments
Desmopressin, gonadorelin, oxytocin, calcitonin	Chemical synthesis	All substantially pure (at least 90%, often higher)
Albumin, factor VIII, factor IX, immunoglobulins, plasma proteins	Human blood	May be less than 10% to more than 95% pure. Free of agents of infection
Gonadotrophins, urokinase	Human urine or placenta	Relatively crude preparations
Aprotinin, corticotrophin, antitoxins, glucagon, insulin, hyaluronidase, trypsin, protamine	Animal tissue	Generally quite impure (e.g. corticotrophin, antitoxins). Insulin is greater than 98% pure
Bacitracin, streptokinase	Bacteria	Partially purified, heterogeneous
Insulin, growth hormone, vaccines, erythropoietin, interferons, tissue plasminogen activator	rDNA technology	Purified by several chromatography steps. Above 90% active principle (often higher). Free from DNA, viruses and host cell proteins
Interferon, monoclonal antibodies, vaccines	Tissue culture	Similar requirement to rDNA technology may apply

endocrinological axes. Like AVP, it has a short half-life which restricts its clinical usefulness. The superpotent long-acting analogue Sandostatin is clinically useful in the symptomatic treatment of acromegaly (caused by growth hormone secreting pituitary tumours) and carcinoid syndrome. A particularly intriguing example of the use of peptide analogues is the superpotent analogues of gonadorelin (gonadotrophin releasing hormone, GnRH). GnRH is a hypothalamic decapeptide whose normal mode of action is to stimulate pituitary gonadotrophin secretion, which in turn stimulates the production of sex steroids from the gonads (Scheme 3.2). Many prostatic cancers depend on testosterone for their growth and in advanced prostatic cancer, castration has often been used to limit the availability of testosterone to the tumour, thus slowing its growth. The use of superpotent GnRH analogues such as Buserelin (Table 3.4) in this situation depends on the observation that GnRH is secreted in a pulsatile manner, and that the pituitary gland requires this pulsatility to respond normally. Continuous application of a superpotent analogue causes receptor down-regulation and desensitization of the pituitary causing the paradoxical effect of inhibiting gonadal steroid production. In a recent review, Vickery (1986) identified 11 analogues of GnRH in clinical use or undergoing evaluation, including true antagonists as well as superpotent agonists. (See Chapter 16).

Table 3.4 — Pharmaceutical peptide analogues

Parent molecule

Cys. Tyr. Phe. Gln. Asn. Cys. Pro. Arg. Gly. NH_2

Vasopressin

Analogue (Superpotency, prolonged action)

S. $(CH_2)_2$. CO. Tyr. Phe. Gln. Asn. Cys. Pro. D-Arg. Gly. NH_2

Desmopressin (Antiduretic)

Parent molecule

Glp. His. Trp. Ser. Tyr. Gly. Leu. Arg. Pro. Arg. NH_2

Gonadorelin

Analogue (Superpotency)

Glp. His. Trp. Ser. Tyr. D-Ser (tBu). Leu. Arg. Pro. NEt

Buserelin (Treatment of prostatic cancer)

Parent molecule

Ala. Glu. Cys. Lys. Asn. Phe. Phe. Trp. Lys. Thr. Phe. Ser. Cys.

Somatostatin

Analogue (Superpotency, long-action)

D-Phe. Cys. Phe. D-Trp. Lys. Thr. Cys. Thr (ol)

Sandostatin (Treatment of acromegaly and carcinoid syndrome)

3.3.2 The use of cytokines in regulation of haematopoiesis

A large number of factors regulating differentiation, growth and function of cells of the erythroid, myeloid and lymphoid lineages have now been identified and are generically referred to as cytokines (Fig. 3.1) (Table 3.5). A detailed account of the biology of cytokines is beyond the scope of this book and the interested reader is referred to an overview of Boxmeyer and Vadhan (1989). Cytokines are being evaluated for clinical potential in a number of situations, one in particular being in the regulation of haematopoiesis. This process, schematically represented in Scheme 3.3, is essentially one of progressive commitment as pluripotent stem cells differentiate first into either erythroid, myeloid or lymphoid lineages, and then progressively into the various cell types. This progress is thought to be regulated by specific haematopoietic growth factors, acting at early, intermediate and late stages, as

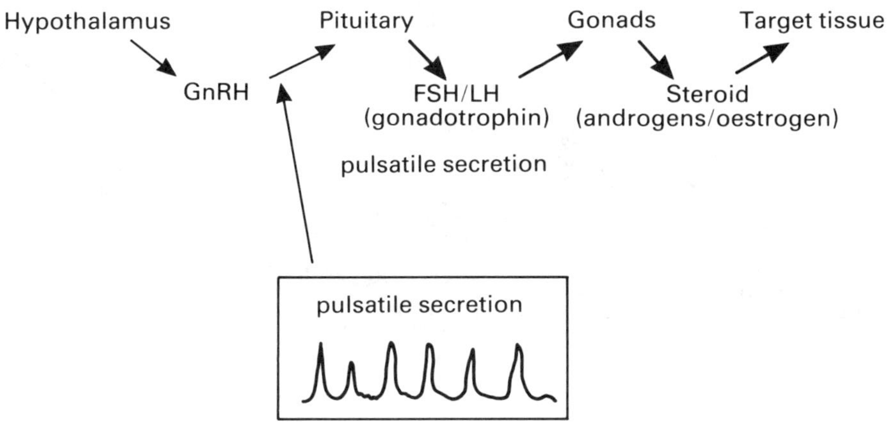

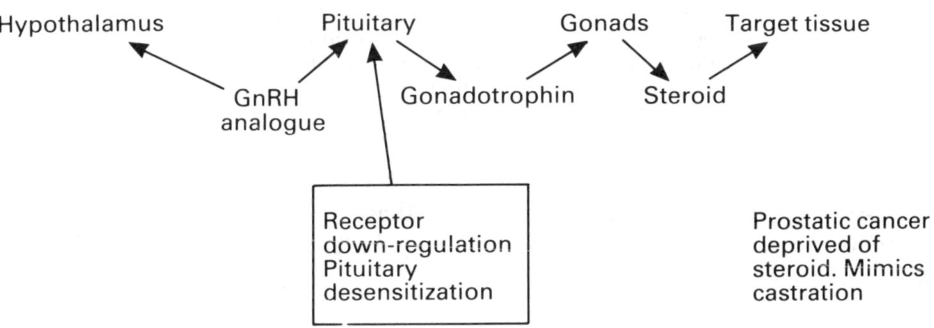

Scheme 3.2 — Clinical application of gonadotrophin-releasing hormone analogues.

illustrated. With the widespread availability of cytokines prepared by recombinant DNA technology, many cytokines are being evaluated in various disorders of haematopoiesis (Table 3.6). It is hoped that they may be of use as proliferative agents on various natural and induced cytopaenias and, in conjunction with chemotherapy, as promoters of differentiation in treatment of undifferentiated leukaemias, and in stimulation of end cell activity (e.g. phagocyte killing activity). While this area

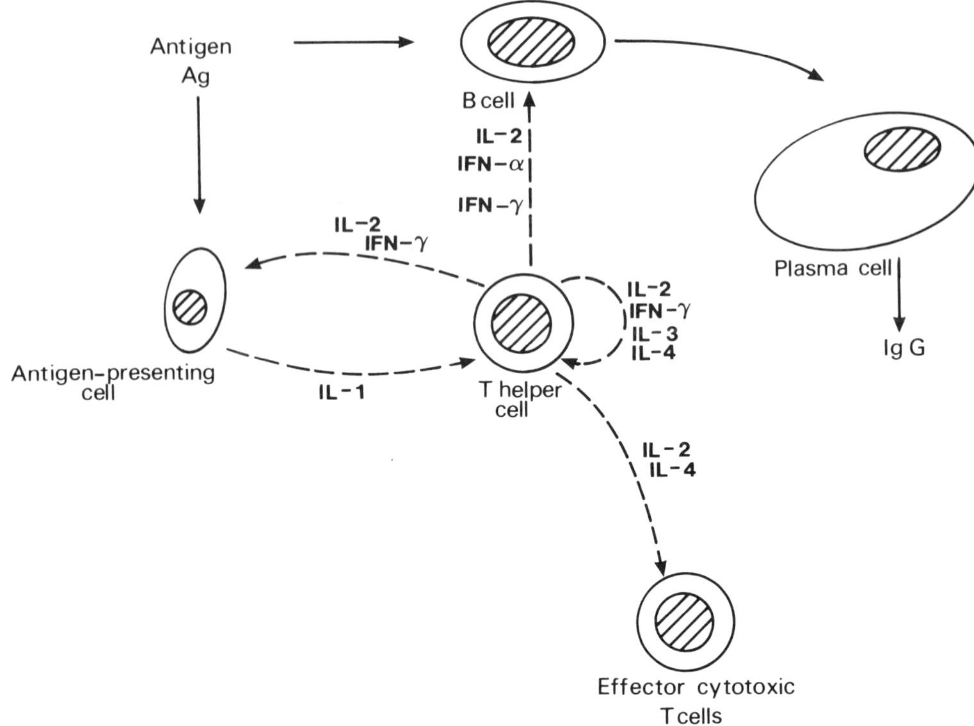

Fig. 3.1 — Involvement of cytokines in the immune response. Stimulation of B-lymphocytes by presentation by antigen-presenting cells of the antigen (Ag) to T-helper lymphocytes. T-helper cells assist in the differentiation of both B-lymphocytes to antibody-secreting plasma cells and effector cytotoxic T-lymphocytes, which have responded to foreign cell antigens. Production of soluble mediators is indicated by dashed arrows.

possesses enormous potential, it is true to say that only in the case of erythropoietin, which has been successfully used to treat anaemia associated with renal failure, has this potential been fully realized.

3.3.3 Use of growth factors in wound healing

A number of connective tissue growth factors have now been identified, cloned and are available in quantity (Table 3.7). Wound regeneration is of great clinical importance in many areas; for instance burns, surgery, ulcers, skin grafts, bone fractures, osteoporosis, and psoriasis. The application of growth factors in this process is an area of considerable activity. The schematic illustration of the wound healing process (Scheme 3.4) illustrates the roles of the various growth factors in this process, and also the reasons for the intense interest. It is noteworthy that transforming growth factor B, epidermal growth factor, platelet-derived growth factor, and fibroblast growth factor have all successfully accelerated healing in animal models, and epidermal growth factor is in clinical trial for wound healing indications. The interested reader is referred to the review by ten Dijke and Iwate (1989).

Table 3.5 — Cytokines

	Mol. wt. (kD)	Glyco-sylated	Biological actions
Interleukin-1	17	−	Pluripotent. Early factor via stromal cell cytokines, lymphocyte activation, inflammation
Interleukin-2	14–16	+	T-cell and B-cell proliferation, NK cell proliferation, macrophage cytotoxicity
Interleukin-3	14.6	+	Early cell growth and differentiation
Interleukin-4	20	+	T-cell growth. Class II MHC production and IgG, IgE production in B-cells
Interleukin-5	50	+	B-cell growth factor
Interleukin-6	22–29	+	Stem-cell growth. B-cell and T-cell function. Hybridoma growth factor
Interleukin-7	25	+	Proliferation of thymocytes, pre-B-cells, T-cells
Interleukin-8	10	−	Neutrophil, T-cell chemotactic factor
Interferon-γ	20–25	+	Class I and II MHC on B-cells macrophages
Tumour necrosis factor (TNF)-α	17	−	Pluripotent. T-cell+B-cell growth factor, cachexia
Granulocyte colony stimulating factor (GCSF)	19.6	+	Growth and differentiation of granulocyte progenitors
Granulocyte macro phase colony stimulating factor (GMCSF)	22	+	Growth factor for granulocytes and macrophages, stimulation of granulocyte and macrophage function
Macrophage colony stimulating factor (MCSF)	40–45	+	Macrophage proliferation, differentiation and cytotoxicity
Transforming growth factor-β (TGF-β)	25	−	Inhibition of T-cell and B-cell proliferation, macrophage and neutrophil function

Stem cells	**Progressively committed cell types**	**End cells**
pluripotent stem cell	myeloblast, neutrophilic juvenile proerythrocyte reticulocyte lymphoblast–prolymphocyte monoblast–promonocyte	neutrophil erythrocyte lymphocyte macrophage
Interleukin-1 **Interleukin-6**	**Interleukin-3** **GM-CSF**	**Erythropoietin** **G-CSF** **M-CSF**
Examples of factors acting directly or indirectly on stem cells	*Examples of intermediate factors*	*Examples of late factors regulating terminal differentiation*

Scheme 3.3 — Growth factors and the control of haematopoiesis.

Table 3.6 — Haematopoietic growth factors: potential clinical applications

1. Stimulation of cell proliferation
   naturally occurring cytopaenia e.g. renal failure — associated anaemia, congenital neutropaenia
   chemotherapy- or radiotherapy-induced cytopaenia
   chemotherapy dose escalation
   enhancement of cytoxic drug sensitivity
2. Induction of cell differentiation
   treatment of leukaemia
3. Stimulation of end cell activity
   increase killing activity of phagocytes

Table 3.7 — Growth factors

Epidermal growth factor (EGF)		5.7 kD
Transforming growth factor-α (TGF-α)		5.7 kD
Platelet-derived growth factor (PDGF)		30 kD
Transforming growth factor-β (TGF-β)		25 kD
Fibrobast growth factor	basic FGF	16.4 kD
	acidic FGF	15.6 kD

	Growth factors
Wounding Haemostatsis (platelet aggregation, clotting)	Release of TGF-β, PDGF and FGF subsequent inhibition of PDGF release by prostaglandin
Inflammatory phase	
Cell recruitment (e.g. monocytes for debridement; neutrophils for infection resistance)	
	Macrophages release TGF-β, PDGF Chemoattractant for monocytes and fibroblasts
Vascularization Cell proliferation Epithelialization	Suppression of macrophage activity (TGF-β)
Granulation, Tissue formation	TGF-β stimulates matrix gene expression (e.g. fibronectin) TGF-β: fibroblastic proliferation, migration, differentiation into myofibroblasts FGF: mitrogen for vascular endothelial cells Stimulation of PA
Wound closure, Extracellular matrix formation Remodelling	FGF: stimulation of extracellular matrix TGF-β: indirect angiogenesis vai chemoattracted macrophages TGF-β: syhthesis of new collagen
Healed wound	PDGF: degradation of old collagen

Scheme 3.4 — Wound healing and growth factors.

3.4 CONTROL OF POLYPEPTIDE PHARMACEUTICALS: CURRENT STATUS

Polypeptide pharmaceuticals, like all others, are required to satisfy criteria of identity, potency, purity, efficacy and safety. The way in which these criteria are being met is, in many cases, undergoing revision as a result of the rapid developments in both production and analytical methodology.

3.4.1 Identity and potency

Traditionally pharmaceutical polypeptides have been defined as a particular biological activity. Thus for instance insulin is still defined as 'the anti-diabetic principle of the mammalian pancreas'. The main, in many cases the only, analytical criterion of identity was some form of bioassay. With the advent of recombinant DNA technology this situation has changed. Almost without exception the physicochemical structure of a protein is known prior to its administration to man, and an increasingly greater emphasis is being placed on physicochemical criteria of identity. During the last decade there has been a dramatic increase in the number of high resolution techniques available to peptide and protein chemists. With the availability of reverse-phase and ion-exchange high-performance liquid chromatography, fast-atom bombardment mass spectrometry and high-field NMR, analysts dealing with peptides up to about 5000–6000 in molecular weight can probably resolve and quantify virtually any change in primary structure. Although at the present time biological activity remains an essential criterion of identity for most proteins, its obligatory role is being increasingly questioned, and has already been relaxed for many smaller peptides. In a similar way, quantitative determination of activity (potency) has always depended on a biological assay in comparison with a standard. However, the increasing sophistication and utility of analytical methodology now presents real alternative procedures for the determination of biologically active protein content for many quite large molecules. Indeed, current developments such as laser desorption and electrospray mass spectrometry (Barinaga 1989) combined with peptide mapping techniques seem likely to extend the analyst's working molecular weight range above 100 000.

3.4.2 Purity

Purity requirements for polypeptide pharmaceuticals have, to a large extent, been technology-driven. Whilst many of the potential contaminants in protein preparations, and the potential consequences of impurity, set out in Scheme 3.5, have long been recognized, it is only recent developments in production and analytical methodology that have made possible their reduction to acceptable levels. Thus when products such as insulin and oxytoxin were first used in man, there was virtually no purity requirement other than a minimum specific activity. Indeed, as recently as 1980, many animal insulins prepared for use in man contained less than 40% insulin, the remainder being the various chemically degraded forms. It is also noteworthy that for all that period of administration of highly impure insulins to man, there were few, if any, clinical sequelae directly attributable to the impurities. Nonetheless, a modern recombinant DNA-derived product would almost invariably be expected to comply with the purity considerations set out in Scheme 3.5.

The product would be expected to contain very low levels of chemically modified forms, possibly to less than 5%, not more than 10 parts per million of host or vector proteins, determined by sensitive immunochemical techniques, not more than 100 pg per therapeutic dose of DNA, determined by hybridization analysis, and to be free of viruses or other agents of infection. The interested reader is referred to recent surveys by the Committee for Proprietary Medicinal Products (1989a) and Anicetti *et al.* (1989) for further consideration of purity testing or recombinant DNA polypeptides.

3.4.3 Safety and efficacy

Evaluation of safety and efficacy in animal models of protein pharmaceutials presents a number of problems which are covered in detail elsewhere in this book. In particular these problems include the structural differences between human and animal forms of proteins which render the human form immunogenic in animals, limiting the usefulness of chronic toxicity testing, and the different biological

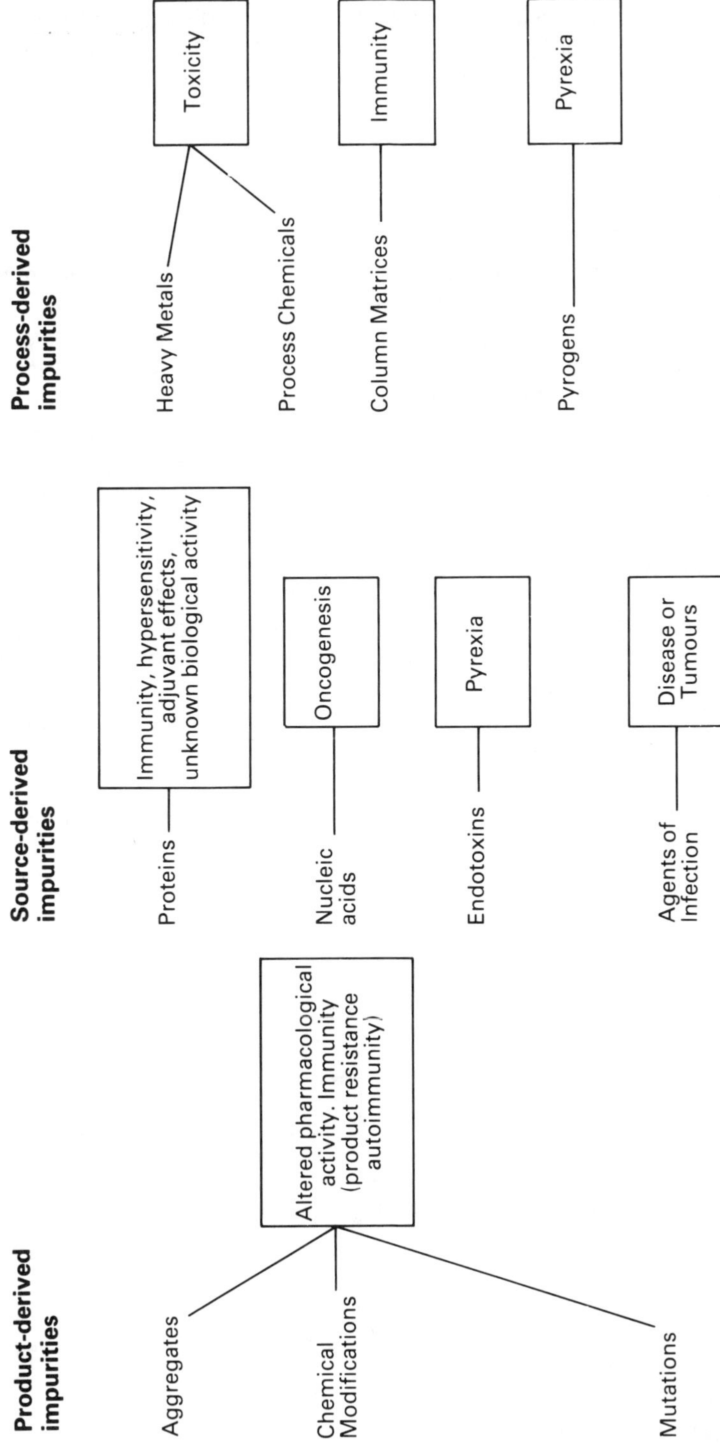

Scheme 3.5 — Polypeptide pharmaceuticals: impurities and potential chemical consequences.

activities of human proteins in animals and which make it difficult to predict human toxicity in animal models and also efficacy in animal models of the disease state (Graham 1987, Committee for Proprietary Medicinal Products 1989b). See Chapter 8.

3.5 FUTURE TRENDS IN THE USE OF POLYPEPTIDE PHARMACEUTICALS

It is not difficult to predict that the current interest in the development and application of polypeptide drugs will continue. The ability to produce in bulk virtually any peptide, protein or glycoprotein that is required has presented the pharmacologist with a potentially unlimited number of drugs to study. It is, however, worth noting that a number of factors will inhibit the development of such pharmaceuticals unless they are overcome. For example, it seems possible that simple parenteral administration will not be appropriate for many of the proteins currently being evaluated. Many of the earlier-used protein drugs were hormones or endocrine in their action, which by definition work by transport in the blood-stream to their site of action. Many polypeptides currently being evaluated such as cytokines and growth factors are probably paracrine or autocrine in action, and act locally, regulating the activity of different cell types or even of the same cell type. Thus, it may not be possible to mimic this situation by simple parenteral administration. Indeed, there is evidence that some growth factors such as insulin-like growth factor-I are actually rendered inactive in the blood. There is also the problem that many cytokines and growth factors have wide-ranging biological activities depending on the target tissue. The activities of tumour necrosis factor for instance (Table 3.8) are extremely diverse, and it must be likely that such compounds will be of limited clinical use unless they can be directed to the intended site of action.

Table 3.8 — Some biological actions of tumour necrosis factor

Tumour cell cytotoxicity (*in vivo* and *in vitro*)
Tumour necrosis (*in vivo*)
Inhibition of lipogenesis (*in vitro*)
Hyperproliferation of normal cells (*in vitro*)
Activation of neutrophils (*in vitro*)
Activation of macrophages (*in vivo*)
Enhanced immune cell/endothelium adhesion (*in vitro*)
Stimulation of arachidonic acid metabolism (*in vivo* and *in vitro*)
Suppression of haematopoiesis (*in vitro*)
Calcium resorption (*in vitro*)
Induction of other cytokines (e.g. haemotopoietic colony stimulating factors)
Antiviral activity

Modification of the structure of a peptide may enhance activity, and in theory the same approach is feasible in proteins. However, for proteins with molecular weights above 2–3 kDa, which is the approximate threshold for immunogenicity, even minor modifications will probability result in the protein being immunogenic in man. This is currently a limiting factor in the design of most novel protein drugs.

Finally, one has to consider the cost of protein drugs produced by biotechnology. For instance, it seems that erythropoietin is a highly successful drug in treating the anaemia associated with renal failure. Like all recombinant-DNA-derived drugs however, it is not cheap. The current cost of maintaining one patient is of the order of £5000 per year and it is clear that its widespread introduction would place strain on health-care budgets no matter what the system of funding. To develop a product of biotechnology and bring it to market is a high-cost venture, and one that drug companies are unlikely to undertake willingly if there is little prospect of the widespread use of a successful product for financial reasons. Indeed, problems of cost may well represent the most serious obstacle to the impact of biotechnology on therapy using peptides and proteins.

REFERENCES

Anicetti, V. R., Keyt, B. A. & Hancock, W. S. (1989) Purity analysis of protein pharmaceuticals produced by recombinant DNA technology. *Trends in Biotechnology* **7** 342–349.

Barinaga, M. (1989) Protein chemists gain a new analytical tool. *Science* **246** 32–33.

Blohm, D., Bollschweiler, C. & Hillen, H. (1988) Pharmaceutical proteins. *Angewandte Chemie* **27** 207–225.

Boxmeyer, H. E. & Vadhan, Raj, S. (1989) Preclinical and clinical studies with the haematopoietic colony stimulating factors and related interleukins. *Immunol Res* **8** 185–201.

British Pharmacopoeia (1988) HMSO Books.

Committee for Proprietary Medical Products: Ad Hoc Working Party on Biotechnology/Pharmacy (1989a) Notes to applicants for marketing authorizations on the production and quality control of medicinal products derived by recombinant DNA technology. *J. Biol. Stand.* **17** 223–231.

Committee for Proprietary Medicinal Products: Ad Hoc Working Party on Biotechnology/Pharmacy (1989b) Notes to applicants for marketing authorizations on the pre-clinical biological safety of medicinal products derived from biotechnology (and comparable products derived from chemical synthesis). *J. Biol. Stand.* **17** 203–212.

Graham, C. E. (Ed) (1987) *Preclinical safety of biotechnology products intended for human use*. A. R. Liss Inc., New York.

ten Dijke, P. & Iwata, K. K. (1989) Growth factors for wound healing. *Biotechnology* **7** 793–798.

Vickery, B. H. (1986) Comparison of the potential for therapeutic utilities with gonadotrophin-releasing hormone agonists and antagonists. *Endocrine Reviews* **7** 115–124.

4

Current chemical strategies in peptide synthesis

Eric Atherton
Cambridge Research Biochemicals, Gadbrook Park, Northwich, Cheshire CW9 7RA

4.1 INTRODUCTION

Over recent years peptides have become increasingly important in such disciplines as molecular biology, immunology and pharmacology. They have been intensively investigated for their clinical potential in the treatment of diseases ranging from growth disorders to AIDS, and are already demonstrated to be commercially viable in such diverse applications as artifical sweeteners (e.g. Aspartame) and anti-cancer agents (e.g. Zoladex, see Chapter 16). Market surveys of the current activity in peptide research suggest that as many as 125 peptide-based products are in some phase of pre-registration development.

As a consequence of this upsurge in interest, a great deal of attention has been focused on the methods available for synthesizing these molecules. Since the first successful synthesis of the peptide hormone oxytocin using classical solution phase methods (du Vigenaud *et al.* 1954), thousands of peptides have been made employing a variety of different techniques. Although the current arsenal of synthetic methods is quite considerable, even the most innocous-looking sequences can still prove troublesome to synthesize. The challenge of peptide synthesis remains constant, and therefore, there is still widespread research into new synthetic methods.

The purpose of this chapter is to review methods available for the chemical synthesis of peptides and to highlight current popular techniques, commenting on their scope and limitations. Spatial restrictions demand that the material presented is selective and reflects a personal viewpoint. It is not within the scope of this article to enter into detailed chemistries regarding the particular methods, but for further information the reader is referred to the excellent series by Gross *et al.* (1979) and to other references cited below.

4.2 METHODS OF PEPTIDE/PROTEIN SYNTHESIS

There are several strategies available for the synthesis of peptides and proteins. They include:

(a) Classical solution-based methods
(b) Solid-phase methods
(c) Enzymatic methods
(d) Recombinant techniques
(e) Semi-synthetic methods.

These different routes are not general for every synthetic objective and the choice of method depends upon the target molecule under investigation. Each of the methods has its advantages and limitations (see for example, Marshak & Liu 1988) and any particular method should be considered judicuously before embarking on a synthesis. In general it may be noted that:

(1) The recombinant, enzymatic and semi-synthetic methods are not *directly* applicable to peptides, so that there is groundwork required in optimizing the synthesis before the best route can be established.
(2) Recombinant techniques are mainly used for large polypeptides and proteins and it seems that none of the other methods will be able to compete routinely in this sphere even in the longer term.
(3) Semi-synthetic methods are likely to be particularly useful in the synthesis of peptides and proteins which incorporate non-natural amino acids or carbohydrate moieties.
(4) Enzymatic methods are very efficient, but require extensive optimization to establish synthetic routes. Once these are established, however, this approach becomes very competitive and the synthesis can readily be scaled up.

The solution-based and solid phase peptide syntheses are more ubiquitous and are discussed in some detail below.

With both types of chemical synthesis, it is necessary that the component amino acids are chemically protected prior to any peptide bond formation: the carboxyl component with an amino protecting group, and the amino component with a carboxyl protecting group. In addition, to allow an unambiguous chemical reaction, it is also necessary to protect any reactive side-chains that may interfere with peptide bond formation (see Fig. 4.1).

Activation of the free carboxyl group, to effect peptide bond formation, can then be achieved through the use of a coupling agent such as dicyclohexylcarbodiimide, or through preformation of an active (e.g. pentafluorophenyl) ester.

In all of the chemical operations involved in the preparation of peptides, it is important that the correct chirality of the amino acids is preserved (see Chapter 1).

The synthesis of peptides can be made using a stepwise addition of single residues or by fragment condensation. The former involves addition of suitably protected and activated amino acids to a growing chain, proceeding almost universally (to avoid problems of racemization) from the carboxyl to the amino terminus. The fragment

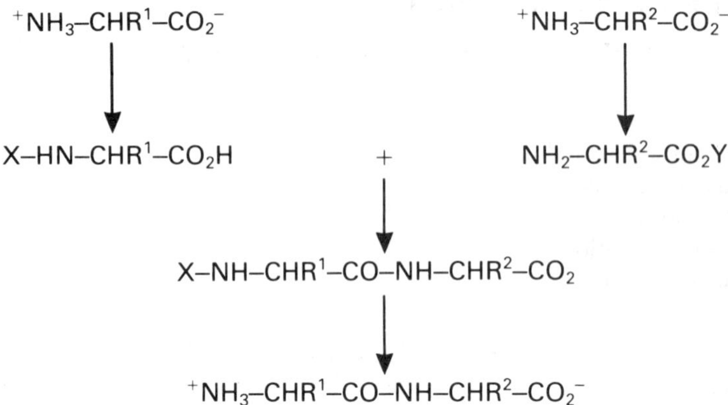

Fig. 4.1 — Formation of a dipeptide. X and Y represent protecting groups.

approach involves the synthesis of judiciously chosen, protected segments of a peptide, which are then coupled together to make the target sequence (for details see Bodansky 1984).

Solution-based methods offer the chemist the attractive opportunity to characterize the compounds formed after each individual chemical reaction. Checking by several different analytical techniques, e.g. NMR, UV, micro-analysis, etc., gives assurance that a synthesis is progressing smoothly. However, these operations are laborious and time-consuming and require the slowly acquired skills of experienced peptide chemists. In contrast, the solid phase technique (invented by Merrifield 1962) is a rapid, cyclic process which lends itself to automation. Although the products of individual chemical reactions cannot be fully characterized, in careful hands the method can produce highly pure peptides.

The difference in effort required between the two methods is aptly illustrated by two early reports describing syntheses of ribonuclease: the one, based on a solution phase approach, was detailed in a series of communications, and involved the efforts of over twenty people (Denekelwalter *et al.* 1969); the other, based on the solid phase procedure was fully described in a single communication, and involved only two authors (Gutte and Merrifield 1969). As a consequence of these differences in effort, the solid phase methodology has become the more widely adopted, and many fully automated commercial instruments are available which utilize this approach.

4.3 SOLID PHASE PEPTIDES SYNTHESIS

4.3.1 Conventional methods

The original Merrifield method employs the use of an apolar solid support (beaded polystyrene), a protecting group strategy based on the mildly acid labile *t*-butyloxycarbonyl group (Boc) for α-amino protection, benzyl-based side-chain protection, a benzyl ester linkage to the resin, and *in situ* activation with dicyclohexylcarbodiimide. At the end of an assembly, treatment with a strong acid (e.g. HF) is required to remove the peptide from the solid support and effect side-chain deprotection.

Fig. 4.2 shows a summary of the procedure, and comprehensive details are given in Barany and Merrifield (1979). This procedure has been refined over the years but for its success the method still depends upon the differential acid lability between the Boc groups and benzyl or modified benzyl protecting groups. The method is carried out in a batchwise manner and the repetitive stepwise addition involving acylation, deprotection and neutralization of individual amino acids can be carried out in one reactor in an automatic manner.

A more recent alternative, milder method (developed by the group led by R. C Sheppard at the Laboratory of Molecular Biology in Cambridge) is rapidly gaining in popularity. The support employed here is a polyamide which swells considerably in dimethylformamide (DMF), a solvent in which the growing peptide chains are likely to be highly solvated and hence accessible for chemical reaction. The chemistry employed is based on the base labile fluorenylmethoxycarbonyl (Fmoc) group for α-amino group protection. Because of the base lability of the Fmoc group, the acid labile *t*-butyl group can be used for side-chain protection and the peptide-to-resin linkage can also be mildly acid labile.† A consequence of this is that repetitive acid treatment is not required and hence the method is milder and more flexible. The method employs DMF for washing and 20% piperidine in DMF to remove Fmoc groups. At the end of the assembly a single trifluoroacetic acid (TFA) treatment is used to cleave the peptide from the support and remove side-chain protecting groups. Fig. 4.3 shows a summary of the procedure, and further details may be found in Atherton and Sheppard (1989). With this method, it is possible to obtain highly pure peptides directly upon cleavage. This is demonstrated in Fig. 4.4, which shows an HPLC trace obtained in a recent large-scale synthesis of substance P.

The polyamide support employed in such syntheses is formed from dimethylacrylamide, using a cross-linking reagent ethylenebisacrylamide, and a functional reagent acryloylsarcosine methyl ester. The polymerized material can be prepared in a beaded form (Pepsyn) which allows for a batchwise mode of operation. Alternatively, the monomers can be polymerized within macroporous fabricated kieselguhr, to provide a column support (Pepsyn K) which enables a low pressure continuous flow method for assembling peptides (Sheppard 1983). Diffusion of reagents, and activated amino acids into and out of the support is rapid and its chemical effectiveness is essentially the same as the beaded support.

Using this column style of operation and the Fmoc method of synthesis, very simple flow machines can be constructed from a couple of valves and a pump (Fig. 4.5). An attractive feature is that if the column effluent is passed through a UV cell the chromophoric nature of the Fmoc group allows the monitoring of the individual chemical reactions as they occur. A typical profile for a complete reaction cycle is shown in Fig. 4.6. During the acylation, the activated Fmoc-amino acid is cycled through the column by configuring the recirculating or waste valve to operate in the recycling mode. This results in an initial sharp rise in the UV trace after the activated Fmoc-amino acid is introduced, followed by a gradual levelling out as the concentration throughout the system becomes constant. When the acylation is complete the amino acid is washed out with DMF and the Fmoc-group removed by

† This type of solid phase chemistry was simultaneously, independently developed using polystyrene supports by Meienhofer (Chang and Meienhofer 1978).

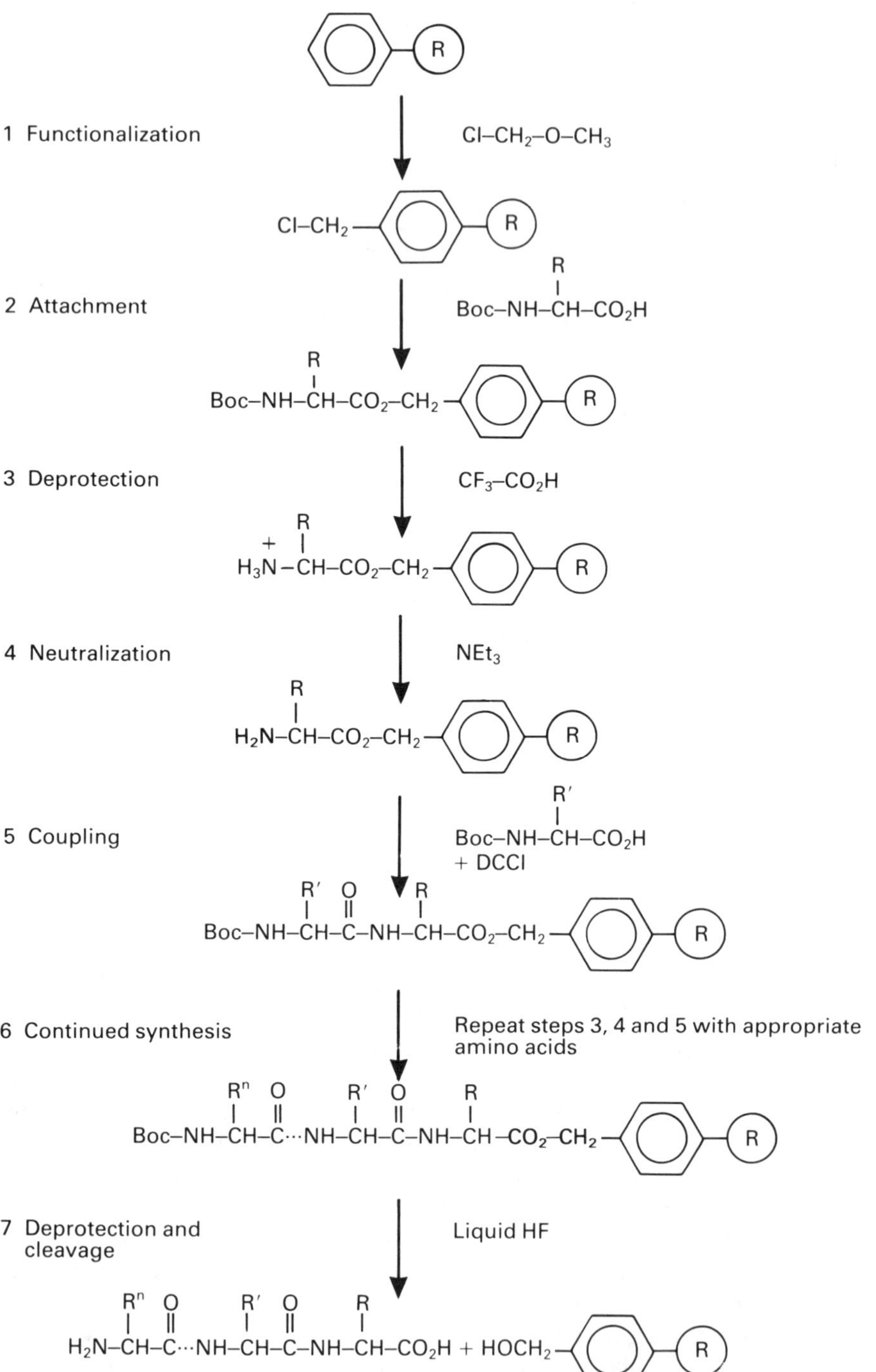

Fig. 4.2 — Merrifield (Boc/Benzyl) solid phase peptide synthesis.

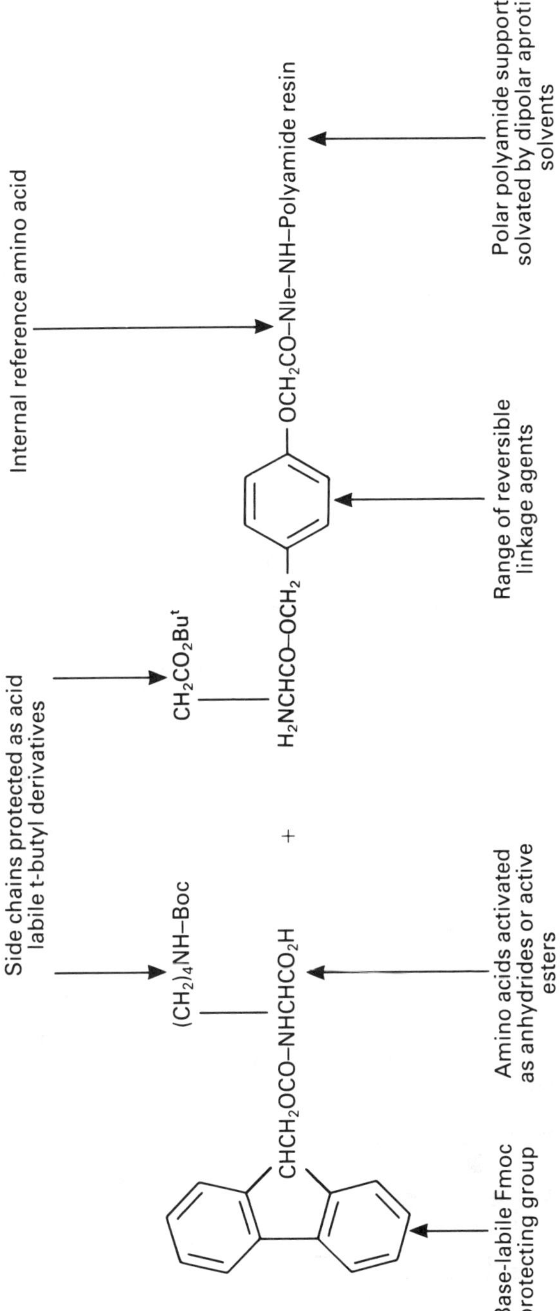

Fig. 4.3 — Summary of Fmoc-polyamide peptide synthesis procedure.

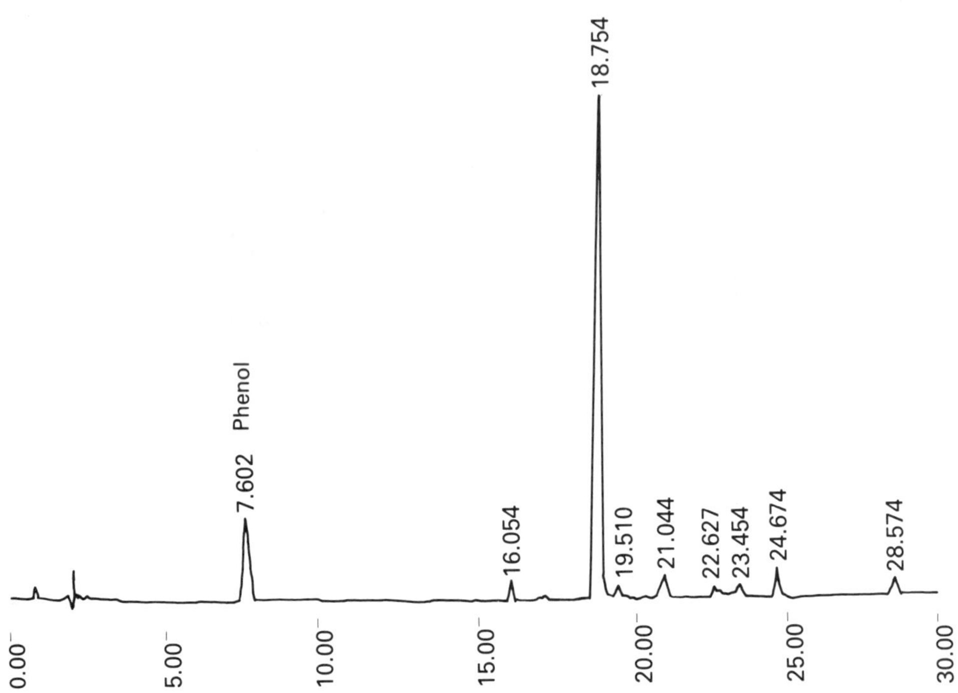

Fig. 4.4 — HPLC of substance P directly cleaved from the resin. HPLC conditions: solvent A 0.1% TFA in water, solvent B 0.1% TFA in CH_3CN; flow rate 1.5 ml/min; gradient 10% to 50% B over 30 min; monitoring at 230 nm.

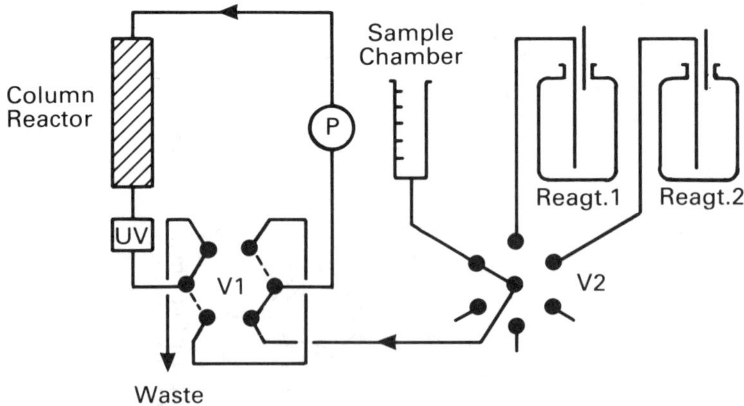

Fig. 4.5 — Simple low pressure continuous flow instrument for solid phase peptide synthesis.

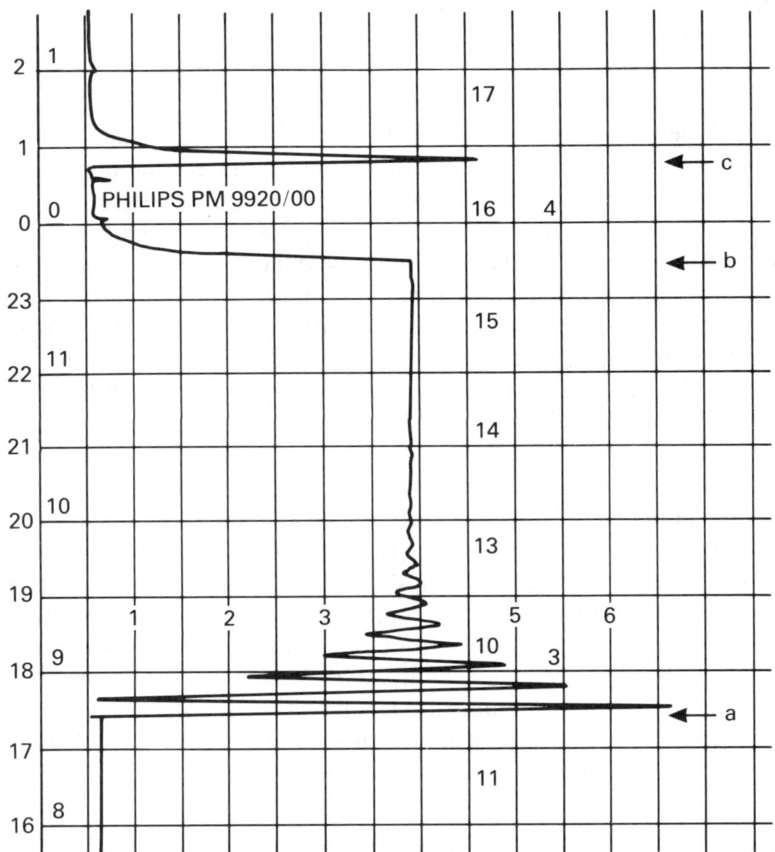

Fig. 4.6 — UV recording of one complete cycle of Fmoc-amino acid addition and deprotection in continuous flow synthesis.

the introduction of 20% piperidine in DMF. The sharpness of the deprotection peak, due to the piperidine adduct of dibenzofulrene, illustrates the rapid nature of the chemical reactions on the support. Although these traces do not give *quantitative* results concerning acylation and deprotection reactions, they do provide the operator with some assurance that the chemistry is progressing satisfactorily. Recently it has become possible to ascertain the completeness of acylation reactions through the use of indicators (Flegel and Sheppard 1990, Salisbury *et al.* 1990), and two fully automated commercial instruments utilizing this style of operation are now on the market.

4.3.2 Multiple, rapid methods of solid phase synthesis

In the field of immunology, and more generally in studies concerned with the structure–activity relationships of peptides, the need has arisen for the rapid production of small quantities of large numbers of peptides for screening purposes. One quite successful approach to this problem (developed by Houghten 1985) involves the simultaneous assembly of peptides in porous 'tea bag' containers. The tea bags are combined during washing and deprotection parts of the cycle and are then separated for acylation reactions.

A recent more novel approach employing Fmoc chemistry involves synthesis on plastic (polyethylene) pins. This method (invented by Geysen *et al.* 1980) has been particularly useful for scanning proteins to determine contiguous parts of the sequence that may be important in immunological recognition (epitopes).

The method involves grafting, by a radiation process, acrylic acid onto the surface of polyethylene pins. After activating the carboxylate groups on the pin surfaces, followed by the addition of hexamethylene diamine and B-alanine spacers, the desired peptide is assembled using Fmoc chemistry. A 96-well micro-titre plate is used for washing the pins and carrying out individual chemical reactions. A protein can be scanned for reactive epitopes by synthesizing a series of different peptides encompassing the whole of the sequence. For instance, the sequence corresponding to residues 1 to 6 can be synthesized on the first pin, residues 2 to 7 on the second, residues 3 to 8 on the third, and so on, until peptides spanning the entire sequence are assembled. The peptides remain covalently attached to the pins, and after deprotection they can be used in an enzyme-linked immunosorbent assay (ELISA) test to determine which peptides react with antibody raised against the native protein. In order to confirm the results obtained, the peptides should then be synthesized by conventional methodology, and their effectiveness ascertained with characterized products. This method is now being extended to cleavable peptides producing large numbers of peptides in small amounts for T-cell epitope work and structure–activity relationships. (Further details of the epitope mapping and the cleavable peptide kits can be obtained from Cambridge Research Biochemicals.)

4.4 PURIFICATION AND CHARACTERIZATION OF PEPTIDES

No matter what chemical method is employed to synthesize a peptide, the challenges of purification and characterization can often be quite considerable. Depending on the efficiency of the preceding chemical reactions involved in assembling peptides on solid supports, or in solution, the task of purification can be made simple, as shown in Fig. 4.4 or extremely difficult. In general, the longer the peptide the more difficult it is to isolate, purify and characterize. Also, in the absence of a sample of the natural material, characterization can be very difficult and sometimes conventional methods can be misleading (Flegel *et al.* 1990).

4.4.1 Purification

There are many accepted, well-practised ways of purifying peptides which are based on the individual characteristics of the target sequence. The purification procedures generally exploit the differences in the peptides' size, shape, solubility, hydrophobicity or affinity. The traditional techniques such as gel chromatography, ion exchange chromatography and partition chromatography are particularly useful.

Recently however, high performance liquid chromatography (HPLC) has played an increasingly prominent role in both the analysis and the purification of peptides. HPLC machines are now standard pieces of apparatus in most laboratories and reverse phase analytical procedures are now in common usage. The technique is extremely powerful and very small amounts of impurities can be readily detected. Scaling up this rapid procedure to a preparative scale has been quite a challenge but machines are now available which can operate at high flow rates with reproducible and programmable gradient systems. Consequently, the technique is becoming widely used and accepted for the purification of peptides and proteins in kilogram amounts (e.g. Kroeft *et al.* 1989).

The chemistry adopted for purification, i.e. eluents and supports, of course depends upon the individual physical characteristics of the peptide being purified. However, the most important feature is the effectiveness of the reverse phase support. This is usually spherical silica of a regular size and porosity, derivatized with an alkane, commonly C_8 or C_{18}, ligand. Silica particles in the size range 8 to 10 μm tend to be the most efficient in obtaining peptide purification, but larger particle sizes can be used with less-demanding purification problems.

As methods of preparing reverse phase supports become more reliable, giving highly defined supports with reproducible performance, this method will become increasingly popular for effecting peptide purifications. To complement this method, which is based on the hydrophobicity of a peptide, it would be advantageous to have an ion exchange material that could be used in this type of rapid procedure. The two methods then used in combination would provide a very powerful purification procedure.

4.4.2 Characterization

There is no one analytical technique which can be used in isolation to characterize a peptide unambiguously. The greater the number of analytical procedures to which a peptide product can be subjected, the more certain the user can be of its purity and authenticity. Common methods of characterization include:

(a) *Amino acid analysis*
This identifies the constituent amino acids in a peptide and quantifies the amount of each present. Note, however, that glutamic acid and aspartic acid are not distinguished from glutamine and asparagine, and that, under conventional hydrolysis conditions, some amino acids are partially destroyed (e.g. serine, cysteine) and some totally destroyed (e.g. tryptophan).

(b) *Analytical reverse-phase and ion-exchange HPLC*
This technique is particularly useful if an authentic sample of the synthetic target is available, but care should be taken when interpreting the results, particularly with longer peptides (Flegel *et al.* 1990).

(c) *Fast-atom bombardment mass spectrometry*
This gives a molecular ion and therefore the molecular weight of the peptide. The experienced user can also determine sequence information.

(d) *Sequencing*
This provides information regarding the linear sequence of amino acid residues, and can be carried out either by conventional Edman degradation methods or by mass spectrometry.

(e) *Peptide mapping*
This involves breaking the molecule into fragments using enzymes and then mapping by analytical HPLC techniques. The authentic material treated with

the same enzyme under the same conditions should give an identical HPLC fingerprint.

(f) *Thin layer chromatography*
This traditional method of analysis, although not as discriminating as HPLC when impurities that have similar physical properties are present, can sometimes highlight impurities that do not show up on an HPLC, particularly highly polar low UV absorbing species.

There are numerous other methods of analysing peptide products that can be potentially useful but those listed above are the ones most often used. Note also that with biologically active peptides, it is important to establish that the peptide elicits the correct biological dose–response curve.

4.5 CONCLUSION

Although there are many techniques that can be employed for the synthesis of peptides, the chemical methods are most general and tend to be used most often for targets up to fifty residues in length. The use of the classical solution phase methods has declined, and the solid phase procedures, which can be automated, have become universally accepted and very popular. The solution phase methods still have the advantage in syntheses of modified peptides, however, the situation here too is changing through the development of solid phase techniques with improved flexibility.

Of the two main chemistries used in solid phase synthesis, i.e. Merrifield's method and that based on Sheppard's Fmoc-polyamide technique, Merrifield's procedure is still probably the system more often used. This method is long-established, and has been proven to be reliable; it uses Boc-amino acids (which are cheap), and is capable of giving high quality peptides. However, the Fmoc chemistry is more flexible, requires fewer reagents, and involves the application of only one acid (TFA) treatment in a synthesis. It is also simpler and easier to use, and when the cost of Fmoc-amino acids is comparable to Boc-amino acids (although this is not the only cost factor) it will become the method of choice.

Although the solid phase method has been used to tackle longer (protein) objectives the method of choice is the recombinant route. In the past a drawback with this technique has been its inability to introduce non-natural, or modified amino acids into a protein. However, in a recent publication (Noren *et al.* 1989) this problem has been overcome at a research level and consequently this methodology may be more widely applied in the future.

REFERENCES

Atherton, E. & Sheppard, R. C. (1989) *Solid Phase Peptide Synthesis — A Practical Approach*. IRL Press, Oxford.

Barany, G. & Merrifield, R. B. (1979) Solid phase peptide synthesis. In: Gross, E. & Meienhofer, J. (Eds). *The Peptides — Analysis, Synthesis and Biology*, Academic Press, New York.

Bodanszky, M. (1984) *Principles of Peptide Chemistry*. Springer-Verlag, New York.

Chang. C.-D. & Meienhofer, J. (1978) Solid phase peptide synthesis using mild base cleavage of fluorenylmethyloxycarbonyl amino acids. Exemplified by a synthesis of dihydrosomatostatin. *Int. J. Peptide and Prot. Res.* **11** 246–249.

Denekelwalter, R. G., Veber, D. F., Holly, F. W., Hirschmann, R., Strachan, R. G., Paleveda, W. G., Nutt, R. F., Vitali, R., Dickinson, M. G., Garsky, V., Deak, J. E., Walton, E., Jenkins, S. R., Dewey, R. S., Lanza, T., Schoenwaldt, E. F., Berkameyer, H., Sondey, J., Varga, S. L., Milkowski, J. D., Joshua, H., Conn, J. B. & Jacob, T. A (1969) Studies on the total synthesis of an enzyme. *J. Amer. Chem. Soc.* **91** 502–508 (5 parts).

Flegel, M., Johnson, T. & Sheppard, R. C. (1990) Fmoc-polyamide solid phase peptide synthesis: Some Recent Developments, Innovation and Perceptives in Solid Phase Synthesis. In: Epton. R. (Ed.) *First International Symposium Proceedings*, SPPC, Birmingham, pp 307–314.

Flegel, M. & Sheppard, R. C. (1990) A sensitive general method for quantitative monitoring of continuous flow solid phase peptide synthesis. *J. Chem. Soc. Chem. Comms.*

Geysen, H. M., Rodda, S. J., Mason, T. J., Tribbick, G. & Schoofs, P. G. (1987) Strategies of epitope analysis using peptide synthesis. *J. of Immunological Methods* **102** 259–264.

Gutte, B. & Merrifield, R. B. (1969) The total synthesis of an enzyme with ribonuclease activity. *J. Amer. Chem. Soc.* **94** 501–502.

Gross, E., Meienhofer, J., Udenfriend, S. & Smith, C. W. (1979 onwards) *The Peptides*: *Analysis*, *Synthesis and Biology*, Academic Press, New York.

Houghten, R. (1985) General method for the rapid synthesis of large numbers of peptides. *Proc. Natl. Acad. Sci.* (*USA*) **82** 5131–5135.

Kroeft, E. P., Owens, R. A., Campbell, E. L., Johnson, R. D. & Marks, H. I. (1989) Production scale purification of biosynthetic human insulin by reverse phase high performance liquid chromatography. *J. Chromat.* **461** 45–61.

Marshak, D. R. & Liu, D. T. (1988) Therapeutic peptides and proteins: assessing the new technologies, Banbury Report 29, Cold Spring Harbour Laboratory.

Merrifield, R. B. (1962) Peptide synthesis on a solid polymer. *Fed. Proc. Fed. Amer. Soc. Exp. Biol.* **21** 412.

Noren, C. J., Anthony-Carhill, S. J., Griffith, M. C. & Schultz, P. G. (1989) A general method for site-specific incorporation of unnatural amino acids into proteins. *Science* **244** 183–204.

Salisbury, S. A., Tremeer, E. J., Davies, J. W. & Owen, D. E. I. A. (1990) Acylation monitoring in solid phase peptide synthesis by the equilibrium distribution of coloured ions. *J. Chem. Soc. Chem. Commun.* 538–541.

Sheppard, R. C. (1983) Continuous flow methods in organic synthesis. *Chem. Brit.* 402–413.

du Vigenaud, V., Ressler, C., Swann, J. M., Roberts, C. W. & Katsoyannis, P. G. (1954) The synthesis of oxytocin. *J. Amer. Chem. Soc.*, **76** 3115–3121.

5

Polypeptide production by recombinant DNA technology

E. Griffiths
National Institute for Biological Standards and Control, South Mimms, Potters Bar, Hertfordshire, EN6 3QG

5.1 INTRODUCTION

The ability to amplify individual DNA sequences, and to purify, break and rejoin DNA molecules at virtually any desired site, has greatly accelerated the progress of research in almost every field of biology and has already led to major medical and commercial applications. It is now possible to identify genes coding for natural (biologically active) proteins, to analyse them in fine detail, and to transfer them from one organism to another, so as to obtain highly efficient synthesis of their products. It is also possible to produce genes coding for modified or entirely new proteins, in order to obtain novel products that have reduced side-effects and/or enhanced biological activity. The construction of novel DNA molecules by joining sequences from different sources leads to what is described as recombinant DNA (rDNA), and the processes involved are often described as genetic engineering.

The first commercially produced biological medicine made by rDNA technology was insulin, which was licensed in the UK in 1982. Since that time an ever-increasing number of new medicinal products derived by genetic engineering have been developed, some to only the pre-clinical stage, but others to the fully licensed product. These include hepatitis B vaccine, interferons, tumour necrosis factor, growth hormone, erythropoietin, various cytokines and tissue plasminogen activator. Such products have extensive implications in the prevention or treatment of very diverse conditions like hepatitis, cancer, diabetes or myocardial infarction.

In order to assess the potential usefulness and expected clinical behaviour of rDNA-derived products, as well as to control their quality and ensure their safety, it is essential to understand the principles involved in the development and production of these molecules. Although there are many excellent texts available which deal in great detail with the molecular biology of the gene and its expression, few approach the topic from the point of view of commercial gene cloning. Generally the methods used by gene cloning companies to obtain rDNA derived products are much the same as those used in basic research (Watson *et al.* 1983, Williams and Patient 1988, Lewin 1990). However, certain issues which might only be of incidental interest to basic researchers may be very important medically and commercially.

This chapter outlines the general principles of polypeptide production by rDNA technology and focuses on the problems encountered and strategies used in producing medically useful products on a commercial scale. In particular it explores some of the aspects which might affect the quality and safety of the product and possibly its behaviour when administered to man. Two basic steps in gene expression need to be clearly differentiated and borne in mind when reading this chapter. The first is *transcription* which refers to the copying of the coding strand of DNA into messenger RNA (mRNA) by RNA polymerase; the second is *translation*, which involves the copying of mRNA into protein and involves ribosomes and transfer RNA carrying the appropriate amino acids, which decode the triplet code in the mRNA to produce an amino acid sequence.

5.2 ISOLATION OF THE REQUIRED DNA SEGMENT

In principle, there are three basic methods for obtaining the specific coding sequence of interest (see Table 5.1):

(i) isolation of genomic DNA;
(ii) reverse transcription of mRNA to give double-stranded complementary (copy) DNA (cDNA);
(iii) chemical synthesis.

In practice, however, these methods may not all be applicable in all situations. For example, the chemical synthesis of genes only becomes an option when the amino acid sequence of the protein is known; this is how rDNA-derived insulin was first produced (Goeddel *et al.* 1979). Often, neither the nucleotide sequence of the gene nor the complete amino acid sequence of the product of interest is known, and the gene has to be obtained from cells. However, a problem with many eukaryotic genes of medical interest is that they are longer than the sequence encoding the polypeptide itself. Genes from bacteria and viruses are relatively simple structures, where all of the information in the mRNA between the initiation codon and the stop codon is translated into polypeptide. In many eukaryotic genes there are additional DNA sequences that lie within the coding region, and which interrupt the sequence that actually encodes the protein product. These extra non-coding regions are called *introns*, and they are removed during the production of the mRNA species which is ultimately translated to give the final protein product; the process of removing introns is called *splicing*. Bacteria are unable to carry out this splicing operation and cannot be used directly therefore to give the correct expression of many of the genes from mammalian cells. One solution to this problem involves the isolation of the already spliced mature mRNA from the appropriate mammalian cell and the conversion of this to cDNA, using the enzyme reverse transcriptase. Where possible, cells which produce significant amounts of the desired product are used as a source of mRNA, because they will be rich in product-specific mRNA; for example, pancreatic cells would be used for isolating insulin-specific mRNA.

Table 5.1 — Different cloning strategies

Cloning method	Advantages	Disadvantages
Shotgun cloning	Relatively easy	Lots of different clones isolated. Needs good screening procedure to isolate desired gene.
		Genes with introns will not be expressed correctly in *E. coli*, the usual cloning host.
		Expression dependent on recognition of foreign transcription/translation signals by *E. coli*.
cDNA cloning	Introns not a problem since mRNA will already be spliced.	Desired mRNA not always abundant, selection may be difficult.
		Cloned genes need to be placed downstream of promoter since natural promoter eliminated by starting with mRNA.
Gene synthesis	Little selection needed after identification of transformants. Sequences of transcription/translation signals can be optimized as can codon choice for desired host cell.	Need to know complete sequence of protein product before synthetic gene can be made.

In the future, it is likely that the *in vitro* polymerase chain reaction (PCR) will find increasing applications in molecular cloning work, including the establishment of cDNA libraries from mRNA extracted from as few as one or two mammalian cells. PCR allows the direct amplification of segments of target DNA provided that the sequences on either side of the region of interest are known; that is, the technique allows the regions between two defined sites to be amplified (see reviews by Erlich *et al*. 1988, Marx 1988, Watson & James 1989).

5.3 CLONING AND MANIPULATION OF THE DNA

A naturally occurring gene, or synthetic nucleotide, which codes for a specific product can be propagated by inserting the DNA into a suitable cloning vector. A vector is a piece of DNA to which other DNA molecules can be attached for the purpose of producing more material or a protein product. The combination of the specific and vector DNA is achieved using highly specific enzymes: restriction endonucleases, which cleave DNA at predetermined sites, and ligases, which join DNA fragments together. The vector is then introduced into a suitable host organism, initially usually *Escherichia coli* (*E. coli*), where it is replicated in parallel with the host cell genome (Table 5.1). Some vectors can direct their own independent replication within a host cell, others can be integrated into the host cell chromosome and maintained in this form during growth and multiplication of the host organism. Individual clones which carry the desired gene can then be selected and grown in mass culture of the same organism or introduced into another more appropriate host cell.

Many vectors are bacterial plasmids, but bacteriophages, viruses or specially constructed plasmids which function both in prokaryotic and in a variety of eukaryotic host cell types have all been used for cloning. Vectors are also constructed to contain powerful expression signals which allow the relevant genes to be transcribed and translated into large amounts of protein. Much genetic engineering has been carried out in bacteria but other systems involving yeast or continuously growing (transformed) cell lines of mammalian origin have also been developed and are already used for production. Various methods can be used to introduce a vector into a host cell, the method chosen depending on the type of vector and host cell being employed. Plasmid DNA can be introduced into bacteria by the process of transformation, where the organisms are treated with a mixture of divalent cations to make them temporarily permeable to small DNA molecules. A similar process works in yeast once the yeast cellulose cell wall is removed enzymically to give yeast spheroplasts. Many procedures have been used to get DNA into mammalian cells. These include transfection mediated by the polycation polybrene, calcium phosphate or DEAE–dextran, and the process of electroporation. Electroporation involves applying brief, high voltage electric pulses to the cells, which is believed to generate nanometre-sized pores in their plasma membranes. DNA is then taken up directly into the cell cytoplasm as a consequence of this pore formation although the precise mechanism involved is unknown.

Even under the best conditions, extraneous DNA enters and becomes stably established in only a small proportion of cells in a given culture. It is helpful, therefore, to home in on these cells rather than to screen the whole culture for cells which have taken up the recombinant gene of interest. Identification of transformants is usually achieved by the introduction into the host cell of a second gene that encodes a selectable marker, such as an enzymic activity that confers resistance to an antibiotic or other drug. The most commonly used selectable markers on plasmids for bacterial systems are genes that confer resistance to antibiotics such as ampicillin, tetracycline, chloramphenicol or neomycin. Virtually all plasmid vectors in common use carry one or more antibiotic resistance genes in addition to the gene encoding the protein of interest. For mammalian cells, however, the genes encoding the protein of interest and the selectable marker do not necessarily need to be on the same vector. Mammalian cells are generally capable of taking up DNA very efficiently with high frequency. Co-transfection obviates the need to construct complex recombinant plasmids and is often used to introduce a selectable marker on one plasmid, and the gene of interest on another plasmid, into mammalian host cells. The best known selectable markers for mammalian cells are:

(1) Thymidine kinase gene (*tk*). Thymidine kinase is an enzyme used in the salvage pathway of pyrimidine biosynthesis. A number of *tk*$^-$ cell lines have been isolated from various mammalian species and these are unable to grow in a selective medium called HAT. HAT medium contains hypoxanthine, aminopterin and thymidine, and only supports the growth of cells expressing the *tk* gene; aminopterin blocks normal conversion of deoxy-CDP to deoxy-TDP and in such conditions the only source of deoxy-TTP for DNA synthesis is via thymidine kinase. By the appropriate use of this medium it is possible to select for or against cells expressing the *tk* gene. Transfection of a vector containing the *tk* gene into a *tk*$^-$ host cell will allow that cell to grow in HAT medium.
(2) Dihydrofolate reductase gene (*dhfr*). Mutant cells that lack the enzyme dihydrofolate reductase are unable to synthesize tetrahydrofolate and thus can only grow in media supplemented with thymidine, glycine and purine. Transfection of such cells with vectors that express a cloned copy of *dhfr* produces clones that can grow in the absence of these supplements. The *dhfr* gene can also be used for amplifying recombinant genes of interest (see section 5.4.1).

Once transformants have been identified, they can be screened for the presence of the gene of interest. This is generally achieved through the use of DNA probes, or by searching for expression products. If the protein of interest has been purified in sufficient quantities to obtain even part of the amino acid sequence, it may be possible to predict the sequence of an oligonucleotide which could be used as a hybridization probe to detect the corresponding DNA clone. Screening for expression is often carried out using immunochemical methods utilizing a suitable antibody to the protein. These two techniques, sometimes in combination, allow the isolation of cDNA clones derived from even low abundance mRNA sequences; relevant clones are then grown up for further study or use. The gene of interest can be separated from the vector DNA sequences by digestion with restriction enzymes, purified by electrophoresis and sequenced using an appropriate strategy. Manipulation of the gene can also be undertaken. For example, codons can be changed so that the amino acid sequence and the properties of the resulting polypeptide product are altered. Insulin analogues, which remain monomeric at pharmaceutical concentration but retain biological activity, have been produced in this way. Normal soluble insulin exists mainly in the form of hexamers at pharmaceutical concentration and the initial delay in absorption has been attributed to the rate of dissolution of hexamer units into dimers and monomers. On introducing charge repulsion into the monomer–monomer interface, by replacement of uncharged amino acids with those that carry a negative charge at neutral pH, insulin can be made which remains monomeric even at pharmaceutical concentrations. Kang *et al.* (1990) show that one such analogue, with B9 serine and B27 threonine (see Fig. 1.5b) replaced by aspartate and glutamate respectively, gives a more rapid early rise in plasma insulin levels in diabetic patients. Much higher plasma insulin levels are obtained with the analogue owing to a threefold faster absorption and a lower metabolic clearance rate.

However, altering amino acid residues in a natural molecule should not be undertaken lightly, since there is always the risk that an analogue might be immunogenic when used in man, especially if used repeatedly. Furthermore, antibodies raised to an analogue might also react with the native protein and lead to considerable therapeutic problems for the individual concerned. Unfortunately, predicting the precise immunogenicity of a specific analogue is impossible (Konrad 1989). There is no substitute for clinical trials and careful immune monitoring; animal experiments are of little use since the human protein is likely to be immunogenic in the animal model anyway.

5.4 MAXIMIZING GENE EXPRESSION

High-level production of the desired gene product is the general goal of all commercial rDNA technology. However, the factors affecting the expression of foreign genes introduced into a new host are complex, and the efficient, reproducible and faithful transcription and translation of cloned DNA sequences is thus an important aspect of research. The rate of synthesis of a protein will depend on the efficiency of transcription of the gene, on the turnover of mRNA and on the rate of translation. The final yield will also depend on the product turnover, i.e. synthesis versus degradation in the host cell or culture used. Some of the factors affecting the expression of a foreign gene in a new host are listed in Table 5.2.

Table 5.2 — Factors affecting the expression of a foreign gene in a new host cell

Copy number of gene
Strength of promoters
Sequences of ribosome binding site and flanking region
Controls on gene expression (constitutive/induced)
Codon choice in cloned gene
Genetic stability of recombinant system
Proteolysis of protein product

5.4.1 Gene copy number

One fundamental factor which influences the amount of product obtained is the number of copies present of the recombinant gene of interest. If the gene is cloned into a plasmid, then the number of copies of the plasmid in the host cell is important. Generally, the more copies of the plasmid, and thus of the gene, the more product is produced. Each type of plasmid is maintained in a bacterial host at a characteristic copy number which is primarily a consequence of the type of replication control mechanism on the plasmid. Different plasmids use different subsets of enzymes to duplicate the bacterial chromosome, and replicate to different extents in their hosts. Some can reach very high copy numbers per cell, whereas others are maintained at the minimum level of one plasmid per host-cell chromosome. Most cloning vector plasmids in common use are maintained in their host cells at 15–20 copies per cell under normal growth conditions.

Foreign genes introduced into mammalian cells can also exist in different copy numbers. DNA introduced into mammalian cells can integrate into the host cell DNA and can be amplified if the vector DNA sequences used contain an amplifiable gene. The principle lies in the fact that when cultured mammalian cells are subjected to certain toxic drugs, variant clones can be selected which are more resistant than the wild type. A common mechanism of resistance involves the overproduction of an enzyme whose activity is inhibited by the drug; this is achieved by increasing the number of copies of the structural gene in each cell. The dihydrofolate reductase gene (*dhfr*) is an example of a gene often used to help amplify rDNA sequences. Dihydrofolate reductase can be inhibited by methotrexate, a folate analogue. Progressive selection of cells that are resistant to increasing concentrations of methotrexate leads to amplification of the *dhfr* gene with concomitant amplification of considerable regions of DNA that flanks the *dhfr* sequences. rDNA that is co-transfected with the *dhfr* gene tends to become integrated into the same region of the host-cell chromosome and therefore can be co-amplified with *dhfr* under methotrexate selection. Cells lacking dihydrofolate reductase activity can also be transfected with a recombinant construction containing the gene of interest linked to the *dhfr* gene. The linked gene is then amplified by selecting with successively higher concentrations of methotrexate. Using these and similar procedures, cell lines, such as Chinese Hamster Ovary cells (CHO cells) can be obtained which express very high levels of the desired recombinant protein product; they can carry 1000–2000 or more copies of the rDNA gene.

Although the *dhfr* gene is a widely used amplifiable marker in CHO cells, a number of others have been developed recently. One system which appears to show considerable promise is that based on the hamster glutamine synthetase gene (GS). Cockett *et al.* (1990) describe the use of a vector containing GS for the high-level expression of rDNA products. Glutamine is a key metabolite in a number of different pathways and must either be provided in the medium or must be synthesized in the cell from glutamate and ammonia by means of GS. Some cell lines are GS-deficient and cannot grow in the absence of glutamine; others do not need glutamine, provided that enough glutamate is present in the medium. Under the latter conditions, GS is an essential enzyme for cell growth and its inhibition by the specific inhibitor methionine sulphoxide is lethal. However, GS genes can be amplified to high copy number by selecting with increasing concentrations of the inhibitor.

5.4.2 Transcription and translation signals

Cloned genes can only be expressed if they are correctly placed in a vector that supplies all of the elements which control transcription and translation in the chosen host cell. Control is mediated by a series of instructions, in the form of nucleotide sequences built into the DNA or mRNA (Table 5.3). Genomic DNA sequences may

Table 5.3 — Common transcription and translation signals

Promoter	Site at which transcription is initiated by RNA polymerase.
Operator	Sequence in DNA at which a repressor protein binds to prevent transcription from initiating at the adjacent promoter.
Enhancer	An element which acts to increase the utilization of some eukaryotic promoters.
Terminator	A sequence of DNA that causes RNA polymerase to terminate transcription.
Polyadenylation signal	Allows the addition of a sequence of polyadenylic acid to the 3′ end of an eukaryotic mRNA after its transcription.
Ribosome binding site	RNA sequence to which ribosomes bind; essential for efficient initiation of translation.
Translation initiation codon	Usually AUG, but can be GUG or UUG in prokaryotes.
Translation termination codon	These are UAG, UAA or UGU.

already carry controlling elements for expression in a mammalian cell, but there is no guarantee that they will work effectively, or at all, in the line of cultured cells available. This may be a particular problem when dealing with genes that are expressed in a tissue-specific fashion. cDNA sequences will lack a natural promoter since they are derived from mRNA. In order to ensure efficient expression, it is usual to build into the vector all relevant controlling elements. For example, mammalian expression vectors now generally contain both prokaryotic sequences that facilitate the propagation of the vector in bacteria and essential eukaryotic transcription and translation signals. Cloned sequences that are expressed in mammalian cells will include the ribosome-binding site and initiation codon found in the natural gene. If the same gene is to be expressed in a prokaryotic expression system, such as *E. coli*, then the ribosome-binding site and initiation codon of the natural gene are invariably replaced by an effective prokaryotic ribosome-binding site which includes the Shine–Dalgarno ribosome-binding sequence and an AUG codon that have been optimally spaced for efficient translation. Bacterial mRNAs contain part or all of a short polypurine stretch known as the Shine–Dalgarno sequence which is complementary to the sequence at the 3′ end of the 16S ribosomal RNA. The number of nucleotides between the Shine–Dalgarno sequence and the initiation codon are critical for optimizing translation. The placing of this sequence closer to or further away from the AUG start codon can greatly lower the translation efficiency of the mRNA.

In some cases, rDNA-derived products are expressed constitutively using unregulated systems. In others, controlled expression is built into the system by including in the vector not only the bacterial promoter but also the operator sequences sensitive to a particular repressor protein. In such cases translation is under the control of an inducing signal, such as a temperature shift or the addition of an inducer.

5.4.3 Codon usage

Most amino acids can be specified by more than one codon in the mRNA. For example, lysine is specified by AAA or AAG, phenylalanine by UUU or UUC, and serine by UCU, UCC, UCG, UCA, AGC or AGU. Sometimes different codons are read by a single tRNA; in other cases different tRNA molecules are involved in reading different codons. It has generally been found that the codons used in mRNA coding for highly expressed genes of several species are not random; a marked preference for particular codons is found (Ernst 1988, Andersson & Kurland 1990). In some cases this preference appears to correlate with the abundance of different tRNA species. Furthermore, the codon context is found to be significantly biased in highly expressed genes; codon context refers to the nucleotide or codons found adjacent to the codon of interest. Since codon preferences of eukaryotic and prokaryotic genes differ, it is thought that the levels of certain tRNAs will affect translational efficiency of these genes in a foreign host cell. Also, bias against certain nucleotide sequences is seen in prokaryotic coding segments. For example, the sequences GAGG and GGAG are rarely used in bacterial coding regions probably to avoid internal translation starts, since these are the sequences found in the Shine–Dalgarno regions. Generally, however, experiments to test the effects of codon usage on translation have led to controversial results. No clear causal relationship between codon usage and gene expression has been found for expression of heterologous genes. However, the data are often difficult to interpret since sequence changes may affect other parameters, such as the secondary structure of mRNA, which could affect ribosome movement. The effect of codon changes in the coding region is therefore not predictable. Nevertheless, a conservative approach is usually taken in designing synthetic genes, and the codons and codon context chosen are those of naturally highly expressed genes in the host-cell system chosen for production. Sometimes, cloned natural genes are altered to give more appropriate codon assignments when bacterial or yeast expression systems are being used; efforts are also made to free the mRNA coding region of secondary structures and potential ribosome binding sites which could interfere with efficient translation.

5.5 CHOICE OF CELL TYPE FOR PRODUCTION

Several important factors influence the choice of host cell used for producing an rDNA protein to be used in human medicine (Table 5.4). Some, such as the cost of fermentation, are purely commercial considerations. Others, however, such as the need for post-translational modification of the protein product, can have important clinical implications. Of course, it would be expected that the nature of the vector into which the gene of interest has been cloned will have been chosen with the final fully scaled-up production process in mind, whether it will be in bacteria, yeasts or mammalian cells. The choice of the actual mammalian cell type to be used also needs to be considered and will influence the choice of a suitable vector. For example, the use of bovine papilloma virus vectors is generally restricted to mouse fibroblast cell

Table 5.4 — Some factors to be taken into consideration when choosing cell type for production

1. Vector choice
2. Available methods of culture and yield required
   - batch culture
   - continuous culture
   - size of culuture
   - cost (mammalian cells grow slower than *E. coli* and are more expensive to culture)
3. Desired/necessary post-translational modification of the expressed polypeptide
   - removal of terminal methionine
   - removal of signal peptide
   - glycosylation
4. Choice of host cell will influence range of potential contaminants in final product
5. Choice of host cell and production method will influence downstream processing and extent of control testing needed.

lines which are 'anchorage-dependent', that is they will only grow when attached to a plastic surface. This may present problems later in development if very large cultures are needed. Anchorage-dependent cells can be grown on microcarrier beads, but the system is complex, expensive and difficult to scale-up to very large volumes (Bliem & Katinger 1988a,b, Looby & Griffiths 1990); the immobilization of cells in porous carrier culture offers some advantages, although it is an area still in its infancy. The choice of cell type can also be crucial for function of the protein product. Sometimes unusual post-translational modification of the target protein is needed for activity. For example, in the case of factor IX a vitamin K-dependent carboxylation of 12 glutamic acid residues is required. This modification is carried out when the factor IX gene is expressed in baby hamster kidney (BHK) cells or rat hepatoma cells, but not in mouse fibroblasts (Busby *et al.* 1985, Anson *et al.* 1985).

Four aspects which have important considerations for the quality of the product and on its expected behaviour in man will be considered in more detail.

5.5.1 The problem of N-terminal methionine

The incorporation of a methionine residue at the N-terminal of each nascent polypeptide is part of the universal translation initiation signal used by prokaryotic and eukaryotic cells alike. However, in bacteria the methionine residue carried by the initiator tRNA is N-formylated prior to incorporation. Once made, many bacterial proteins are subjected to post-translational modification reactions which remove sequentially the formyl group and the terminal methionine residue so that in *E. coli* only a portion of the polypeptide chains found in the cytoplasm retain their methionine. The deformulation step is more tightly coupled to the translation process than is the excision of methionine. Recent work has shown that the extent of the cleavage of methionine depends on the length of the side-chain of the amino acid in the penultimate position in the polypeptide chain (Hirel *et al.* 1989; Dalboge *et al.* 1990). Of importance for genetically engineered proteins is the fact that products of foreign genes expressed in *E. coli* generally do not have the methionine removed. *E. coli* could, therefore, provide a product with an amino acid sequence identical to the native human sequence except for an additional methionine. One such product, methionyl (human) growth hormone made by Kabi (see Chapter 14) has been used clinically. However, in view of the potential immunogenicity of modified human proteins, considerable efforts have been made to produce therapeutic products without the terminal methionine, and a number of different strategies have been devised.

One trick is to employ a vector construction which uses methionine as a link that can be specifically cleaved between the desired recombinant polypeptide chain and a bacterial carrier peptide, producing a recombinant protein lacking an N-terminal methionine. This was the method used to produce recombinant insulin. Oligonucleotides coding for the A and B chains of insulin were synthesized chemically and the codon ATG added to the 5′ end of each. These oligonucleotides were then cloned downstream of the β-galactosidase promoter and part of the β-galactosidase protein coding sequence, in a common plasmid vector called pBR322. When the plasmid was introduced into *E. coli*, hybrid proteins were produced, consisting of the N-terminal portion of the β-galactosidase protein (followed by a methionine residue from the ATG) fused to either the A or B chain of insulin. The β-galactosidase protein portion was removed by treatment with cyanogen bromide, which cleaves polypeptide chains at a methionine residue, thus liberating free A or B insulin chains. The two chains were finally combined to give biologically active insulin. Unfortunately, this particular strategy is very limited since it can only be used for products that contain neither methionine nor tryptophan (also cleaved by cyanogen bromide) in their natural sequence.

Another approach is to borrow part of the *E. coli* protein export system to produce an rDNA-derived product that is not only free of N-terminal methionine, but also exported into the periplasmic space. Many bacterial secreted proteins are synthesized as precursor molecules which contain an additional hydrophobic N-terminal signal sequence of between 15 and 30 amino acids; the signal peptide is cleaved during passage through the membrane. This is the system used to produce human growth hormone (hGH) without a terminal methionine using *E. coli*. In the system used by Kabi Vitrum, the hGH gene is preceded by codons corresponding to the 23-amino-acid signal sequence for *E. coli* enterotoxin ST2. This polypeptide is secreted through the membrane and the signal peptide is removed by signal peptidase as normal; the hGH is then released into the periplasmic space with no terminal methionine (Frykland 1988). In this particular system, expression is under the control of the alkaline phosphatase promoter so that the gene for the precursor protein is not turned on until the phosphate content of the medium is low, thus allowing for much better control of production (Gray *et al.* 1985). Since the expression of hGH is limited to the latter part of the fermentation cycle, there is an added advantage for hGH itself in that it is exposed to potential proteolytic attack for a shorter period of time.

5.5.2 Intracellular or extracellular production

Once high level expression of recombinant protein has been achieved it is important that the product is isolated and purified rapidly, because both lysed *E. coli* and eukaryotic cells release proteases that may 'clip' the desired product. Traces of protease are a major concern in the purification of recombinant proteins because these can be extremely difficult to remove and may significantly affect the stability and quality of the final product. One approach would be to develop *E. coli* strains with low levels of protease activity, but rapidity of purification, at least at the initial stages, is very important. Minimizing the release of protease is also crucial. Generally, secretion of the desired product into the culture medium has a number of advantages: purification is simplified, the medium is likely to have fewer proteases to degrade the recombinant protein, and there is a large space for the accumulating product.

When transformed mammalian cells are used as hosts for the production of rDNA-derived proteins, a fully functional recombinant molecule is normally synthesized and, if the protein is naturally secreted, the product can be recovered directly from the culture medium. This is in contrast to the expression of many heterologous proteins in *E. coli* where post-translational modification such as glycosylation (see section 5.5.3), phosphorylation, amidation, carboxylation, proteolytic cleavage, assembly and secretion cannot usually be performed. Furthermore, the reducing environment of the *E. coli* cytoplasm leads to incomplete or incorrect formation of disulphide bonds, and the gene product can be unstable and easily degraded. However, early observations indicated that high level expression of some cloned genes in *E. coli* leads to the intracellular formation of insoluble proteinaceous aggregates, or granules, which are sedimentable by low speed centrifugation. These granules are readily visible under the microscope. Not all recombinant proteins form these so-called inclusion bodies and the factors which might affect the state of recombinant proteins in *E. coli* and lead to aggregate formation are discussed by Kane and Hartley (1988). A model for inclusion body formation is shown in Fig. 5.1.

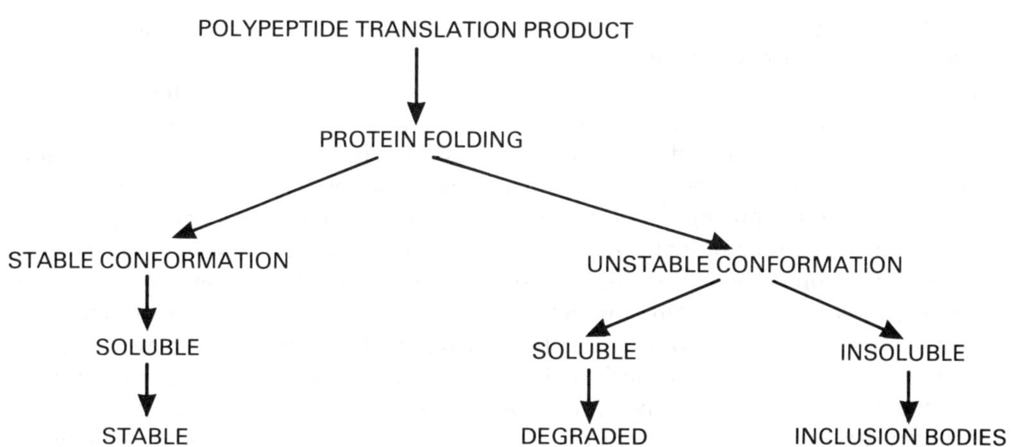

Fig. 5.1 — Model for formation of inclusion bodies (from Kane and Hartley 1988).

and although the conditions required are not entirely understood, it is clear that the process is affected by parameters relating to the host cell, growth conditions and the properties of the particular protein concerned. There are in fact two major advantages to the formation of an aggregated recombinant protein product. First, the initial isolation step is relatively simple: physical disruption of the bacterial cell and centrifugation can yield concentrated product of relatively high purity (30% or more of the cellular protein forming inclusion bodies). Second, there is evidence that intracellular proteases do not attack aggregated forms of the recombinant proteins. However, there is also a major problem when recombinant proteins form aggregates. Once the inclusion bodies have been partially purified there is a need to solubilize them in order to achieve further purification, and then to refold them to give the final native active form of the molecule (Chapter 2). Harsh solubilizing agents are required for this purpose, followed by complex renaturing processes, and only certain proteins can be refolded to give an active product once denatured. Another disadvantage of intracellular production of aggregates is that it tends to lead to cell death or lysis.

In view of the difficulties encountered in renaturing proteins correctly, some workers have sought to engineer bacterial systems to allow secretion of the gene product out of the reducing environment of the cytoplasm; correct folding of polypeptides containing disulphide linkages is then expected to occur. Secretion also has the advantage of simplifying the recovery of the product, especially if it occurs directly into the culture medium. Significant purification can then be achieved merely by separating the unruptured cells from the medium. Until recently, the secretion systems which had been developed for *E. coli* allowed secretion of the protein product only into the periplasmic space. An example of such a system, which uses the signal peptide of *E. coli* enterotoxin ST2 for the production of human growth hormone, has been described earlier (see section 5.5.1). Although the hormone is correctly folded, soluble and easily isolated (the product being released by osmostic shock), the ideal is still to get recombinant proteins secreted directly into the culture medium. One interesting method for achieving this has recently been developed by Hsiung *et al.* (1989) and involves the use of the *E. coli* bacteriocin release protein. Bacteriocin release protein is a small protein consisting of 28 amino acids synthesized as a precursor with a signal peptide of 21 amino acids. It is secreted across the cytoplasmic membrane where it is *N*-acetylated and inserted into the outer membrane. Here it activates phospholipase A, resulting in the formation of permeable zones in the cell envelope through which proteins can pass and be released into the culture medium. By using *E. coli* containing two vectors, Hsuing and colleagues were able to produce recombinant human growth hormone which was secreted into the culture medium. One vector carried the human growth hormone gene proceded by the signal sequence for outer membrane protein A (OmpA) which directed the secretion of the gene product into the periplasm; the other vector encoded the bacteriocin release protein derived from plasmid pCloDF13. Expression of both products was regulated by the *E. coli lac* repressor, so that in the presence of low concentrations of isopropyl-1-thio-β-D-galactopyranoside (IPTG) both the bacteriocin release protein and the OmpA–human growth hormone hybrid protein were expressed and mature growth hormone was released into the medium. The excreted hormone was found to contain the correct amino acid terminus (no extra methionine) corresponding to the N-terminus of growth hormone derived from human pituitary glands showing that the correct processing of the OmpA signal peptide had taken place.

5.5.3 Glycosylation

Although high level expression, and in some cases secretion into the culture medium, of the active recombinant protein is possible in a prokaryotic organism like *E. coli*, such cells are unable to effect many of the post-translational modifications which normally occur to proteins from higher organisms. Of particular significance are the variety of post-translational modifications involving carbohydrate moieties and there is increasing awareness of the biological significance of glycosylation. Indeed, many of the human proteins of interest for therapeutic use are glycosylated, including erythropoietin, tissue plasminogen activator, and various cytokines. *N*-Glycosylation is the most common form of glycosylation, where the oligosaccharides are attached covalently to the polypeptide through an *N*-glycosidic bond to the NH_2 group of asparagine; only an asparagine in the specific Asn-X-Ser/Thr is glycosylated. *O*-Glycosylation can also occur and this involves covalent attachment of an oligosaccharide to the hydroxyl group of serine or threonine (Rademacher *et al.* 1988).

Most glycosylated proteins have two or more *N*-glycosylation sites and each site is usually associated with several oligosaccharides. A glycoprotein therefore really exists as different sets of molecules, called glycoforms, that share a common polypeptide backbone but differ in the number, location or structure of *N*-linked oligosaccharides. From the point of view of producing recombinant products it is important to note that the amino acid sequence and the cell type influence glycosylation, but that under constant physiological conditions the pattern of glycosylation is highly reproducible. Given the known influence of cell type it would be expected that an rDNA-derived glycoprotein would not have the same set of glycoforms as the native human form. Also, it would be expected that products expressed from the same recombinant gene in yeast, CHO cells or other cell line, although all glycosylated, would differ not only from the native human material, but also from each other with respect to the nature of the glycosylation. Bacterial cells do not glycosylate protein.

Knowing about the diversity of glycoforms of native proteins it is important to know how significant might be the differences between a native glycoprotein and its recombinant counterpart, as well as the differences between different glycoforms of the recombinant molecule when made in different host cells. Accumulating evidence points to the fact that glycosylation can have important consequences (Parekh *et al.* 1989, Stanley 1989). The presence of oligosaccharides affect not only the physical properties of a glycoprotein but also influences its biological activity, its distribution *in vivo*, its circulation half-life and its immunogenicity (Parekh *et al.* 1989). It must be accepted, therefore, that, in most cases, recombinant glycoproteins will be analogues of the natural material even though they may have identical amino acid sequences. When used as pharmaceutical products, the problems expected will be those encountered with any analogue, namely that the product may have a different spectrum of biological activity, different *in vivo* distribution, different circulatory half-life, and also perhaps novel biological properties. It is difficult to generalize, however, and each product must be considered on a case-by-case basis. For example, the recombinant erythropoietin produced in *E. coli* is not glycosylated and is inactive in man because of its extremely short lifetime in circulation; only glycosylated erythropoietin can be used clincially. In contrast, all forms of recombinant granulocyte macrophage colony stimulating factor (GM-CSF) are active *in vivo* (see Gribben *et al.* 1990). However, interesting observations have recently been reported on the effect of glycosylation of these molecules on their immunogenicity in man (Gribben *et al.* 1990).

Recombinant hGM-CSF is available in several forms:

(1) rhGM-CSF, which has the identical amino acid sequence to the natural protein produced in a mammalian cell line, and is variably glycosylated on both *O*-linked and *N*-linked sites. This gives rise to a heterogeneous population of proteins of 18–30 kDa.
(2) rhGM-CSF, which is produced in *E. coli*, is not glycosylated and has a molecular weight of about 14 kDa.
(3) rhGM-CSF, which is produced in yeast. This differs from the natural product in the substitution of a leucine for a proline at position 23 of the amino acid sequence to remove a protease-sensitive site. The molecule produced by yeast is glycosylated at *N*-linked sites which results in three species of 15.5, 16.8 and 19.5 kDa.

Gribben *et al.* (1990) have found that a high proportion of patients given the yeast-derived rhGM-CSF developed antibodies to the product within seven days of its administration. These antibodies reacted with sites on the native protein backbone which are normally protected by *O*-linked glycosylation, but which are exposed in the rhGM-CSF produced in yeast and *E. coli*. In one patient the presence of antibodies to GM-CSF was associated with a decline in biological effect and a marked decrease in serum levels of GM-CSF despite continued infusion; this suggests that in this case the antibodies were accelerating serum clearance of the active substance. Clearly, the antigenicity of recombinant human proteins due to a failure to glycosylate appropriately may well have relevance to the choice of host system for the production of rDNA-derived proteins for clinical use.

Antibodies directed to the protein backbone of a recombinant molecule might not be the only problem. Antibodies directed to certain carbohydrate determinants also need to be considered. *N*-Glycosylation is species-specific and oligosaccharide determinants can define barriers to inter-species exchange of glycoproteins. There are naturally occurring antibodies against certain carbohydrate groups that are not found on the circulating glycoproteins of the host but are present in other species. A good example might be the Gal-(α 1$\rightarrow$3)-Gal determinant; this is present on certain cells from a variety of species such as mouse, rat, guinea-pig, rabbit, dog and cow. If a recombinant molecule was to be made in a cell which resulted in the presence of the determinant on the product, then it might be rejected in man by naturally occurring antibodies: up to 1% of all circulating antibodies in humans are directed against this sugar residue (Galili *et al.* 1987). Certain oligosaccharide epitopes, such as D-xylose linked β 1→2 to the β-mannosyl residue, or L-fucose linked α 1→3 to the reducing terminal of *N*-acetylglucosamine, are highly immunogenic in man. Both are common on plant glycoproteins, implying that the therapeutic use of human glycoproteins made in plant cells might be associated with adverse immunological reactions. Products made from such cells would need to be clinically evaluated very carefully indeed. Better still would be to analyse the nature of the carbohydrate groups on recombinant proteins early in product development, so that an informed choice might be made between different expression systems in order to avoid potentially problematic oligosaccharide epitopes.

5.5.4 Influence on nature of potential contaminants

Quite apart from the various issues discussed above relating to the efficacy and safety of the recombinant molecule itself, the choice of host cell and manufacturing process will also influence the nature and range of potential contaminants which might be present in the final product. Examples of potential contaminants of rDNA-derived products are listed in Table 5.5. Some, such as solvents, column matrices, and

Table 5.5 — Potential contaminants of rDNA-derived biological products

Small moleuclar weight materials	Antibiotics
	Solvents
	Column matrices
	Specific chemicals (e.g. methotrexate)
Macromolecules	Host-cell proteins
	Non-host-cell proteins (e.g. serum proteins)
	Product variants (breakdown/modification aggregates)
	Endotoxins
	Lipids/carbohydrates
	DNA
Microbiological	Bacteria
	Mycoplasma
	Yeast
	Cells
	Viruses

product variants will be common both to products derived from prokaryotic and eukaryotic cells, and indeed some might be encountered in conventionally manufactured biological substances. Others relate specifically to the nature of the cell chosen for production and some of these may have serious and undesirable consequences when administered to man. Clearly, endotoxin (lipopolysaccharide) should be considered as a contaminant if the recombinant protein is made in *E. coli* and measures must be taken to ensure its removal during purification. Residual host cell proteins are a problem whether bacteria, yeast or mammalian cells are used, but the nature of those proteins will of course be host-cell specific. There are, however, two major concerns which relate specifically to the use of transformed mammalian cells for production. These are the presence of contaminating cellular DNA and viruses.

Nucleic acid contamination from transformed (continuously growing) mammalian cells is a particular concern because of the possible presence of DNA of oncogenic potential. A Study Group convened by the World Health Organization concluded that there is no need to exclude transformed cell lines for producing recombinant proteins or other products, provided that steps are taken to minimize DNA contamination in the final product (World Health Organization 1987). Unfortunately, there is no direct evidence for a level of DNA which can be considered safe, or indeed unsafe. Estimates of a safe level of DNA are based on the extrapolation of data derived from experiments using highly oncogenic viral DNA in susceptible animals. After consideration of the data available and of calculations of the tumour-inducing potential of residual DNA concentrations, the Study Group concluded that there is negligible risk from heterogenous contaminating DNA at a concentration of 100 pg or less per parenteral dose.

Considerations of the problems associated with viral contamination of recombinant proteins are on surer grounds. Some potential contaminants are known to be highly pathogenic in man; others may give problems in immunosuppressed individuals. Furthermore, there is evidence that in the past several biologicals administered to humans have been contaminated with viruses with dire consequences for the recipients. HIV-contaminated blood products are a recent example. Potential viral contamination of recombinant products arises from two main sources. First, the host cells used in production may be contaminated with a virus indigenous to the species of origin. Cells may have a latent or persistent infection which may be transmitted vertically from one cell generation to the next since the viral genome persists within the cell, e.g. retroviruses; this may be expressed unexpectedly as an infectious virus. Second, adventitious viruses may be introduced by the use of contaminated animal products used in the manufacturing process. For example, mammalian cells might become contaminated with bovine viruses through the use of bovine sera during culture. It should also be borne in mind that any recombinant product, of bacterial or eukaryotic origin, might be contaminated with a murine virus if affinity chromatography utilizing a murine monoclonal antibody has been used during downstream processing of the product.

Clearly, the nature of the host cell used in manufacture, as well as the production method itself, greatly influences the nature and range of potential contaminants expected in the product. It is the purpose of the purification process to remove these completely or to ensure that they are kept below defined limits. A knowledge of the potential contaminants arising from the choice of host cell is essential in order to design effective purification strategies as well as appropriate quality control testing procedures.

5.6 QUALITY CONTROL

From the above it is seen that the production of therapeutic or prophylactic proteins by rDNA technology is a very sophisticated process, and one that is open to innumerable variable. Recombinant proteins, like insulin, growth hormone or tissue plasminogen activator, can be and often are produced by very different processes, in prokaryotic or eukaryotic cells, and can be made intracellularly or secreted into the medium. Such substances may therefore have subtle, but possibly very important, differences in product characteristics and purity, as well as clinical action. It is essential therefore that quality control strategies are developed to take account not just of the product itself, but also of the production process chosen. It is also important to bear in mind the scale-up which is generally required for commercial production. Extensive scale-up is usually needed as laboratory developments progress to full scale production and this may have considerable consequences for the quality of the product and implications for control testing. Unintended variability in the culture during this transition, or during production, may lead to changes which favour the expression of other genes in the host–vector system, or which lead to alterations to the polypeptide product. Such variations might result in decreased yield of product, quantitative or qualitative differences in impurities, or to subtle changes in the quality of the desired biologically active molecule. For instance, glycosylation may be affected by minor changes in cell culture technique; protease production by the host cell may also vary with culture conditions, resulting in unexpected nicking of the recombinant molecule. Controls therefore need to ensure the consistency of the production conditions as well as the final product. Emphasis is thus placed on ‘in-process’ controls and on the validation of certain aspects of the manufacturing process such as the ability of the purification system to remove unwanted molecules like DNA and viruses. The object of validation is to estimate quantitatively the overall levels of clearance of DNA or viruses achieved by the various stages of purification and/or any virus inactivation step. This is done by the deliberate addition (spiking) of significant amounts of say a virus to the crude material to be purified and to different fractions obtained during the various purification stages, and measuring its removal or inactivation during the subsequent stage of purification. The results provide a theoretical estimation of the removal of these contaminants and provide added assurance concerning the elimination of viruses which might be present below the threshold of detection of laboratory methods used for testing and selecting the source material. As far as viruses are concerned the most attractive option is to aim for freedom from contamination rather than relying on methods for their removal. Evidence that the host–vector cells used in manufacture, the seed pool, are free from adventitious agents is essential. However, it is recognized that certain cell lines contain endogenous viruses, e. g. retroviruses, which may not be readily eliminated, and it is for this reason that quality assurance strategies must include evidence for the effective removal of viruses which may inevitably be present in the source material as endogenous agents.

Given the need to ensure the safety, efficacy and consistency of the medicinal products made by rDNA technology, guidelines for control and standardization of these substances destined for use in man have been developed by several countries and organizations. Such guidelines are intended to indicate appropriate methods for the manufacture and testing of rDNA-derived products, as well as the information which might be submitted to national control authorities in support of clinical trial or market authorization applications.

Evaluation procedures are generally considered under three broad headings:

(1) starting materials,
(2) manufacturing process,
(3) control of final product.

The detailed quality control strategies recommended are discussed in papers by Duncan *et al.* (1987), Garnick *et al.* (1988), Cuthbert & Griffiths (1988) and Anicetti *et al.* (1989) and also in the recent guidelines from the Commission of the European Communities (1989a,b,c).

REFERENCES

Andersson, S. G. E. & Kurland, C. G. (1990) Codon preferences in free-living microorganisms. *Microbiological Reviews* **54** 198–210.

Anicetti, V. R., Keyt, B. A. & Hancock, W. S. (1989) Purity analysis of protein pharmaceuticals produced by recombinant DNA technology. *Trends in Biotechnology* **7** 342–349.

Anson, D. S., Austen, D. E. G. & Brownlee, G. G. (1985) Expression of active human clotting factor IX from recombinant DNA clones in mammalian cells. *Nature* **315** 683–685.

Bliem, R. & Katinger, H. (1988a) Scale-up engineering in animal cell technology: Part I. *Trends in Biotechnology* **6** 190–195.

Bliem, R. & Katinger, H. (1988b) Scale-up engineering in animal cell technology: Part II. *Trends in Biotechnology* **6** 224–230.

Busby, S., Kumar, A., Joseph, M., Halfpap, L., Insley, M., Berkner, K., Kurachi, K. & Woodbury, R. (1985) Expression of active human factor IX in transfected cells. *Nature* **316** 271–273.

Cockett, M. I., Bebbington, C. R. and Yarranton, G. T. (1990) High level expression of tissue inhibitor of metalloproteinases in Chinese Hamster Ovary cells using glutamine synthetase gene amplification. *Biotechnology* **8** 662–667.

Cuthbert, M. F. & Griffiths, E. (1988) Medicines from new biotechnologies. *Journal of the Royal Society of Medicine* **81** 126–128.

Dalboge, H., Bayne, S. & Pederson, J. (1990) *In vivo* processing of N-terminal methioninein *E. coli*. *FEBS Letters* **266** 1–3.

Duncan, M. E., Charlesworth, F. A. & Griffiths, J. P. (1987) Regulatory requirements for licensing medical products of biotechnology. *Trends in Biotechnology* **5** 325–328.

EEC Committee for Proprietary Medicinal Products: Ad Hoc Working Party on Biotechnology/Pharmacy (1989a) Notes to applicants for marketing authorizations on the production and quality control of monoclonal antibodies of murine origin intended for use in man. *Journal of Biological Standardization* **17** 213–222.

EEC Committee for Proprietary Medicinal Products: Ad Hoc Working Party on Biotechnology/Pharmacy and Working Party of Safety of Medicines (1989b) Notes to applicants for marketing authorizations on the pre-clinical biological safety testing of medicinal products derived from biotechnology (and comparable products derived from chemical synthesis). *Journal of Biological Standardization* **17** 203–212.

EEC Committee for Proprietary Medicinal Products: Ad Hoc Working Party on Biotechnology/Pharmacy (1989c) Notes to applicants for marketing authorizations on the production and quality control of medicinal products derived by recombinant DNA technology. *Journal of Biological Standardization* **17** 223–231.

Erlich, H. A., Gelfand, D. H. & Saiki, R. K. (1988) Specific DNA amplification. *Nature* **331** 461–462.

Ernst, J. F. (1988) Codon usage and gene expression. *Trends in Biotechnology* **6** 196–199.

Frykland, L. (1988) Development of high quality recombinant protein pharmaceuticals: in *Biotechnology Pharmaceuticals — A Regulatory Challenge*. BIRA Publications pp 9–14.

Galili, U., Clark, M. R., Shohet, S. B., Buehler, J. & Macher, B. A. (1987) Evolutionary relationship between the natural anti-Gal antibody and the Gal α1→3 Gal epitope in primates. *Proceedings National Academy Sciences* (*USA*) **84** 1369–1373.

Garnick, R. L., Solli, N. J. & Papa, P. A. (1988) The role of quality control in biotechnology: an analytical perspective. *Analytical Chemistry* **60** 2546–2557.

Goeddel, D. V., Kleid, D. G., Bolivard, F., Heyneker, H, Yansura, D., Crea, R., Hirose, T., Kraszewski, A., Itakura, K. & Riggs, A. (1979) Expression in *Escherichia coli* of chemically synthesized genes for human insulin. *Proceedings National Academy Sciences* (*USA*) **76** 106–110.

Gray, G. L., Baldridge, J. S., McKeown, K. S., Heyneker, H. S. & Chang, C. N. (1985) Periplasmic production of currently processed human growth hormone in *E. coli*: natural and bacterial signal sequences are interchangeable. *Gene* **39** 247–254.

Gribben, J. G., Devereaux, S., Thomas, N. S. B., Keim, M., Jones, H. M., Goldstone, A. H. & Linch, D. C. (1990) Development of antibodies to unprotected glycosylation sites on recombinant human GM-CSF. *Lancet* **335** 434–437.

Hirel, P. H. H., Schmitter, J. M., Dessen, P., Fayet, G. & Blanquet, S. (1989) Extent of N-terminal methionine excision from *Escherichia coli* proteins is governed by the side-chain length of the penultimate amino acid. *Proceedings National Academy Science* (*USA*) **86** 8247–8251.

Hsiung, H. M., Cantrell, A., Luirink, J., Oudega, B., Veros, A. J. & Becker, G. W. (1989) Use of bacteriocin release protein in *E. coli* for excretion of human growth hormone into the culture medium. *Biotechnology* **7** 267–271.

Kane, J. F. & Hartley, D. L. (1988) Formation of recombinant protein inclusion bodies in *Escherichia coli*. *Trends in Biotechnology* **6** 95–101.

Kang, S., Owens, D. R., Vora, J. P. & Brange, J. (1990) Comparison of insulin analogue B9 AspB27Glu and soluble human insulin in insulin-treated diabetes. *Lancet* **335** 303–306.

Konrad, M. (1989) Immunogenicity of proteins administered to humans for therapeutic purposes. *Trends in Biotechnology* **7** 175–178.

Lewin, B. (1990) *Genes IV* Oxford University Press, Oxford.

Looby, D. & Griffiths, B. (1990) Immobilization of animal cells in porus carrier culture. *Trends in Biotechnology* **8** 204–209.

Marx, J. L. (1988) Multiplying genes by leaps and bounds. *Science* **240** 1408–1410.

Parekh, R. B., Dwek, R. A., Edge, C. J. & Rademacher, T. W. (1989) *N*-Glycosylation and the production of recombinant glycoproteins. *Trends in Biotechnology* **7** 117–122.

Rademacher, T. W., Parekh, R. B. & Dwek, R. A. (1988) Glycobiology. *Annual Review of Biochemistry* **57** 785–838.

Stanley, P. (1989) Glycobiology: a new dimension in diversity. *Trends in Biotechnology* **7** 47–49.

Watson, S. P. & James, W. (1989) PCR and the cloning of receptor subtype genes. *Trends in Pharmaceutical Sciences* **10** 346–348.

Watson, J. D., Tooze, J. & Kurtz, D. T. (1983) *Recombinant DNA — a Short Course*, Scientific American Books.

Williams, J. G. & Patient, R. K. (1988) *In Focus–Genetic Engineering*. IRL Press, Oxford.

World Health Organization (1987) Acceptability of cell substitutes for production of biologicals. *WHO Technical Report Series* **747** 3–29.

6

High-pressure liquid chromatography of proteins

S. S. Bansal
Department of Pharmaceutical Chemistry, King's College London

6.1 INTRODUCTION

Recent advances in chemical synthesizers and recombinant DNA technology has provided a means of preparing virtually unlimited quantities of therapeutic peptides and proteins. Whilst this new technology can provide the initial peptide or protein product, the product must be extensively purified and analysed, prior to formulation. Analytical methods must be developed to determine the homogeneity of the protein product. Trace impurities such as proteins, nucleic acids, antibiotics, surfactants, and proteases must be determined. Special attention must also be paid to altered amino acid sequences derived from DNA mis-coding or amino acid misincorporation. Chromatographic techniques, particularly high-pressure liquid chromatography (HPLC), are a powerful tool in the analysis and separation of peptides and proteins. In this chapter, the range of applications of analytical and preparative HPLC are illustrated by example of the hormone insulin.

6.2 HIGH-PRESSURE LIQUID CHROMATOGRAPHY

Traditionally, proteins were separated using semi-rigid or non-rigid gels which have relatively low mechanical strength requiring low flow rates, and large particle sizes resulting in poor resolution and long separation times. Consequently protein analyses were dominated by electrophoresis techniques. With the development of high-pressure pumping systems providing constant flow rates and pressure-stable microparticulate packing materials, HPLC has become available to protein chemists. The applications of HPLC (Table 6.1) have had a great impact on protein chemistry.

The fact that proteins vary enormously in their behaviour can be utilized for their separation in HPLC. The diverse biological functions of proteins are largely due to

Table 6.1 — Some applications of HPLC in protein chemistry

Analyses
- Amino acid analysis
- Purity control of synthetic peptides
- Identification of multiple or mutant forms of proteins
- Molecular weight determinations
- Analytical peptide mapping
- Sequence analysis
- Metabolism studies — identification of prohormones

Preparative
- Isolation of natural products from extracts
- Purification of products obtained from chemical and gene technology
- Isolation of protein fragments from sequencing studies

different amino acid side-chains which govern the properties of the proteins. Separation of proteins can be achieved by exploiting the differences in molecular size, ionic properties, and polar and non-polar character as well as specific interaction with other molecules.

6.2.1 Size-exclusion HPLC

Interaction between the side-chains of a protein or peptide results in a specific tertiary structure for the protein and hence a difference in size which can be utilized for HPLC (Regnier & Gooding 1980). Size-exclusion (SE))HPLC became operational in 1980 with the development of uniform rigid particles which possessed relatively uniform pores, sufficiently large to be permeated by polypeptide molecules. The pore size is selected such that small molecules readily pass in and out of the pores essentially without hindrance while large molecules do not penetrate the pores and elute from the column first at the void volume. The elution volume available for separation thus lies between two limiting volumes: the volume of liquid between the particles (V_0) and the total diffusion volume which is the sum of V_0 and the pore volume (V_i). The elution volume (V_e) of a substance peak is given by:

$$V_e = V_0 + K_d \times V_i$$

where K_d is the distribution coefficient. In general there is a linear relationship between the elution volume or the distribution coefficient and the logarithm of the molecular weight of the protein, in the range K_d=0.15–0.85 (Regnier 1983). Therefore, SE-HPLC can be used not only to purify proteins but also to determine their molecular weight with an accuracy of 10%.

One important criterion in the choice of SE-HPLC column is that there should be no interaction between the polypeptide and the surface of the stationary phase. The problem is especially common with silica-based materials but can be minimized by bonding the surface with modifiers. Bonding of silica is often incomplete for steric reasons, so these chromatography materials exhibit residual ion-exchange properties. The effect of the slight negative charge on the residual silanols can be compensated by adding salts to the mobile phase. However, care should be taken not to use too high salt concentrations as such action can increase hydrophobic interactions. Should these occur, an organic modifier can be added to the mobile phase. Clearly, in SE-HPLC the choice of mobile phase is extremely important. Selection is also rendered critical by the possibility of the presence of denaturing agents which can influence the effective radius of the polypeptide, thereby moving its exclusion range to lower molecular weights. Columns containing support materials having a pore size from 5 to 400 nm are readily available and several different bonded silica materials find routine use. In particular, diol-modified silicas have found application for separation of proteins, although the elution behaviour can be influenced by ionic interaction due to unreacted silanols (Roumeliotis, & Unger 1979). Glycerylpropyl bonded surfaces are available in different pore diameters enabling the separation of a wide range of proteins and peptides. (Regnier 1983). They have good capacity for proteins and can be used for semipreparative separations. Other column supports are TSK-PW, which is a hydroxylated polyether copolymer and Superose, which is a cross-linked agarose derivative. Since these latter two column supports are not silica-based, they are stable over a much wider pH range. The polymeric supports allow very efficient protein separations, particularly in the high molecular weight range.

6.2.2 Ion-exchange chromatography

Because proteins possess a wide range of net charges, ion-exchange chromatography provides an ideal for the separation of proteins (Dizdaroglu 1984). The retention of proteins is dependent on the ionic interaction between the support surface and the charged groups in the protein. The strength of interaction is determined by the number of cooperatively effective ionic groups on the protein, the charge density of the ion-exchanger and the ionic strength of the mobile phase. Ion-exchange chromatography consequently provides a wide variety of variables for influencing retention and selectivity and this is one of the reasons for its popularity as an HPLC-based technique. The best choice for most protein separations is a pore size of 30 nm; however, for molecules with a molecular weight larger than 150 000 a pore size of 100 nm or more is preferred (Gooding & Schmuck 1985). Silica gel supports have been more frequently used, being available with a wide variety of different chemistries. Recently, however, hydrophilic polymers such as hydroxylated polyether-based supports (TSK-PW), have become available and provide good results (Nakamura & Kato 1985), in both analytical and preparative separations. The advantage of polymeric supports is their stability at high pH, enabling the use of a wider range of buffers and more stringent column cleaning procedures, hence increasing the working lives of the columns.

Conventional HPLC systems are made of stainless steel, which may corrode by the long-term use of high concentrations of chloride ions. In addition some proteins may be inactivated by metal ions leaching from the stainless steel. Hence systems based on inert materials have been developed (e.g. the Pharmacia-LKB FPLC and titanium HPLC systems). However a much more cost-effective approach is to use a conventional HPLC system with a mobile phase containing metal complexing agents such as EDTA or nitriloacetic acid (Venkatram & Rahman 1987).

Weak anionic and cationic ion-exchange materials are frequently used for peptides and proteins. Bonding silica with diethylaminoethyl (DEAE) or carboxymethyl (CM) moieties gives materials which are stable without unwanted hydrophobic interactions. The TSK-PW range is available in 13 nm pore size for small proteins. For preparative HPLC, a 50 nm weak cation-exchanger (Accell CM) is available (Millipore UK Ltd). Of the polymeric supports available for ion-exchange chromatography of proteins, the mono-Q and the Mono-S resins (Pharmacia Fine Chemicals) are the most prolific. These are macroporous, hydrophilic particles produced from polystyrene–divinylbenzene copolymer as monosized, 10-μm particles. The low dispersion allows high flow rates with low back pressure. They have high mechanical stability and do not suffer from swelling.

6.2.3 Hydrophobic interaction chromatography

The chromatographic methods available for the separation of proteins by virtue of their hydrophobicity are reverse-phase HPLC (RP-HPLC) and hydrophobic interaction HPLC (HIC-HPLC) (Fausnaugh *et al.* 1984). In both cases, hydrophobic functional groups are bound to the substrate and used as the functioning stationary phase.

However, in RP packings the functional groups are very densely distributed, producing a strong hydrophobic interaction between the protein and the stationary phase. As a result, organic solvents must be used to elute the proteins and this may cause loss of activity. By contrast, the functional groups in HIC packing material are more thinly distributed, so that elution can be accomplished with buffers that do not appreciably denature the protein. Until recently, hydrophobic packing materials for HIC-HPLC were not readily available. Early columns utilized silica as the support matrix, but dissolution of the silica under alkaline conditions severely limits the use of such columns. Recently polymer-based HIC columns have been developed. These are 10-μm porous (100 nm pore size) hydroxylated polyether support which is stable between pH 2–12, with a low density of covalently bound phenyl groups (Kato *et al.* 1984).

HIC provides a powerful means of separation which is applicable to the purification of most proteins. The diversity of potential eluting conditions can enable the resolution of even complex mixtures of proteins which would be difficult to separate by other techniques. Predicting the best conditions for separation is, however, difficult. The effectiveness of HIC is generally reduced by the presence of hydrophobic contaminants. An additional problem encountered in the use of HIC is the strong conditions which are frequently required to elute the protein from the hydrophobic matrices. This may involve the use of non-ionic detergents or ethylene glycol which can increase the likelihood of protein denaturation.

6.2.4 Reverse-phase chromatography

RP-HPLC has become the preferred method of HPLC because of its wide scope of applications (Regnier & Gooding 1980, Janssen *et al.* 1984; Hancock *et al.* 1981). It is characterized by the use of silica particles derivatized with alkyl functionalities (C_2–C_{18} alkyl chains) and an aqueous mobile phase containing an organic solvent such as methanol, propanol, acetonitrile or ethanol. It is basically a variant of HIC as discussed above. In contrast to HIC, significantly more non-polar, i.e. more heavily alkylated, stationary phases are employed so that proteins bind to the matrix even in aqueous buffer at low ionic strength.

The retention of the protein depends on the ratio of polar to non-polar amino acid residues on part of the surface that is involved in binding to the stationary phase. In the case of small peptides (fewer than 30 residues), it is possible to calculate the retention time time accurately from the sum of the hydrophobicities of the individual amino acids (Sasagawa *et al.* 1982). However, predictions are difficult for some proteins owing to their tertiary structure.

Almost all the packing materials available for RP-HPLC are silica-based. Some of these silicas are irregular in size, but the majority are spherical of 5, 7, and 10 μm mean diameter. Stationary phases with a pore size of 30 nm are best suited for proteins with a molecular weight greater than 20 000 with respect to loading capacity, resolution, and recovery. Advantage can be taken of the fact that salt-free eluents can be used in RP-HPLC to facilitate their use in preparative HPLC. A solvent frequently used is 0.1% trifluoroacetic acid which is readily available at the high purity required for UV monitoring at 215 nm. In addition to being a strong acid and highly volatile, it is an excellent solubilizing agent. Even at low concentration, it will protonate the carboxylate function of proteins, increasing the affinity of the protein for the reverse-phase surface (Mohoney & Hermodson 1980). RP-HPLC is unquestionably one of the most efficient methods of separation. Its greatest potential lies in the separation of closely related proteins and protein derivatives.

6.3 INSULIN PRODUCTION

Insulin is produced by the β-cells of the pancreas, and is critically involved in the regulation of glucose levels in the blood. The insulin molecule consists of two chains: the A-chain of 21 amino acid residues and the B-chain of 30 amino acid residues. The insulin molecule is stabilized by disulphide bonds linking the chains at A7–B7 and A20–B19. Within the A-chain there is a disulphide between A6–A11 Fig. 1. Consequently, insulin is a relatively rigid structure (Blundell *et al.* 1972). Insulin is insoluble at its isoelectric point at pH 5.4, it dissociates at acidic pH and forms dimers and hexamers at neutral pH; the hexamers are stabilized by zinc. Separation above the isoelectric point of insulin is only successful in the presence of zinc chelating agents and/or dissociating agents such as ethylenediaminetetra-acetic acid (EDTA), alcohols (Brange, 1987), urea (Sato *et al.* 1983), or detergents (Lougheed *et al.* 1983).

Insulin is an immensely important pharmaceutical, and an ideal candidate to demonstrate the scope and limitations of analytical and preparative HPLC. Bovine, porcine, and human insulin preparations are currently used in therapy. Bovine insulin differs from human insulin by two amino acid residues in the A-chain and one amino acid residue in the B-chain, whereas in porcine insulin the last amino acid residue in the B-chain is substituted (threonine and alanine).

6.3.1 Semi-synthesis of insulin

Of the four routes to human insulin which have been employed, extraction from pancreas (Burchermeister *et al.* 1975), chemical synthesis (Meienhoffer *et al.* 1963, Katsoyannis 1964), conversion of procine insulin (Morihara *et al.* 1979, Obermeier & Seipke 1984), and fermentation of suitably encoded micro-organisms (Goeddel

1979, Chance *et al.* 1981), only the latter two are currently used in the production of insulin.

Crude porcine insulin is reacted with a threonine ester, the reaction being catalysed by porcine trypsin in a mixture of water and organic solvents. This yields human insulin ester (Morihara *et al.* 1979) with the concomitant release of alanine (B30) (Fig. 6.1). The reaction can be described as a transpepidation reaction, in that

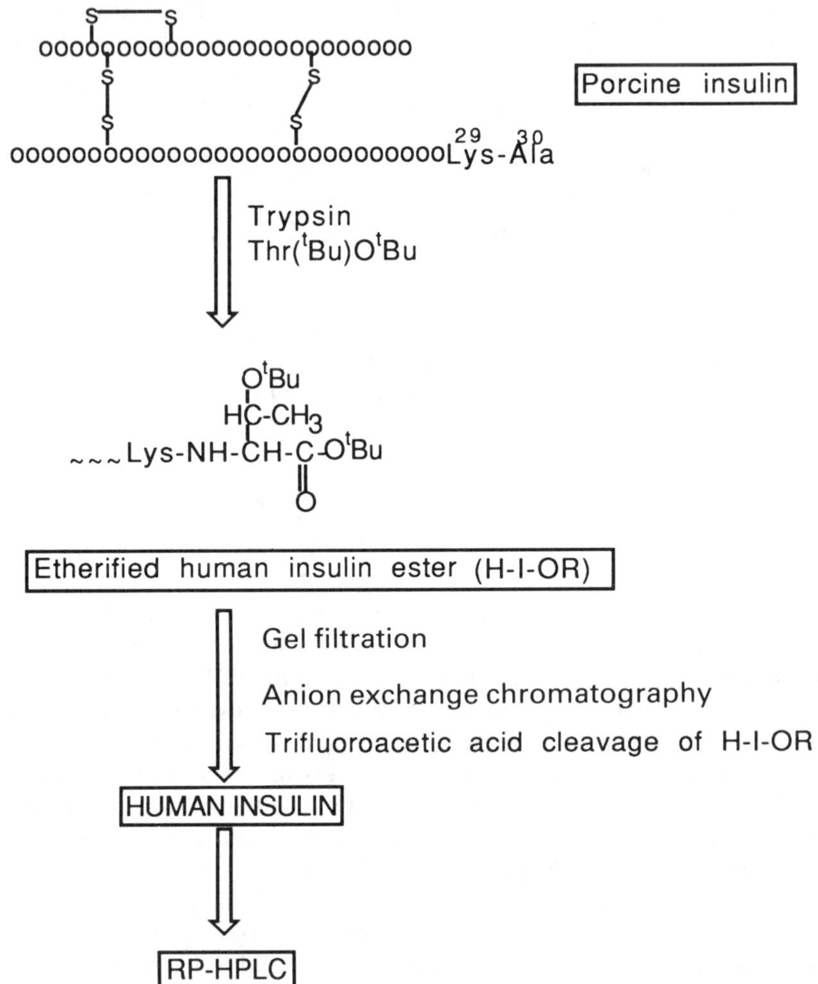

Fig. 6.1 — Semi-synthesis of insulin.

the acyl group of lysine (B29) is transferred from the amino group of the alanine (B30) to the amino group of the threonine ester. HPLC analysis of this procedure (Fig. 6.2) shows no des(B30) insulin is formed, however, some des(B22–30) insulin

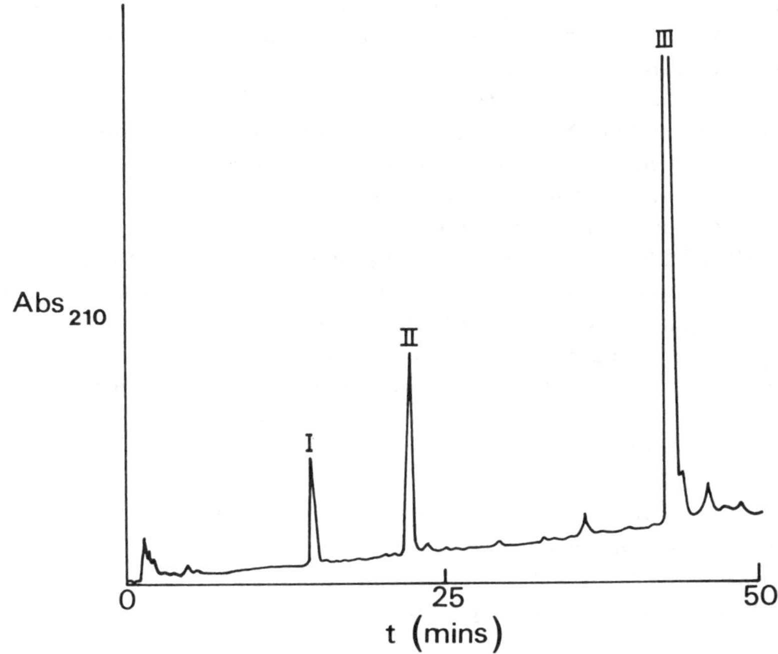

Fig. 6.2 — HPLC chromatogram of the crude reaction mixture obtained in the semi-synthesis of insulin. HPLC system: Rad Pak C8 10 μm. (see text) (Seipke *et al.* 1986).

(I), unconverted porcine insulin (II) and the desired side-chain etherified human insulin ester (III) are in evidence. The des(B22–30) insulin is formed by proteolysis at arginine (B22). The insulin product must be extensively purified and processed before it can be formulated into the final dosage form. As shown in Fig. 6.1, the semi-synthetic insulin route employs a number of purification steps. The etherified insulin ester is initially purified by size-exclusion chromatography in order to remove any trypsin, and non-converted porcine insulin and des-Ala(B30) insulin are removed by anion exchange chromatography. The side-chain and ester-protecting groups on threonine (B30) are hydrolysed with trifluoroacetic acid, and finally the semi-synthetic insulin is purified by RP-HPLC.

6.3.2 Production of human insulin by recombinant DNA technology

Insulin obtained from the pancreas is in its biologically active, correctly folded form. Gene technology on the other hand yields inactive and poorly soluble precursors of insulin. For instance in the *E. coli* cells, the precursors are deposited in the form of large inclusion bodies. Insulin in this form must be renatured and folded by chemical means (Chapter 2).

The production of biosynthetic human insulin (BHI) has been accomplished by gene technology in two ways. In the first method the A-chain and the B-chains are separately fermented, the chains are isolated as the *S*-sulphonate and then combined to yield BHI (Chance *et al.* 1981) (Fig. 6.3(a)). In the second method proinsulin is obtained in a single fermentation, and then isolated as the *S*-sulphonate, which is then folded to yield the proinsulin. The proinsulin is enzymatically converted into BHI (Fig. 6.3(b)) (Chance *et al.* 1981, Frank *et al.* 1981). Recently an RP-HPLC system has been developed (Snel 1987) for the quantification of recombinant DNA insulin precursors in yeast fermentation broth. HPLC monitoring can be achieved without interference from substrate components or proteins stemming from the host cell, thus serving to determine the optimum harvesting time.

Once the BHI is obtained from either the single-chain fermentation or the two-chain fermentation process the correct folding pattern must be established. This can be carried out using fingerprint analysis (Johnson 1983, Frank & Chance 1983, Grau 1985a). The principle of fingerprint analysis is staightforward. The protein is digested by the action of a protease into peptide fragments, the size and nature of these fragments being characteristic of the parent protein. The usefulness of the method depends on a number of factors:

— the formation of a reasonable number of well-sized fragments of interest;
— a minimum of non-specific reactions;
— a rapid efficient separation and detection method;
— the identification of individual fragments.

For insulin the *Staphylococcus aureus V8* protease is ideal for this purpose as it cleaves C-terminally at glutamate. Four of the eight glutamates are strategically positioned within the insulin sequence. The theoretical fragmentation pattern for insulin is shown in (Fig. 6.4).

Fragments I (A[5–8]–B[1–13]) and II (A[18–21]–B[14–21]) are particularly important for analysis since they contain the disulphide bridges. By-products from incorrectly folded product would give rise to new peaks at the expense of fragments I and II. An HPLC chromatogram of the fingerprint analysis adapted from Grau (1985a) is shown in Fig. 6.5.

In the next step of the single-chain fermentation method, the proinsulin is enzymatically transformed into BHI (Kemmler *et al.* 1971) by digestion with a mixture of trypsin and carboxypeptidase B. Proteolysis with trypsin occurs mainly at the six arginines and lysine. Cleavage at arginine (A0) at the N-terminal end of the A-chain, at arginine (B0) at the end of the N-terminal end of the B-chain, and at arginine (B32) are kinetically favoured. Cleavage at lysine (B29) yields des-Thr(B30) insulin in small amounts, while cleavage at arginine (B22), which would yield des(B23–B30) insulin, does not occur for steric reasons. Trypsin-catalysed proteolysis yields Arg-insulin which is further converted to insulin with carboxypeptidase B.

HPLC analysis is extremely useful for monitoring the processing of proinsulin into insulin. Time-dependent HPLC analysis of proinsulin digested with trypsin has demonstrated that both Arg-insulin and $(Arg)_2$-insulin occur as intermediates while no des(B23–30)-insulin is not observed (Seipke *et al.* 1986). Successful resolution of des Thr(B30)-insulin, human Arg-insulin and BHI is largely dependent on the level of perchlorate in the eluent. Perchlorate-containing eluents also tend to afford sharper peaks.

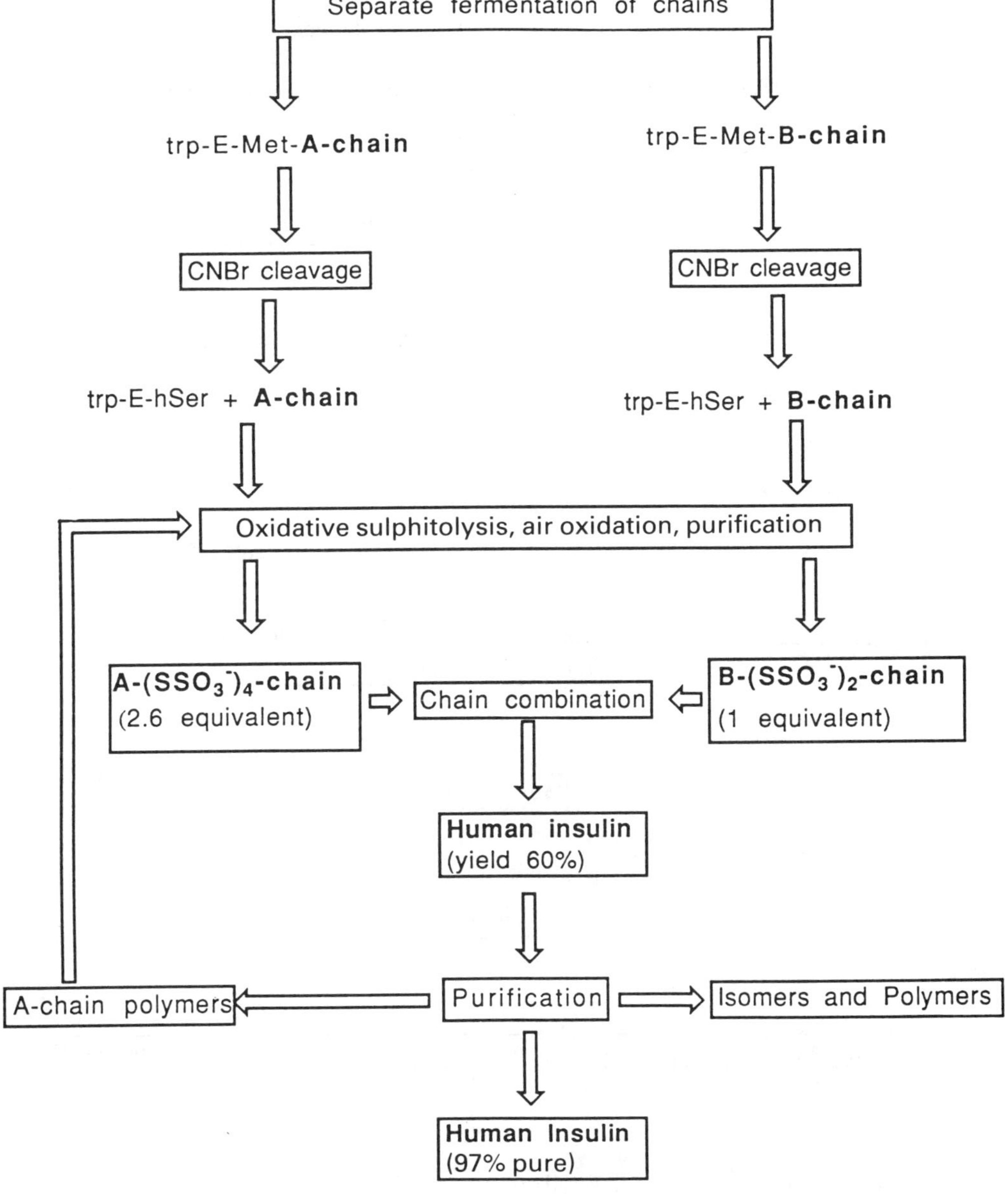

Fig. 6.3. — (a) Preparative steps in the chain combination strategy from chimerically bound A- and B-chains to human insulin (Chance *et al.* 1981a).

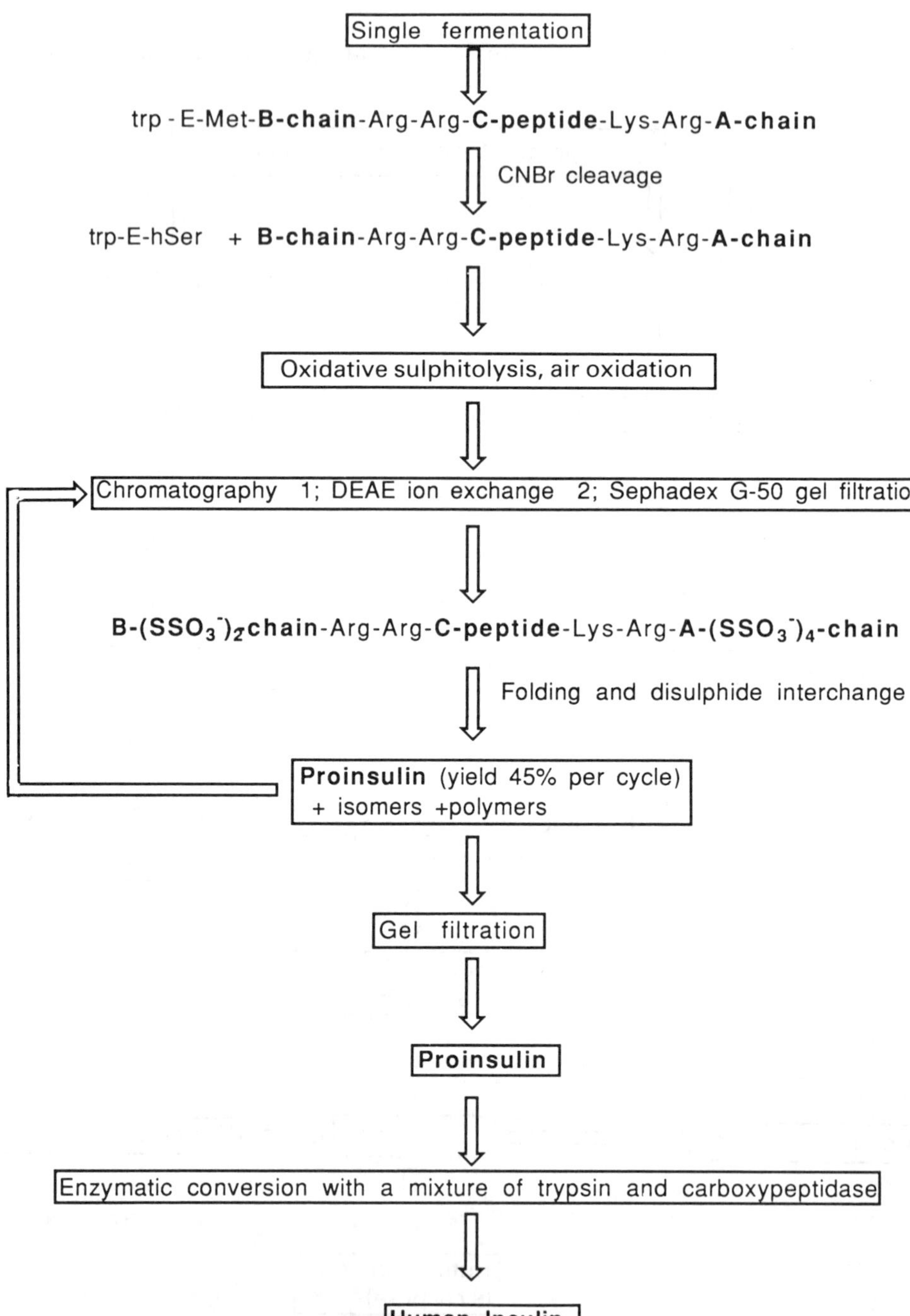

Fig. 6.3. — (b) Preparative steps in the proinsulin strategy from the chimerically bound linear sequence of proinsulin to BHI (Chance *et al.* 1981, Frank *et al.* 1981).

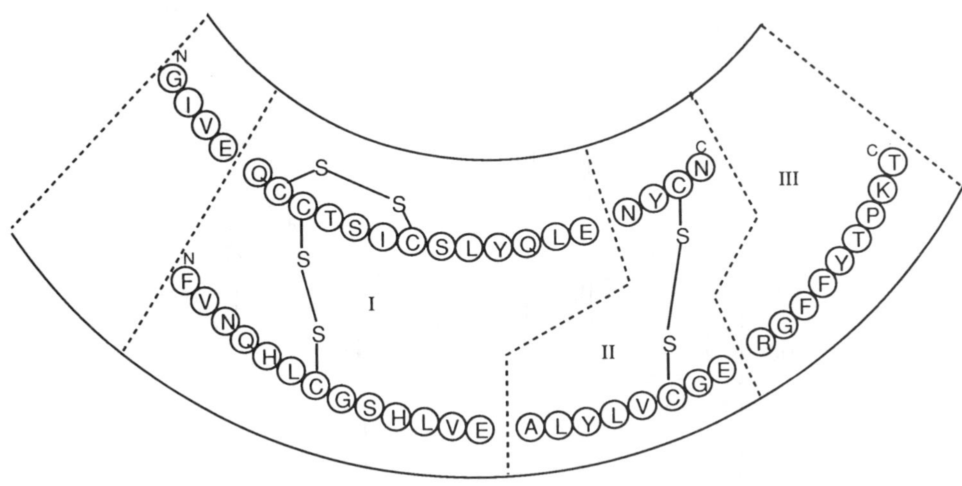

Fig. 6.4 — The theoretical *S. aureus* protease V8 peptide fragments of human insulin.

The purification of BHI is more complex than the procedure used for semi-synthetic insulin. The proinsulin route, which is the current method of choice for producing BHI, includes RP-HPLC as a routine step in the overall production scheme. In this procedure the *S*-sulphonate derivative of proinsulin is isolated and chromatographically purified. The *S*-sulphonate is then folded to form proinsulin, which is further purified before being enzymatically transformed to BHI. The resulting BHI is purified by ion-exchange chromatography and is then crystallized. The redissolved crystals serve as the charge material for RP-HPLC. Following RP-HPLC the insulin is further purified by size-exclusion chromatography and crystallization. In the proinsulin procedure a 10-μm Zorbax Process grade C8 silica was used as the stationary phase to purify kilogram quantities of BHI by RP-HPLC. (Kroeff *et al.* 1989). By inserting the RP-HPLC step late in the overall scheme the majority of the *E. coli*-derived impurities, which arise from the bacterial host system, have been removed by the preceding steps.

The BHI is identical to that produced in humans as determined by HPLC, gel electrophoresis, isoelectric focusing, circular dichroism, amino acid analysis, radioreceptor assay, radioimmunoassay, and rabbit hypoglycaemic assay. Furthermore, the purified product appears to have negligible contamination by foreign proteins (e.g. *E. coli* endotoxin) as determined by HPLC and the *Limulus* amoebocyte lysate assay (Chapter 9).

6.4 QUALITY CONTROL OF INSULIN

To ensure optimal potency and minimal immunogenicity, all insulins currently used in therapy are of high purity. When insulin was first introduced into therapy, preparations only contained 5 units per milligram. In contrast present day insulins possess an activity of 28 units per milligram. HPLC is currently proving to be a particularly useful tool for the determination of the insulin purity. RP-HPLC is used

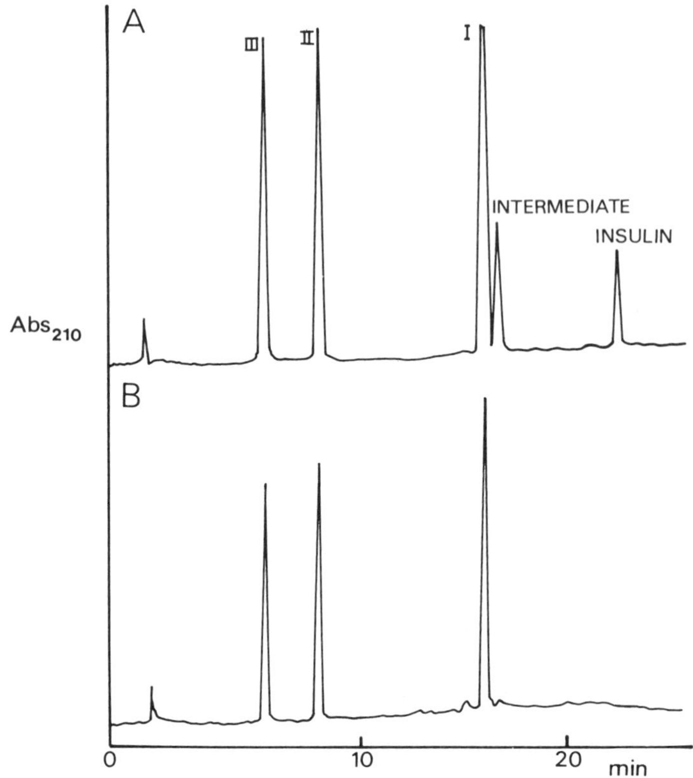

Fig. 6.5 — HPLC pattern of *S. aureus* V8 fingerprint of human insulin(A) after 60 min and (B) after 2 h. Intact insulin elutes at 22 min (see text) Grau 1985a). I, II and III are given in Fig. 6.4.

for fingerprint analysis of the proteolytic enzyme digest of production batches of insulin. Each batch is digested and chromatographed in parallel with an insulin standard. This is demonstrated with proinsulin (Fig. 6.5), and can be similarly performed with bovine, porcine, and human insulin. The resolving power of RP-HPLC for proteins of molecular weight 50 000 can be demonstrated by the assay developed for the determination of monomeric related substances in insulin (Smith *et al.* 1985). Human insulin can be separated from bovine and porcine insulins and their principle degradation products, desamido-(A21)-insulins (Farid *et al.* 1989) (Fig. 6.6).

RP-HPLC assays may also be used to detect other substances which are closely related to insulin, for example, des-Thr-(B30)-insulin, a by-product in the semi-synthesis of insulin. The two peptides are resolved using Zorbax C8 eluted with 70% 0.1 M phosphate buffer, pH 3.5 and acetonitrile containing 0.58% of the ion pairing agent pentane sulphonic acid (sodium salt), (Farid *et al.* 1989).

SE-HPLC is now routinely used for quality control and stability tests. In this method high molecular weight insulin derivatives, such as dimers originating as a result of transamidation, can be quantified. The high molecular weight derivatives

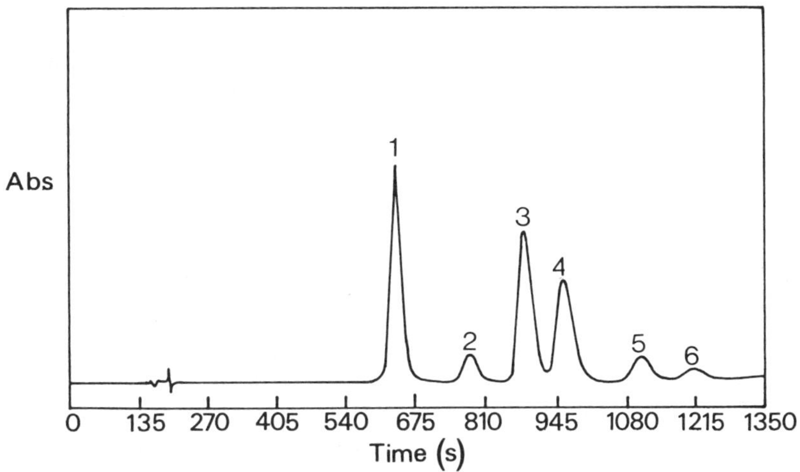

Fig. 6.6 — HPLC chromatogram of insulins: 1, Bovine insulin; 2, Bovine A21 desamido insulin; 3, Human insulin; 4, Purcine insulin; 5, Human desamido insulin; 6, Porcine A21 desamido insulin (Farid *et al.* 1989).

are separated using Zorbax GF-250 with a mobile phase of 70%, 0.4-M ammonium bicarbonate and acetonitrile. The method is equally applicable to bovine, porcine, and human insulins and to their formulations.

The potency of insulin has traditionally been determined by a rabbit biological assay. It is now widely recognized that appropriately developed RP-HPLC assays are superior to biological assays in many applications (Fisher & Smith 1986). RP-HPLC assay for insulin gives an estimate of the potency that is in excellent agreement with that obtained by the blood glucose method. Indeed, the HPLC assay gives more precise and more reproducible results (Table 6.2) (Smith *et al.* 1985).

Table 6.2 — Comparison of HPLC potency and rabbit bioassay results of human insulin zinc crystals (Smith *et al.* 1985)

	X	RSD	Range
Biopotency (U/mg)	28.06	2.08%	26.4–29.0
HPLC Potency (U/mg)	28.33	1.00%	27.8–29.0

With insulin (drug substance) and insulin injections subjected to accelerated stability tests, the HPLC assay method can detect decomposition that cannot be detected by either blood glucose assay or the immunochemical assay (Fisher & Potter, 1981). A decrease in the immunochemical assay will only be caused by chemical decomposition of the antigenic determinant, whereas decomposition of the other parts of the molecule might leave the antigenic determinant intact.

Throughout the 1980s there has been increasing pressure on regulatory authorities to replace the animal response assay by an HPLC assay. The United States Pharmacopoeia has already included an HPLC method of assay in its insulin monograph, (US Pharmacopeia) and the British and European Pharmcopoeias are about to make the change.

6.5 STABILITY OF INSULIN IN ARTIFICIAL DOSAGE DEVICES

In healthy individuals the pancreas excretes a small amount of insulin as a basal supply and larger amounts as blood glucose increases after meals. This induces the blood glucose levels to fall to normal (5 mmol/l). However, normal secretion of insulin in healthy individuals can be achieved by means of artificial infusion systems. Unfortunately, during the development of these systems, a severe problem emerged. Insulin has a tendency to denature, and aggregate, leading to precipitation. In external infusion pumps, catheter blockage is a significant source of clinical complication and infusion of altered insulin has been implicated as a source of severe problems (Lougheed *et al.* 1983, Irsigler and Ktiz 1979). Difficulties associated with implantable pumps are even greater, because residence times in the reservoir may be relatively long, thermal exposure is greater, and there is long-term contact with metallic surfaces. To overcome these problems various insulin preparations have been considered, including the addition of serum (Albisser *et al.* 1980), bicarbonate (Schade *et al.* 1981), calcium (Brange & Havelund 1983), glycerol (Blachshear *et al.* 1983) and even the application of sulphated insulin (Normura *et al.* 1983). HPLC methodology has been critical for these studies. Encouraging *in vitro* and *in vivo* results have been reported with neutral insulin stabilized by the addition of the surface-active polyethylenepolypropylene glycol (HOE 21 PH) (Grua 1985b). In acidic solution (pH 3), insulin undergoes rapid deamidation at asparagine (A21); however, in the implantable devices deamidation can occur at several of the six asparagines and glutamines. The HPLC chromatogram of acidic porcine insulin in the dosage device shows a number of modified insulin products (Grua 1985b). These effects are minimized with the HOE 21 PH insulin solution.

6.6 SUMMARY

Within the last few years there has been an explosive growth in the application of HPLC in protein chemistry. Newly developed support materials made of extremely small non-porous particles (smaller than 2 μm), (Seipke *et al.* 1986) promise even further improvements. The successful high resolution separation of kilogram quantities of insulin is very encouraging for the preparative separation of proteins in general (Kroeff *et al.* 1989). A crucial problem is the biocompatibility of stationary phases and mobile phases. While SE- and ion-exchange HPLC are carried out under approximately physiological conditions, separations using reverse-phase may involve the risk of denaturing the protein. In principle HIC-HPLC can avoid this problem since it is based on the same criteria as reverse-phase but requires much milder conditions. The introduction of polymer supports is advantageous as a wider range of buffers may be used to carry out separations under milder conditions (Kato *et al.* 1984).

It is clear that HPLC is an invaluable tool in the development of peptide and protein pharmaceuticals. HPLC is essential in the characterization of protein and peptide products obtained from either chemical synthesis or gene technology. It is a powerful tool in the preparation, purity determination and stability analysis of these products.

In the future we may expect all proteinaceous drug substances to be assayed by HPLC methods. The future also holds great potential for proteins being profiled by tryptic mapping/HPLC/mass spectroscopy (Borman 1987). This combined technique is capable of providing a fingerprint of the protein where the slightest modification can be detected.

REFERENCES

Albisser, A. M., Lougheed, W. D., Perlman, K. & Bahoric, A. (1980) Nonaggregating insulin solutions for long-term glucose control in experimental and human diabetes. *Diabetes* **29** 241–243.

Blachshear, P. J., Rohde, T. D., Palmer, J. L., Wigness, B. D., Rupp, W. M. & Buchwald, H. (1983) Glycerol prevents insulin precipitation and interruption of flow in an implantable insulin infusion pump. *Diabetes Care* **6** 387–392.

Blundell, T. L., Dodson, G. G., Hodgkin, D. C. & Mercola, D. A. (1972) Insulin: the structure in the crystal and its reflection in chemistry and biology. In: *Advances in Protein Chem.* **26** 379–402.

Borman, S. (1987) Analytical biotechnology of recombinant products. *Anal. Chem.* **59** 969–973.

Brange, J. (1987) *Galenics of insulin. The Physio-chemical and pharmaceutical properties of insulin and insulin preparations.* Springer Berlin, Heidelberg, NY, Tokyo.

Brange, J. & Havelund, S. (1983) Insulin pumps and insulin quality — requirements and problems. *Acta Med. Scand. Suppl.* **671** 135–138.

Burchermister, W., Enzmann, H. H. and Schone, H. H. (1975) The isolation of insulin from the pancreas. Bruchhausen, F. V. & Hasselblatt, A. (eds) In: *Handbach der experimental Pharmakologie*. Springer, Berlin. p. 715.

Chance, R. E., Kroeff, E. P., Hoffmann, J. A. & Frank, B. H. (1981) Chemical, physical and biologic properties of biosynthetic human insulin. *Diabetes Care* **4** 147–154.

Dizdaroglu, M. (1984) Separation of peptides by high-performance ion-exchange chromatography. In: Hancock, W. S. (ed.) *CRC Handbook of HPLC for ther Separation of Amino acids, Peptides, and Proteins,* CRC Press, Boca Raton, Vol. 11, pp. 23–46.

Farid, N. A. Atkins, L. M., Becker, G. W., Dinner, A., Heiney, R. E., Miner, D. J. & Riggin, R. M. (1989) Liquid chromatographic control of the identity of potency of biomolecules used as drugs. *J. Pharmac. & Biomed. Anal.* **7** 185–188.

Fausnaugh, J. L., Pfannkoch, E., Gupta, S. & Regnier, F. E. (1984) High-performance hydrophobic interaction chromatography of proteins. *Anal. Biochem.* **137** 464–472.

Fisher, B. V. & Potter, P. B. (1981) Stability of bovine insulin. *J. Pharm. Pharmacol.* **33** 203–206.

Fisher, B. V. & Smith, D. (1986) HPLC as a replacement for the animal response assay for insulin. *J. Pharmaceut. & Biomed. Anal.* **4** 377–387.

Frank, B. H. & Chance, R. E. (1983) Two routes for producing human insulin utilising recombinant DNA technology. *Munch Med. Wochenschr* **125** [Suppl 1] 14–20.

Frank, B. H., Pettee, J. M., Zimmermann, R. E. & Burke, P. J. (1981) The production of human proinsulin and its transformation to human insulin and C-peptide. In: Rich, D. H. & Gross, E. (eds) *Peptides. Proceedings of the Seventh American Peptide Symposium*, Pierce Chemical Company, Rockford, Illinois.

Goeddel, D. A., Kleid, D. G., Bolivar, F., Heyneker, H. L., Yansura, D. G., Crea, R., Hirose, T., Kraszewski, A., Irtakura, J. & Riggs, A. D. (1979) Expression in *Escherichia coli* of chemically synthesised genes for human insulin. *Proc. Natl. Acad. Sci.* **76** 106–110.

Gooding, K. M. & Schmuck, M. N. (1985) Comparison of weak and strong high-performance anion exchange chromatography. *J. Chromatogr.* **327** 139–146.

Grau, U. (1985a) Fingerprint analysis of insulin and proinsulin. *Diabetes* **34** 1174–1180.

Grau, U. (1985b) Chemical stability of insulin in delivery system environment. *Diabetologia* **28** 458–463.

Grau, U. & Saudek, C. D. (1987) Stable insulin preparations for implanted insulin pumps. *Diabetes* **36** 1453–1459.

Hancock, W. S., Capra, J. D., Bradley, W. A. & Sparrow, J. T. (1981) The use of reverse-phase high-pressure liquid chromatography with radial compression for the analysis of peptides and proteins. *J. Chromatogr.* **206** 59–70.

Irsigler, K. & Kritz, H. (1979) Long term continuous intravenous insulin therapy with a portable insulin dosage regulating apparatus. *Diabetes* **28** 196–203.

Janssen, P. S. L., van Nipsen, J. W., Hamelinck, R. L. A. E., Melgers, P. A. T. A. & Goverde, B. C. (1984) Application of reversed-phase HPLC in some critical peptide separations. *J. Chromatogr. Sci.* **22** 234–238.

Johnson, I. S. (1983) Human insulin from recombinant DNA technology. *Science* **219** 632–637.

Kato, Y., Kitamura, T. & Hashimoto, T. (1984) New support for hydrophobic interaction chromatography of proteins. *J. Chromatogr.* **292** 418–426.

Katsoyannis, P. G. (1964) The synthesis of insulin chains and their combination to biologically active material. *Diabetes* **13** 339–344.

Kemmler, W., Peterson, J. D. & Steiner, D. F. (1971) Studies on the conversion of proinsulin to insulin. *J. Biol. Chem.* **246** 6786–6791.

Kroeff, E. P., Owens, R. A., Campbell, E. L., Johnson, R. D. & Marks, H. I. (1989) Production scale purification of biosynthetic human insulin by reversed-phase HPLC. *J. Chromatogr.* **461** 45–61.

Lougheed, W. D., Albisser, A. M., Martindale, H. M., Chow, J. C. & Clement, J. R. (1983) Physical stability of insulin formulations. *Diabetes* **32** 424–432.

Meienhoffer, J., Schnabel, E., Bremer, H., Brinkhoff, O., Zabel, R., Sroka, W., Kloslermeyer, H., Brandenberg, D., Okuda, T. & Zahn, H. (1963) Synthese der Insulinktten und ihre Kombination zu insulinaktiven Praparten. *Z. Naturforsch.* **18B** 1120–1121.

Mohoney, W. C. & Heremodson, M. A. (1980) Separation of large denatured peptides by reverse-phase HPLC-trifluoroacetic acid as a peptide solvent. *J. Biol. Chem.* **255** 11 199–11 203.

Morihara, K., Oka, T. & Tsuzuki, H. (1979) Semi-synthesis of human insulin by trypsin catalysed replacement of Ala B30 by Thr in porcine insulin. *Nature* **28** 412.

Morihara, K., Ueno, Y. & Sakina, K. (1986) Influence of temperature in the enzymic semi-synthesis of human insulin by coupling and transpeptidation methods, *Biochem. J.* **240** 803–810.

Nakamura, K. & Kato, Y. (1985) Preparative high-performance ion-exchange chromatography. *J. Chromatogr.* **333** 29–40.

Normura, M., Zinmam, B., Bahoric, A., Marliss, E. B. & Albisser, E. M. (1983) Intravenous infusion of sulphated insulin normalises plasma glucose level in pancreatectomized dogs. *Diabetes* **32** 788–792.

Obermeier, R. & Seipke, G. (1984) Enzyme catalysed semi-synthesis with proinsulin. *Process Biochem.* **19** 29–32.

Regnier, F. E. & Gooding, K. M. (1980) High-pressure liquid chromatography of proteins. *Anal. Biochem.* **103** 1–25.

Regnier, F. E. (1983) High-performance liquid chromatography of biopolymers. *Science* **222** 245.

Roumeliotis, P. & Unger, K. K. (1978) Structure and properties of *n*-alkyldimethylsilyl bonded silica reversed-phase packing. *J. Chromatogr.* **149** 211–214.

Sasagawa, T., Okuyama, T. & Teller, D. C. (1982) Prediction of peptide retention times in reverse-phase high pressure liquid chromatography during linear gradient elution. *J. Chromatogr.* **240** 329–340.

Sato, S., Ebert, C. D. & Kins, S. W. (1983) Prevention of insulin self-association and surface adsorption. *J. Pharm. Sci.* 228–232.

Schade, D. S., Eton, R. P. & Spencer, W. (1981) Implantation of an artificial pancreas — current perspectives. *J.A.M.A.* **245** 709–710.

Seipke, G., Mullner, H. & Grau, U. (1986) High-pressure liquid chromatography of proteins. *Angew. Chem. Int. Ed. Engl.* **25** 535–552.

Selam, J. L., Zirinis, P. & Mirouze, J. (1987) Stable insulin for implantable delivery systems. *Diabetes Care* **10** 343–347.

Smith, H. W., Atkins, L. M., Binkley, D. A., Richardson, W. G. & Miner, D. J. (1985) A universal HPLC determination of insulin potency. *J. Liq. Chromatogr.* **8** 419–439.

Snel, L. (1987) HPLC quantification of rDNA polypeptides like insulin precursors produced in yeast. *Chromatographia.* **24** 329–333.

United States Pharmcopeia XX. The United States Pharmacopeia/The National Formulary NF XV US Pharmacopeial Convention.

Venkatram, S. & Rahman, V. E. (1987) High-performance liquid chromatographic analysis of des ferrioxamine using a metal free system. *J. Chromatogr.* **411** 494–497.

7

The stabilization of proteins by freeze-drying and by alternative methods

S. F. Mathias, F. Franks and R. H. M. Hatley
PAFRA Ltd., Biopreservation Division, 150 Science Park, Milton Road, Cambridge CB4 4GG

7.1 INTRODUCTION

In the past decade significant progress in isolation and separation techniques has led to a dramatic increase in the number of biomolecules (mainly proteins) which can be made available to the bioindustries. The world market value of products which find their end uses as biochemical reagents, therapeutic preparations (pharmaceutical or veterinary) and diagnostic reagents is put at more than $300 billion. The European market in some products, e.g. anti-infective drugs, is expected to grow by some 30% per annum in the next few years (Frost & Sullivan 1989).

A major problem facing the producer, distributor and user of such valuable products is that they are in many cases inherently unstable and lose their activity more or less quickly when isolated from their natural environment. A brief overview of the mechanisms of instability and stabilization is given in Table 7.1 — deterioration under discussion here is generally due to conformational instability. The biological functions of isolated proteins are closely linked with their unique structures, and in most cases this structure is only stable within quite narrow ranges of temperature, pH, pressure, solvent composition, etc. If conditions deviate from these limits then activity is lost. The inactive protein is said to be denatured, and in practice this inactivation is usually irreversible. However, even if conditions remain within the limits necessary to maintain the native state, the activity of enzymes, hormones and blood factors tends to decline with time, for reasons which are not clear. It may still be that gradual denaturation is caused by changes in the environmental conditions with time, or possible that trace amounts of proteolytic enzymes are always present in the isolate (Hatley *et al*. 1987).

Proteins achieve maximum stability when conditions restrict interaction with other molecular species and conformational changes leading to unfolding are

Table 7.1 — Factors which influence instability and stabilization of proteins

Protein instability	
Chemical (covalent)	
Hydrolysis	S–S interchange (disulphide isomerase)
Oxidation	Thio/S–S exchange
β-Elimination	Maillard reaction
Isopeptide formation	(NH_2 + reducing sugar)
Racemization	
Conformational	
pH	Metal binding
Hydrophobic (detergent)	Temperature
Pressure	Shear
Sorption	Lyotropism
Immobilization	
Mutant	
Sequence change	
Protein stabilization	
In solution	Chemical modification
Immobilized	Solid state (glass)
Cofactors/substrates	Frozen
Undercooled (subzero temperature)	Mutant
Cross-link	Heat shock protein

prevented. Therefore proteins in crystalline form, or in a glassy amorphous matrix with extreme viscosity (the product from successful freeze-drying or spray-drying, for example) are very stable, as are proteins in liquid formulation kept at sufficiently low temperature.

The aim of any protocol for stabilization of such products must be to ensure an acceptable shelf-life. Convenience of use and cost-effectiveness are sometimes thought to be of secondary importance given the high value of many of these products, but must surely become more important as the industry matures. The scope of this review is to consider briefly the methods and their relative advantages currently available for the stabilization of biomolecules.

They are:

I 'Traditional methods'
- — freeze-drying
- — stabilization by additives
- — spray-drying

II 'Novel methods'
- — undercooling
- — Permazyme

Freeze-drying is very commonly used as the preferred method of stabilization for labile biomolecules, and as such will be given a more detailed consideration than the alternative methods.

7.2 TRADITIONAL METHODS

7.2.1 Freeze-drying

Surveys of the relevant literature and of current industrial practice show that the optimization of freeze-drying protocols is frequently regarded as an empirical activity. However, there is little doubt that an appreciation of the basic scientific principles and the relationships which determine the efficiency of the freezing and drying processes can result in substantially improved products in terms of yield and stability (Franks *et al.* 1987, Franks 1990a,b). Freeze-drying (lyophilization) is commonly the final step in a sequence of carefully controlled processes for the production of valuable biomolecules. When performed correctly a uniform, dry, lightweight product with an extended shelf-life at ambient temperatures should result. That suppliers frequently recommend storage at 4°C or even −20°C shows that their products are not stable at ambient temperatures, and indicates that freeze-drying protocols have not been optimized.

The physical parameters which govern the behaviour of aqueous solutions during freezing and drying are summarized in Scheme 7.1 whereas the chemical changes which can accompany freezing of aqueous solutions containing proteins are shown in Scheme 7.2. Of paramount practical importance in designing the freeze-drying cycle is the glass/rubber transition of the formulation, as described below. However, if products of quality and stability are to be achieved, all the factors discussed below should if possible be known and taken into account. A more detailed review of the freeze-drying process, including economic considerations and calculation of cycle times, has been discussed by Franks (1990b).

The solution stability of the active component (protein), i.e. possible destabilizing effects of concentration, temperature, pH, salt type and/or stabilizing effects of protectant additives, should be known, as significant changes will occur during the freezing stage as water is progressively removed as ice (see Scheme 7.2). Similarly, precipitation/crystallization/reactions of the various components during freeze concentration should be taken into account. However, it has been shown that during freeze-drying these components (phosphate buffers for example) do not necessarily behave in the way predicted from phase diagrams of salt solutions freezing under equilibrium conditions (Murase & Franks 1989).

Consideration should be given to achievement of optimum freezing conditions to produce a uniform ice crystal dispersion and a freeze concentrate of uniform composition and texture. This can be difficult with the limitations of large scale equipment.

The glass temperature of the freeze concentrate (T_g') is of fundamental importance. At T_g', ice crystallization in the formulation becomes so slow as to be undetectable and the freeze concentrate is said to be glassy. Fig. 7.1 shows a state diagram for a sucrose/water mixture which has T_g' at −32°C. T_g' is characteristic for each formulation, and its importance lies in the fact that at this temperature the

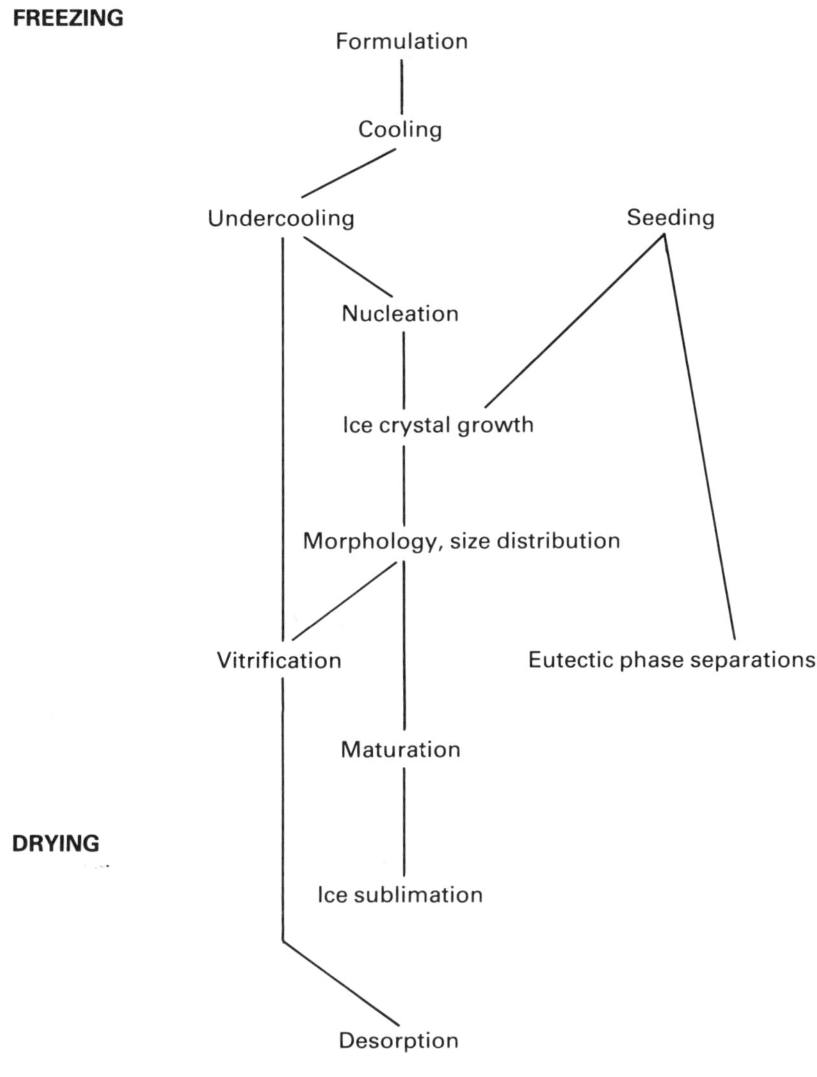

Scheme 7.1 — Important physical parameters which control freezing and drying behaviour (redrawn from Franks 1990b).

formulation is maximally frozen and the freeze concentrate exists as an amorphous glass. Thus the maximum amount of water possible can be sublimed as ice, and the remaining material is glassy and mechanically strong. T_g' therefore represents the

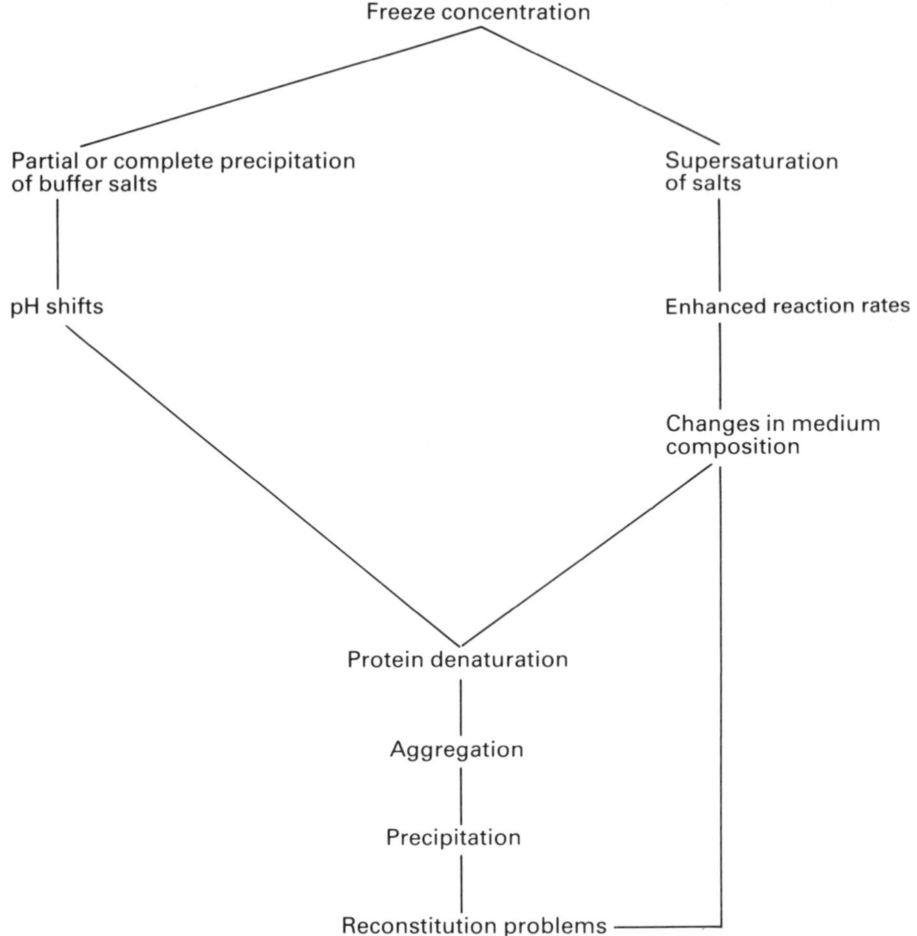

Scheme 7.2 — Chemical changes which can accompany freezing of proteins in aqueous solutions (redrawn from Franks 1990b).

temperature to which the product should be frozen initially and at which the product should be held during primary drying. If the product temperature rises too far during primary drying, then partial melt-back of ice into the product can occur, leading to collapse as the freeze-concentrate becomes rubbery. T_g' can be conveniently measured by differential scanning calorimetry. Fig. 7.2 shows a typical example of the glass transition seen during warming a previously frozen therapeutic proteinaceous product (Hatley *et al.* 1989).

The moisture content of the freeze concentrate (W_g'), is also of vital importance, as it represents the amount of water which will remain after primary drying and which will need to be removed by heating during secondary drying. Fig. 7.1 shows the W_g' for a sucrose/water mixture. Usually, W_g' should be low in order to shorten the

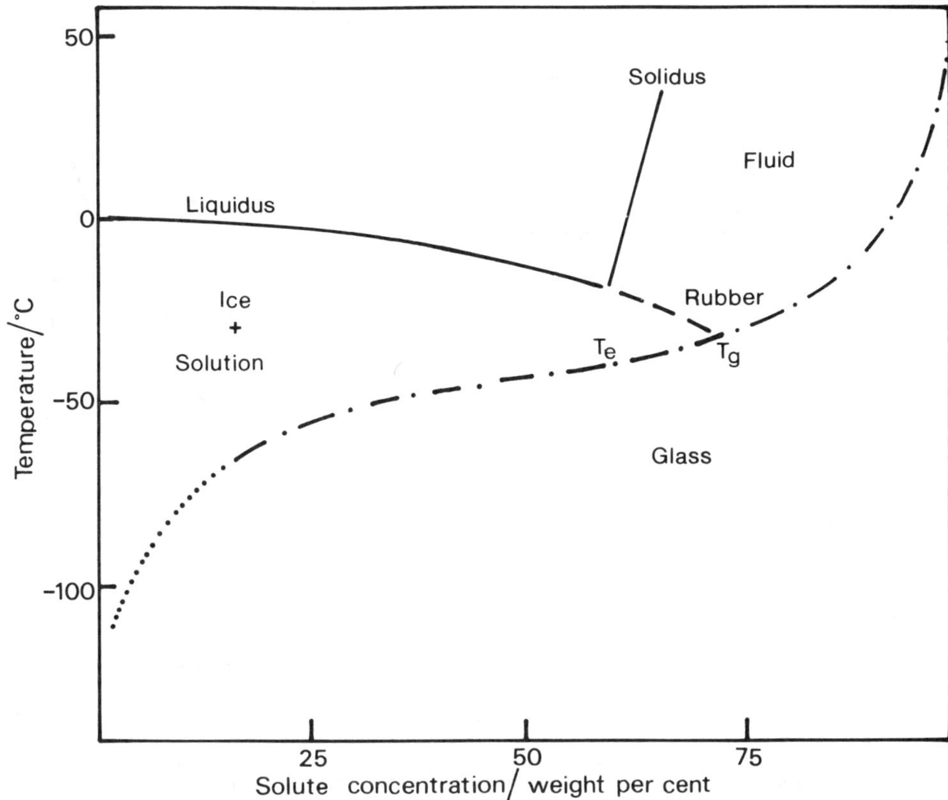

Fig. 7.1 — Solid/liquid state diagram of the sucrose water system. Solid lines correspond to the equilibrium coexistence curves, the broken line indicates supersaturation, and the dotted and broken/dotted curve is the T_g isoviscosity curve (10^{14} Pa s), below which the homogeneous phase is a glass. T'_g is the intersection of the liquidus (T'_g) curve and the glass transition curve and T_e is the (probable) eutectic point. The exact course of the T_g curve below -50°C is uncertain. Redrawn from Franks (1990b).

secondary drying cycle. Table 7.2 shows T'_g, T_g of the anhydrous product and W'_g for various commonly used excipients (Levine & Slade 1988, Finegold *et al.* 1989). Knowledge of these factors can influence the choice of excipients in the final formulation of a product, as the 'correct' choice can help to shorten drying times and increase the likelihood of a stable product. However, the total formulation needs to be taken into account as it is the ratio of salt:protein:excipient which finally determines T'_g. Therefore standard buffer/excipient mixtures with different concentrations of protein added will each have a different T'_g and the optimized freeze-drying protocol will be different for each.

The secondary drying cycle is designed to remove the residual moisture (unfrozen water), in order to render the dried product stable at ambient temperatures. The plot of T_g is an isoviscosity curve: at temperatures above T_g the viscosity of the mixture drops and the material becomes a deformable rubber. At and below T_g the

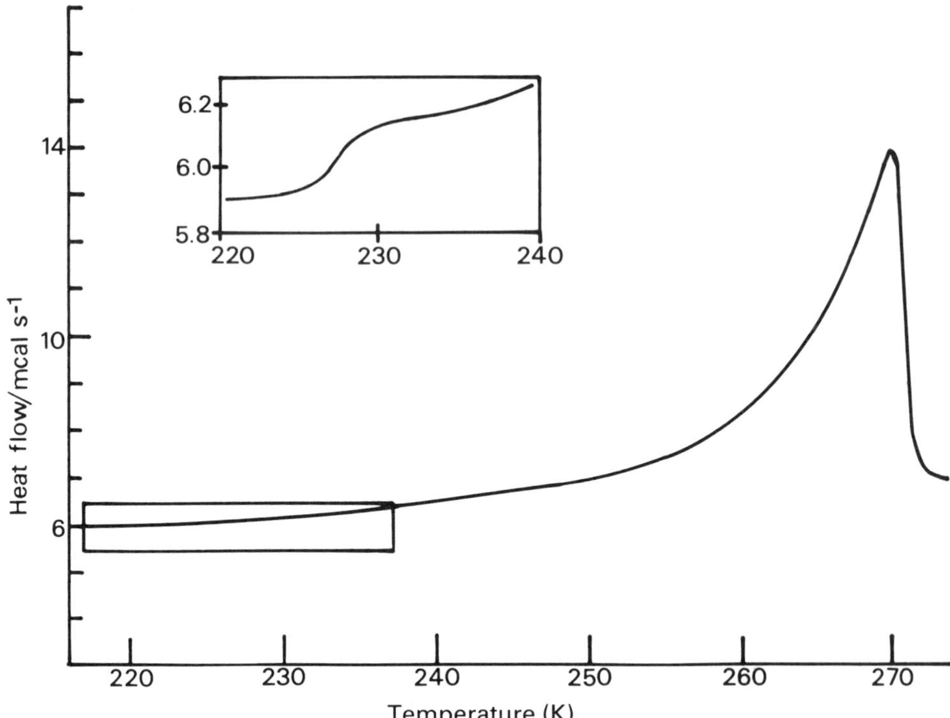

Fig. 7.2 — Typical DSC heating scan of a frozen therapeutic proteinaceous product, showing the usual broad ice-melting endotherm characteristic of ice melting into a concentrated solution. The inset corresponds to the magnification of the section marked by the rectangle; T'_g is clearly visible, centred on 227 K. Data from Hatley *et al.* (1989). Figure redrawn from Franks (1990b).

Table 7.2 — Glass/rubber transitions of anhydrous and freeze concentrated carbohydrates; data selected from Levine and Slade (1989) and Finegold *et al.* (1989)

	T'_g (°C)	T'_g (°C)	Wt% water
Glycerol	− 93	− 65	46
Glucitol (sorbitol)	− 3	− 43.5	18.7
Glucose	39	− 43	29.1
Mannose	30	− 41	25.9
Fructose	13	− 43	49
Maltose	43	− 29.5	20
Trehalose	77	− 29.5	16.7
Sucrose	57	− 32	35.9
Glucose/fructose (equimolar)	20	− 42.5	48
Maltotriose	56	− 23.5	31

viscosity remains high (above 10^{14} Pa s) and the material remains glassy. If the product is to remain stable during storage then the storage temperature must remain below T_g. Knowledge of the glass transition curve as a function of moisture content enables drying to be controlled such that excursions from this $T_g(W_g)$ isoviscosity curve are never excessive. For a stable high quality product, W must be $< W_g$ at ambient temperatures.

The shelf-life of the product depends not only on the amount of residual moisture, and thus T_g, but on the distribution of the residual moisture within the 'dry' product.

In practice, the quality of a freeze-dried bioproduct is judged in terms of the yield of active product, its shelf-life and rehydratability. The yield largely depends on the details of the formulation, i.e. the relative concentrations of the components, the choice of buffer and excipients, etc. Shelf-life and rehydratability on the other hand are largely determined by processing details, such as the filling volume, the cooling rate, the sublimation temperature and the time/temperature characteristics of the drying cycles.

7.2.2 Stabilization by additives

Certain salts and non-electrolytes are able to stabilize proteins against the denaturing effects of high temperature or low pH (Schleich & von Hippel 1969), and conversely, other additives destabilize the native state. These observed effects result from subtle changes in the structural nature of the aqueous solvent medium produced by salts and non-electrolytes (Eagland & Franks 1975), and have been applied in efforts to extend the shelf-lives of commercially produced proteins. Thus many preparations are supplied in high concentrations of ammonium sulphate (above 3 M) for example. In this case the protein crystallizes from the solution, and is therefore present in a very stable form. Storage at refrigerator temperatures is nonetheless generally recommended for these products. Alternative additives include glycerol (c. 50%) which also enhances the stability of the native (active) state (Lee & Timasheff 1974), but at such high concentrations the function of glycerol is to prevent freezing, and the stabilizing effect is achieved by storage at low temperature (− 20°C).

These methods are useful for some products, expecially those for which freezing is not an option (e.g. restriction enzymes). The major disadvantage of this method is that for many applications the stabilizing additives must be dialysed from the active material before use. Further, the active material must be greatly diluted before use, as it is rarely possible for it to be supplied in stable form at the low protein concentration necessary for typical assays. Wastage and uncertainty are introduced with the dialysis and dilution steps.

7.2.3 Spray-drying

Spray-drying has been used for many years as an economical method for drying such items as egg, milk and coffee to give granular, easily dissolved products. More recently the application has widened to include more thermally sensitive materials, e.g. food grade bacteria and washing powder enzymes, and is gaining favour as a possible means of stabilizing the more labile products of biotechnology (Pope 1985).

The factors influencing the final stability of a spray-dried product are similar to those mentioned for freeze-drying — the residual moisture content and distribution and the glass temperature (T_g) of the dried product being of particular importance. Advantages of spray-drying over freeze-drying may lie in the shorter processing time, low energy input and relatively low cost of capital equipment. The disadvantages can be seen in the exposure of labile material to the high air temperatures in the drier necessary to achieve rapid drying (say inlet temperature 150°C, outlet 80°C), plus the shear stress as the liquid formulation is sprayed through a nozzle into the hot air stream. However, when the active material enters the drier it is in dilute solution and is cooled by the rapid evaporation of water. The product temperature apparently only approaches that of the air temperature at the outlet (80°C), so exposure to high temperatures is very short, and only when the product is already dry (typically 4% moisture).

7.3 NOVEL METHODS

7.3.1 Undercooling

Undercooling (Franks 1984) is a novel technique for maintaining the shelf-life of proteins at low concentrations without additives. Low temperature is used to retard deterioration, but freezing (separation of ice) which is undoubtedly injurious is avoided (Hatley *et al* 1987 & Mathias *et al.* 1985). The process depends on the principle that under certain conditions liquid water can be cooled to temperatures substantially below its equilibrium freezing point. Freezing in most cases is initiated at sites of particulate matter (dust, pyrogens, etc.) which catalyse ice nucleation. This type of nucleation can be avoided by dividing the aqueous phase into small droplets, for example, as a mist or as a water-in-oil dispersion. Thus the particulate matter is confined to a few droplets from the total volume, and if even these droplets do freeze, ice crystals cannot propagate any further and the bulk of the aqueous matter remains unfrozen.

In practice, to achieve undercooling, the product in aqueous solution is dispersed in the form of microdroplets throughout an inert oil phase carrier. The carrier is mobile at the mixing temperature, but solid at the storage temperature (usually −20°C) so that the aqueous droplets are locked in, and remain unfrozen for the duration of storage (Franks *et al.* 1987). Reconstitution of the products is achieved by warming the preparation until the oil phase becomes mobile again and the two phases separate. Gentle centrifugation may assist separation. The aqueous phase containing the active product is then collected for use and found to be completely oil-free. One advantage of this method is that proteins can be stored at any required concentration, and that no additives such as glycerol etc. are needed to assist stabilization. Thus once recovered from the carrier phase, the product is ready to use.

This simple, cost-effective method has been shown to be successful for many enzymes and proteinaceous biomolecules. It is particularly suitable for biomolecules known to be susceptible to freezing. Fig. 7.3 shows data for lactate dehydrogenase (LDH) stored at several temperatures frozen or undercooled. Lactate dehydrogenase has been stored at −20°C for four years without loss in activity.

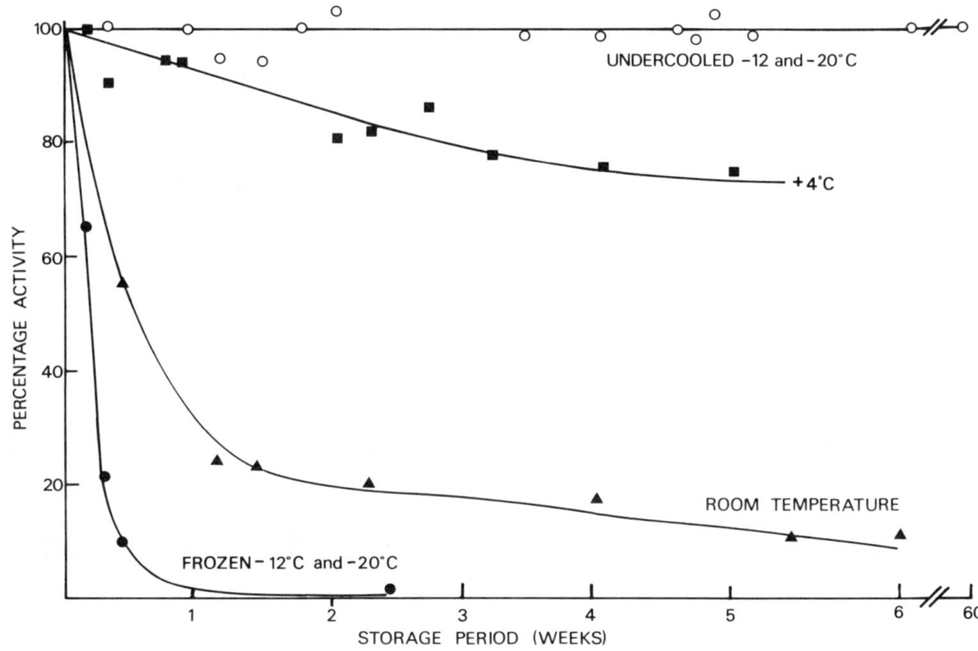

Fig. 7.3 — Maintenance of enzyme activity by LDH stored under different conditions; LDH concentration: 10 μg ml^{-1} in phosphate buffer (pH 7). Redrawn from Hatley *et al.* (1987).

7.3.2 Permazyme†

A further technique for stabilization of labile proteinaceous materials is currently under development (Franks & Hatley 1989). In essence the active material is stabilized by incorporation into an aqueous solution containing a glass-forming substance, which is then dried to its glassy state. Factors affecting the final stability of the product are therefore similar to those mentioned under freeze-drying.

Glass formers (usually carbohydrates or their derivatives) have well-defined glass/rubber transition temperatures (T_g) which depend on the molecular weight and the molecular complexity of the substance (see Table 7.1 for some examples). T_g is depressed by the addition of diluents, and the glass/rubber transition temperature is adjustable by the addition of water or aqueous solutions. This method of stabilization depends on the evaporation of water from a solution containing the glass former and active substance so that the T_g of the resultant product is above ambient, and the product remains glassy throughout storage. This can be achieved by drying at 30–40°C under vacuum for between 24 and 36 h, followed by a short exposure to a higher temperature when already dry. The high viscosity of glasses (10^{14} Pa s and

† Trade mark applied for.

above) means that diffusion is in the order of micrometers per year, and that chemical and biochemical reactions are practically inhibited. Fig. 7.4 shows the glass

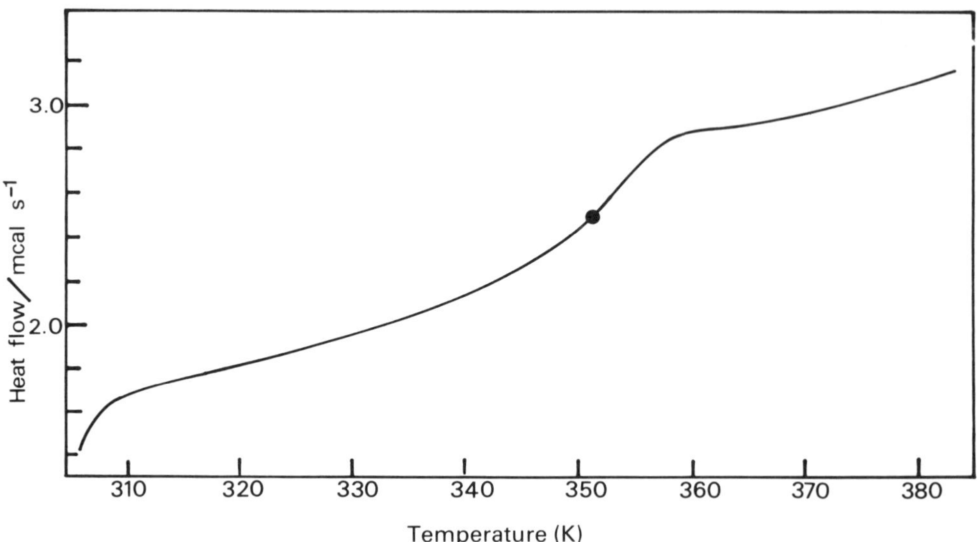

Fig. 7.4 — Typical DSC heating scan of LDH processed by the Permazyme method. The glass temperature T_g of the dried product is clearly visible, centred on 351.5 K, well above room temperature.

transition of a Permazyme product (active ingredient: lactate dehydrogenase) determined by differential scanning calorimetry. The T_g is seen to be 78°C, well above room temperature. After storage at room temperature for three months the sample retained full activity.

The Permazyme product typically contains 3–4% moisture, evenly distributed, which compares favourably with moisture levels and distribution commonly found in freeze-dried products. Recovery of the active material is simply by rehydration. This process is anticipated to be cost-effective in terms of time and energy compared with freeze-drying, and should find a wide application in the stabilization of freeze-labile proteins.

REFERENCES

Eagland, D. & Franks, F. (1975) The role of solvent interactions in protein conformation. *C.R.C. Crit. Rev. Biochem.* **3** 165–219.

Finegold, L. X., Franks, F. & Hatley, R. H. M. (1989) Glass/rubber transitions and heat capacities of binary sugar blends. *J. Chem. Soc. Faraday Trans.* **85** 2945–2951.

Franks, F. (1984) Preservation by refrigeration. European Patent 84 305524.

Franks, F. (1990a) Improved freeze-drying. Process Biochemistry **24** Probiotech. Suppl. iii–vi.

Franks, F. (1990b) Improved freeze-drying: from empiricism to predictability. *Cryo-Letters* **11** 93–110.

Franks, F. & Hatley, R. H. M. (1989) Storage of Materials European Patent Application No 89 03593.5.

Franks, F., Wakabayashi, T. & Mathias, S. (1987) Nucleation kinetics of ice in undercooled yeast cells: long term stability against freezing. *J. Gen. Microbiol.* **133** 2807–2815.

Frost and Sullivan (1989) European market for anti-viral drugs and vaccines, Report E911; Frost and Sullivan, 4 Grosvenor Gdns, London SW1 WODH.

Hatley, R. H. M., Franks, F. & Mathias, S. F. (1987) The stabilization of labile biochemicals by undercooling. *Process Biochemistry* **20** 169–172.

Hatley, R. H. M., Franks, F. & Green, M. (1989) A novel data acquisition retention and examination system (DARES) for differential scanning calorimetry. *Thermochimica Acta* **156** 247–257.

Lee, J. C. & Timasheff, S. N. (1974) Partial specific volumes and interactions with solvent components in guanidine hydrochloride. *Biochemistry* **13** 257–265.

Levine, H. & Slade, L. (1988) Principles of cryostabilization technology from structure/property relationships of carbohydrate/water systems — a review. *Cryo-Letters* **9** 21–63.

Mathias, S. F., Franks, F. & Hatley, R. H. M. (1985) Preservation of cells in the undercooled state *Cryobiology* **22** 537–546.

Murase, N. & Franks, F. (1989) Salt precipitation during the freeze-concentration of phosphate buffer solutions. *Biophysical Chemistry* **34** 293–300.

Pope, J. M. (1985) Achieving the right granulometry, *Food Journal* 1985 42–43.

Schleich, T. H. & von Hippel, P. H. (1969) Specific ion effects on the solution conformation of poly-L-proline. *Biopolymers* **7** 861–877.

8

Toxicity and safety: the testing of genetically engineered pharmaceuticals

A. D. Dayan
DH Department of Toxicology, St Bartholomew's Hospital Medical College, Dominion House, 59 Bartholomew Close, London EC1

8.1 INTRODUCTION

For proteins and polypeptides being developed as drugs, as for more conventional new chemical entities (NCEs), preclinical studies have two complementary aims — to demonstrate potential utility and to rule out useless or harmful compounds — with efficiency and economy. Both decisions must be based on experimental evidence, but extrapolation from the laboratory to human patients or other target animal species also requires less certain expert judgement and experience, and it involves basic knowledge and clinical beliefs, too. The scientist who works in this area must do so in close association with molecular biologists and clinicians, as the best decisions are based on integration of their findings and judgements with his procedures. At the same time, the general precautions appropriate to genetic engineering technology must also be applied (Dayan *et al.* 1988).

So far, the majority of their genetically engineered pharmaceutical products on the market, or undergoing clinical trial, have been developed because they were already known as physiologically active agents, and the therapeutic goal has been either replacement of a deficiency or enhancement of a known activity. The introduction of new molecules with entirely novel properties is only now being attempted; examples of licensed medicines include recombinant insulin, growth hormone, erythropoietin, tissue plasminogen activator and interferon-α; development candidates include various human interleukins and other pituitary hormones; materials under consideration include novel hybrid and consensus interferons, and interleukins and growth factors with modified structures to improve stability or prolong activity.

The pre-clinical development scientist or toxicologist must consider this broad classification, because so far it has been reasonable for his work to be concentrated on the relatively simplistic exploration of the physiological effects of high doses, and a broad screen for toxic actions based on the long-used general catch-all approach developed for conventional NCEs. An entirely novel substance or preparation may require very unconventional procedures to explore its effects, but that can only be decided on an empirical, case-by-case approach.

Studies of monoclonal antibodies, or of deliberate immunogens, such as vaccines, must also be quite different, because of the nature of the preparations and their intended actions, and the characteristics of the recipients; for example, a novel vaccine against, say, dental caries, a common cold virus or malaria, would have to be considered from the viewpoint of its life-long effects on healthy young individuals in the community rather than seriously ill patients in hospital. Again, case-by-case assessment and planning would be essential, but taking into account the need to seek effects both in the treated individuals and the population if a live agent were employed.

Proposals based on science must be aligned with what is possible (e.g. if candidate AIDS vaccines had to be tested in the chimpanzee, could they be obtained?) and cost-effective (in animal, human and monetary terms), and what is required by regulatory agencies. All the latter have accepted the empirical approach so far, within a general structure of requirements to ensure broad coverage of activities. Industrial, academic and regulatory experts must keep each other aware of developments if this valuable scientific flexibility is to be maintained.

8.2 REGULATORY GUIDELINES

Rapidly changing requirements, often in the form of draft or provisional proposals rather than formal guidelines, have been issued in the EEC by the Committee on Proprietary Medicinal Products (CPMP, 1987), by the US FDA (Weissinger, 1989) and by the Japanese Pharmaceutical Manufacturers' Association (1989). In the EEC the suggestions have become closely intermingled with the specific requirements of the Expert Report System, which, in this instance at least, forces a critical assessment of all the pre-clinical and clinical information. Behind these concepts lie vital documents from WHO (Petricciani & Hennessen 1987) and OECD (1986) about the use of mammalian cell lines, including those of human origin, to prepare medicines. There has been concern about the safety of such production substrates and these reports describe prudent working practices and precautions, which have been adopted in all countries.

The guidelines have deliberately been written in so general a fashion that they are of use to scientists only in indicating the broad types of information required, rather than the detailed procedures required to obtain them. It would be foolish to start a development programme before ensuring either that it followed the pattern of requirements, or that if there were good reasons for divergence, discussing the planned work with the agencies concerned.

These provisos apply to cytokines, hormones, growth factors and similar, systemically active materials. The position of antibodies (mono- and polyclonal) is different, because the nature and purpose of their administration is different —

usually short-term administration (often only a single dose), and most commonly as diagnostic imaging agents, although cancer chemotherapeutic combinations with cytotoxicants and radio-nuclides are being developed for multi-dose use.

Vaccines, as isolated antigens and killed or living organisms, raise questions that are so specific to each preparation that no comprehensive general guideline has been devised.

Further discussion of antibodies and vaccines is presented in sections 8.6 and 8.7, and Chapters 10 and 11, as they must be distinguished from the other types of product.

8.3 TESTS

8.3.1 Efficacy testing

The range of substances and potential actions is so extensive that only a general account can be given of the principles underlying efficacy testing. Some discussion can be found in Marshak & Liu (1988) and Balkwill (1989).

Like an ordinary NCE, the first need is to define the specific biological activity that is considered or believed to represent the mechanism of the therapeutic goal. This might be as precise as a change in a receptor or an intracellular effector (e.g. receptor occupancy or binding affinity for an hormone, or the intracellular level of a second messenger), a clear-cut biological response in cells *in vitro* (e.g. resistance to a viral infection, or proliferation or differentiation of a particular cell type), or a more complex response in an animal (for example, heightened resistance to an infection or an implanted heterologous or xenogeneic tumour), or an endpoint as ill-defined as growth. Like any bioassay, it is important always to choose the most precise and reproducible assay, provided that it is of adequate sensitivity, but a compromise may have to be made between a reference technique which is close to the analytical ideal but is awkward or expensive to use, and a secondary method of greater ease and simplicity, but perhaps not so directly or specifically indicative of the particular activity in question; for example, contrast an ELISA method to assay IFN-α, which would be simple and rapid, with a transplanted tumour rejection model *in vivo*, which is inherently slow and difficult. The former measures an antigen (epitope) and the latter the net result of various biological proceses. The two may be related, but very possibly not directly, so that, as discussed elsewhere in this book, any procedure of convenience used most of the time may need periodically to be checked by some other, more appropriate but much more costly technique.

Most genetically engineered products developed so far have been identified and characterized because of a specific biological activity, so the initial assay systems have also largely been the bases of predictive efficacy test systems too. The main limitations have been those of species differences in specific responses, well seen with IFN-α, and in the effects of certain hormones (e.g. somatropin mainly promotes growth in primates but is essential for milk production in cattle), and the common problem in biological cancer therapy of paralleling the complexity of the human immune system and tumour behaviour in animals (see, for example, general discussion in Balkwill, 1989, as well as Marshak & Liu, 1988).

It is necessary to explore the principal effect of the candidate drug and the mechanism of that action in whatever test systems are available, because that knowledge is needed both for extrapolation to the target species, and as a guide to the general and specific activities that should be monitored in therapeutic trials.

Overall, efficacy testing can only be based on knowledge of the particular physiological activities of the substance and so on techniques devised to indicate those effects. This truism must be qualified by distinguishing between artificially simplified test systems, usually of an *in vitro* type, and often based on isolated biochemical or immunological responses, and effects seen *in vivo*, which will be the net results of those responses plus any additive or counter-regulatory changes produced by other body mechanisms. Precision and sensitivity are important, but so is their obverse of generality to indicate the range of activities and effects; for example, an IFN-γ might be developed for its effect on macrophages or lymphocytes, which could be assayed by measures of cell activation, surface receptors, numbers of a specific cell type at rest or after an immune stimulus, or in terms of, say, a humoral or cellular immune response to a defined challenge, or host resistance to standardized infection or tumour challenge. They are all valid as indicators of 'efficacy', but in different ways, and each should be explored in its own right. For modified, non-natural molecules, the breadth of investigation of the biological activities is even more important, if adequate understanding is to be gained of the range of their unpredictable desired and incidental activities.

8.3.2 Toxicity testing

This covers at least as broad an area of biology as the assessment of efficacy. The need is to detect and characterize potentially harmful actions in such a way that an informed judgement can eventually be made between the likely therapeutic benefit and risk of treatment in deciding whether and how to undertake clinical trials, or to apply for a product licence.

As described below, toxicity tests are sub-divided into several categories because of the need to seek distinct types of effects by different means. The principles underlying the toxicity testing of recombinant drugs are comparable to those on which the testing of NCEs is based, but with modifications imposed by the properties and manner of use of those substances (Tables 8.1 and 8.2).

In addition to the practical problems listed in Table 8.2. there may be uncertainty about the proper scope of toxicity testing. In the past there has been confusion between its use to indicate and explore biological hazards (its proper role), as a form of extended survey of biological activities (valuable supplementary data can be obtained from prolonged, multidose experiments, but not if those estimations interfere with the evaluation of toxicity), and as a means to exclude harmful extraneous materials, e.g. reactive chemicals used in preparation of the final dosage form, or extraneous organisms and contaminants arising from the producing culture system or manufacturing process.

The latter are almost always better dealt with by biochemical, chemical or bioassay quality control (QC) procedures, as discussed elsewhere in this book, and not by varieties of *in vitro* or *in vivo* toxicity tests. They are very important factors in the determination of hazard, but they should not be regarded as part of toxicity testing, regardless of the name of the department that undertakes the work. The factors to be considered here might include chemical analyses for process chemicals and column leachates, immunological or other assays for contaminating proteins,

Table 8.1 — Principles of toxicity testing of recombinant DNA drugs

Requirement	Reason
Knowledge of physiological and pharmacological activities	Test nature must be adapted to effect Possibly explore multidose and high dose actions
Information about species specificity of actions	What species is suitable for toxicity tests?
Information about pharmacokinetics	To decide dosing patterns
Information about metabolism and excretion	To decide dosing pattern
Decision about clinical uses — doses — duration — effects of disease — co-therapy	To decide dosing pattern and duration, types of tests; interaction studies

and diverse techniques to measure or rule out contaminating DNA, organisms and endotoxin.

In practice, the most important difficulties to overcome in toxicity testing have been those arising from species specificity, practicable dosing pattern, and antigenicity and abnormal physiology.

8.3.2.1 Species specificity

The interferon-αs are an extreme example of this problem, as they are all virtually devoid of antiviral activity in other than higher primates (Manning & Deloria 1986; Graham 1987; Balkwill 1989). This means that studies in such conventional species as the rat or dog cannot reveal the pharmacodynamic or other consequences of high doses. As a result, only limited investigations have been possible other than in man (Trown *et al.* 1986). Fortunately, IFN-α is very unusual in its narrow specificity, and other cytokines, hormones and enzymes have sufficient actions in animals to make it reasonable to do relatively more conventional toxicity tests on them.

8.3.2.2 Practicable dosing pattern

In toxicity tests it is common to mimic the route and to try to emulate the temporal pattern of dosing in man, having regard for comparative pharmacokinetics. But, in the case of cytokines especially, they are administered systemically but act in specialized local environments, the circulating lifespan in animals is usually only minutes (Manning & Deloria, 1986), whereas there is likely to be continuous or repeated release stimulating a plateau in man, produced by a constant i.v. infusion over several days, or frequent small injections.

Table 8.2 — Problems of toxicity testing of recombinant DNA drugs

Problem	Explanation
Supply and drug characterization	Is a sufficient quantity available of proven purity?
Formulation	Is it formulated and stable; i.e. is there a proven analytical technique?
Blood level or biomarker	Method for pharmacokinetic and metabolism studies Monitor target effect in toxicity tests
Species specificity	Are the species suitable for toxicity testing responsive to specific biological action?
Antigenicity	As a heterologous protein, is there a method to detect antibodies to drug and when are they produced?
Clinical use pattern	Can it be paralleled in toxicity tests. e.g. frequency and duration of dosing; effects of disease?
Nature of physiological pharmacological effects	Can/should any be monitored? Will excess interfere with toxicity tests? Are there counter-regulatory responses?
Phasing preclinical work	Preliminary tests only? Formal tests for major efficacy trials? Extension tests for special patients, e.g. babies or the pregnant?

Although it is feasible to place indwelling cannulae in rats and larger animals for continuous infusions, that is a major procedure and it carries some morbidity and pathological consequences. If the persistence of injected drug is brief, repeated dosing or an infusion may be needed, or perhaps treatment can be given with an implanted minipump, provided that there is a proven formulation. However, the toxicologist may have to accept that the pattern of clinical therapy cannot always be mimicked.

An implication of this problem is that there may be uncertainty about the biochemical or biological endpoint to be followed as a marker of effect; for example, should it be the blood level of the agent, circulating cells carrying an activation marker, or some intracellular index of activity? The time-span of each of these after a single dose will differ, but any might be appropriate under some circumstance.

8.3.2.3 Antigenicity

This is probably the most serious problem in prolonged testing. Any foreign protein is likely to excite antibodies on repeated injection by any route, which may or may not be capable of neutralizing the administered substance, but which will almost certainly greatly truncate its half-life. It is essential, therefore, to make repeated assays for the development of antibodies during dosing, and to stop the test as soon as they appear. Continued treatment may produce serious but quite misleading tissue damage if immune complexes are formed, and in any case, the kinetics of the drug will be so altered as to make the experimental results of little value. In practice, this requires early development of assays for the administered substance and antibodies to it in plasma.

It is sometimes suggested that if antibody formation (or restricted species sensitivity) prevents adequate testing, then why not prepare the homologous species-specific substance and study it in animals of that species? There are considerable weaknesses in the argument, however, especially the fact that many substances do not have the same actions in different species (especially hormones, but consider also the differences in the immune system between, say, the mouse and man) and the enormous practical burden of cloning the gene and manufacturing a pharmaceutically adequate preparation. The idea is best left as an interesting possibility that will rarely be worthwhile following.

8.3.2.4 Abnormal physiology

Many of the recombinant products now being investigated are naturally occurring cytokines, etc., with important regulatory functions in normal life. If large doses are given in a toxicity study, or for a therapeutic purpose, the normal functions are likely to be seriously disturbed — directly by massive stimulation, by down-regulation of responder mechanisms, or by counter-regulatory responses. Equally, if an antibody were injected, or if one were produced in response to the test substance, the corresponding normal function might well be affected.

It would be very important toxicologically to know of and to understand such effects. It is just as important to realize that they might not necessarily indicate toxic effects in man exposed to different doses under different circumstances. Areas of particular concern here would include gross polycythaemia due to an erythropoietin, severe hypoglycaemia and secondary metabolic effects produced by insulin, and actions due to disturbance of the normal roles in embryonic development of IFN-α, IFN-γ and TNF (Balkwill, 1989).

8.4 TOXICITY TESTS

It is both reasonable and a regulatory requirement for the basic nature of individual tests to be very similar whether the substance is an NCE or a product of biotechnology. Differences are likely to arise because of the need for additional investigations to examine particular actions (Dayan, 1990).

The general nature of the studies is of a set of broad screens containing multiple investigations able to reveal a wide range of actions. They contain core procedures of great generality, such as clinical biochemistry and histopathology tests, which are likely to reflect directly or indirectly the consequences of almost any type of tissue damage, and which may often suggest the pathogenetic mechanisms involved. Onto them can be grafted additional, more specific techniques to examine types of action and reaction known or expected to occur, provided that it is worth looking for them, e.g. changes in immune cell populations, in the electrophysiological or metabolic activities of affected cells, or in neural function, i.e. whatever appears appropriate in relation to the substance and the need to seek likely activities and to assess known effects.

8.4.1 Acute toxicity test

The limit-test approach is preferred, in which a single high dose is injected into a small number of one or two species of rodent, which are followed clinically for up to 14 days before being autopsied. In a very few non-rodents, a rising dose type of test should also be done, just to seek clinical and physiological effects, but not to kill animals.

8.4.2 General (safety) pharmacology

A broad screen is required for effects on the nervous, cardiovascular and respiratory systems and on various types of smooth muscle, as for a conventional NCE.

8.4.3 Repeated dose toxicity tests

What can be and is worth doing depends on species sensitivity and antigenicity, and on the intended duration of administration to man and the nature of the disease to be treated.

In general, it is probably worthwhile planning to do full tests for one month in two species, and then eliminating or modifying the experiments if there is evidence of their impossibility or futility for good biological reasons.

As for an NCE, each test should comprise at least daily administration for 30 days of the vehicle (control) and three dose levels to groups of 10 male and female rats and three dogs/monkeys (as appropriate). Additional reversibility groups should be included. The high dose level should be a small multiple of that intended for man. The usual extensive clinical, haematological and biochemical (blood and urine) tests, and autopsy and histological studies should be done. In addition, appropriate measures of the biological effects of the active agent should be made. It is important to monitor the blood level of the active agent and/or antibodies to it.

The duration of these experiments can only be decided pragmatically in the terms of the intended period of human treatment, the feasibility of really prolonged dosing of animals, and the customary practices relating the duration of treatment of animals and man.

8.4.4 Reproduction toxicity tests

Whether these are necessary at all depends on the type of patient being treated, and then on their experimental feasibility, especially in relation to any physiological role of the test substances and the consequences of interference with it.

8.4.5 Mutagenicity tests

The rationale for doing mutagenicity tests on proteins and other substances that occur naturally in the body is unclear, but they are often half-suggested in a despondent fashion as 'valuable' (undefined). Ames and other point mutation tests appear valueless for more or less conventional polypeptides and proteins. Cytogenetic tests may be sensitive to agents that affect the cell cycle, but the relationship between *in vivo* and *in vitro* activities in this instance is at best disputable.

8.4.6 Local irritancy

This should certainly be explored for every injected or topical preparation.

8.4.7 Sensitization

Sometimes this is examined in specific studies of dermal sensitizing potential, e.g. the Magnusson and Kligman and similar tests. It may be reasonable to do so for cutaneous applications, but such information is unlikely to be of value for other preparations. Instead, attention should be focused on the development and nature of antibodies on repeated administration to animals, and on any evidence suggesting a cell-mediated immune response. Interpretation of the findings must reflect any antigenic differences between the human protein being studied and the form native to the species in which it is being tested.

8.4.8 Pharmacokinetics and metabolism

As already mentioned, knowledge of pharmacokinetics (disposition and clearance) is very importnat. It may be investigated using radio-labelled material, or by cold assays. Whatever method is employed it is necessary to show if the radioactivity counted, or the antigen measured by, say, ELISA, represents the biologically active molecule or an inert fragment. As the action of many cytokines and hormones depends on a continuing cellular response, there may be a considerable difference between the circulating half-life and the biological half-life.

Whether there is also need to study general effects on drug metabolism, and hence on drug interactions, is uncertain. Some cytokines do affect the P450 oxidase system (Manning & Deloria, 1986), but that probably does not represent a general phenomenon requiring universal investigation.

The metabolism of small proteins of various sources is well-known. Part is rapidly taken up by the liver and is catabolized there, and a major fraction is likely to be excreted in the glomerular filtrate, and to be resorbed and catabolized in the renal tubules. Simple studies with radiolabelled material, probably *in vitro*, should show whether this is the case, or if more extensive investigation is required.

8.5 PROGRAMME OF TOXICITY TESTING

Although dependent on the intended clinical use, there is a general plan of testing (Table 8.3) which must be adapted to each specific instance.

The pattern of investigations appropriate for an interferon to treat cancer would be very different from those, for example, for a hormone to stimulate ovulation prior to *in vitro* fertilization. In the former instance acute and subacute effects on the major physiological systems and organs would be important, whereas for the latter, there should be concentration on the quality of the ova produced. In every procedure the GLP regulations should be followed.

Table 8.3 — Programme of toxicity testing

Test type	Explanation
General pharmacology Acute toxicity Pharmacokinetics Two- or four-week repeat dose toxicity test in one or two species Local irritancy	Basic set of data before any clinical work (N.B. QC procedures to exclude DNA, extraneous agents and other contaminants) Adequate specification
Antibody formation	During any multidose test
Second species multidose test; perhaps prolonged tests Complete kinetics and metabolism mutagenicity	If feasible (anti-bodies not produced) and if clinical therapy prolonged
Reproduction toxicity	Feasible; only if appropriate patients to be treated
Special procedures; pharmacodynamics; endocrine effects, state of immune system, mechanism of toxicity	As appropriate to known actions and results of other tests Rolling programme

A preliminary categorization of the general classes of biotechnological medicinal products and tentative guidance about their testing has been proposed by several regulatory agencies (Table 8.4). It is too early to make it into a set of definitive requirements, but it does display the flexible relationship between the nature of a product, the pattern of use and the range of investigations that should be considered.

8.6 ANTIBODIES

The toxicological exploration of antibodies is difficult and should be approached on an entirely empirical basis, as recognized even by interventionist regulatory agencies (FDA 1987) and more academic groups (UK CCCR, 1986).

The preparations may be polyclonal, monoclonal, fragments obtained by conventional proteolysis (Fab, Fc, etc.) or chemically reconstructed to minimize antigenicity (modification of constant regions, heavy chains, etc). Biochemical and chemical procedures are required to verify the nature (isotype, etc.) of the preparation and to exclude DNA, organisms and other contaminants.

The toxicologists's job is to determine whether administration of the antibody itself affects physiological systems (e.g. by blocking receptors, or removing normal circulating hormones, etc.). It is also important to show whether the preparation

Table 8.4 — Tentative guide to classes of biotechnology drugs and appropriate toxicity testing

	Product type									
	Hormone, cytokine and other regulatory factors			Blood product			Antibody		Vaccine	
Biochemical group	I	II	III	I	II	III	Polyclonal	Monoclonal	Antigen	Live vector
Toxicity test										
Single dose	?	×	×	×	×	×	×	×	×	×
Repeated dose	0	×	×	×	×	×	C	C	?	?
Reproduction	C	C	×/?	C	C	C/?	?	?	0	?
Genotoxicity	0/?	0/?	0	0	0	0	0/?	0/?	0	0
Carcinogenicity	0/?	0/?	?	0/?	0/?	0/?	0	0	0	0
Local irritancy	×	×	×	×	×	×	×	×	×	×
General pharmacology	×	×	×	×	×	×	×	×	0	0
Kinetics and ADME	×	×	×	×	×	×	×	×	0	0
Immunotoxicity	C	C	C	0	0	0	C	C	0	0

I = Identical to normal substance; physiological dose.
II = Slightly different structure that might affect activity or immunogenicity; high dose.
III = Very different structure.
× = Undertake.
0 = Unnecessary.
C = Consider.
? = Doubtful value/feasibility.

might mimic an autoimmune disease by combining with normal cell components leading to their immunological damage directly or by exposing neoantigens, and whether dangerous immune complexes might be produced by combination of the antibody with its intended antigen. A further remote possibility, but one that has sometimes been detected, is the potentially harmful response of the body by formation of an anti-idiotype antibody.

A serious practical problem is that the body is likely to respond to such highly antigenic substances by the production of anti-antibodies, which will greatly restrict animal experiments.

In practice, it seems reasonable, in addition to carrying out appropriate experiments to demonstrate efficacy, to explore acute general pharmacological effects, and to do a limit acute toxicity test in one species of rodent. Additional biochemical, haematological and pathological tests may sometimes usefully be added to the last study, depending on the nature and properties of the preparation. An *in vitro* screen, best using immunofluorescence, should be done to check for binding to a wide range of normal human or antigenically similar tissues (thyroid, gastric mucsoa, and liver). As part of the efficacy study, or in addition to it if binding is seen, the state of tissues to which binding occurs should be investigated (CPMP Guidance Note, 1987 and UK CCCR, 1986).

8.7 VACCINES

The discovery and preparation of immunogens in a form suitable for use in the community of healthy individuals requires even more stringent testing of safety than the development of other pharmaceuticals, which are usually given to people with some form of illness. Their toxicity testing is peculiarly difficult, because they are normally administered once or on a very few occasions but they are selected to produce a long-lasting change in a particular response of the immune system. The general problems have been reviewed by the Steering Committee on Future Health Scenarios (1988).

For a killed vaccine, or isolated antigen, which is usually combined with an adjuvant, the principal efficacy and safety considerations are the local irritancy of the formulation, the nature, magnitude and duration of the immune response on isolated or repeated challenge, whether that immune response is appropriate to prevent or stop the infection (an inappropriate response may enhance an infection), and whether any autoimmune response occurs.

For toxicity testing, a species is required in which the human immune response is emulated and in which the other factors can also be examined.

A live vaccine poses more problems. In addition to assessing immunogenicity, empirical procedures must be devised to check pathogenicity (including any question of autoimmunity) to the recipient. Then, the likelihood of passage of live organisms to the external world must be considered and explored, and an assessment made of the stability of the vaccine strain in case reversion to wild-type virulence occurs in secondarily infected hosts. The core in all these procedures is case-by-case consideration of the known properties of the immunogen and the empirical design of experiments to explore those actions under conditions that make it possible to seek other untoward effects.

In addition, many QC-type procedures are required to determine the nature, composition and stability of the preparation.

8.8 CONCLUSIONS

Studies of biotechnology products to demonstrate their efficacy and toxicity and so to permit prediction of their safety may appear complex, but that reflects the many powerful physiological and pathological mechanisms involved in bodily responses to them.

The first concern in planning such experiments is to be sure that there are appropriate molecular, biochemical and other data about the nature and composition of the preparation and the system in which it has been produced to ensure consistency and to exclude contamination by extraneous chemicals, organisms and DNA, and other materials derived from the producing cells.

The next type of information required is an account of the known properties and activities of the substance. Tests must be adapted to investigate the responses to high doses and prolonged treatment.

Then, formal toxicity studies can be devised, based on standard methods (acute to chronic, reproduction, etc.), modified in the light of the pharmacological actions of the substance. These experiments should be performed in species capable of responding to the substance. If no suitable animal is available, then only a limited set of experiments is worth doing.

Any immunological response to the substance must be taken into account, because it is likely to nullify the value of further testing. It is important to assess the kinetics of the substance, and to consider both its metabolism and the possibility that it may affect the metabolism of xenobiotics. The toxicity testing must also reflect the intended pattern of clinical use.

Antibodies and their fragments should also be investigated in empirically designed procedures able, within the general structure of toxicity tests, to reveal the consequences of the primary action of the immunoglobulin. Additional procedures are required to demonstrate the lack of autoimmune-like binding and damage to normal tissues, and perhaps to seek anti-idiotype generation.

Vaccines, too, require empirical exploration designed to screen for local effects due to the administered preparation, and the systemic consequences of the immune reaction and of the adjuvant.

Above all, the testing of biotechnology products must be firmly based on knowledge of their physiological and pharmacological actions and a general screen for unanticipated actions. As such, it requires a careful blend of toxicology, cell biology and immunological understanding applied to the product, on a background of careful biochemical and other procedures done to rule out other hazards arising from the producing system and the production process.

REFERENCES

Balkwill, F. R. (1989) *Cytokines in cancer therapy*. Oxford University Press, Oxford.

Committee of Proprietary Medical Products (CPMP) (1987) Guidelines on the production & quality control of medicinal products derived by recombinant

DNA technology. Commission of the European Community. *Trends Biotechnol.* **5** G1–G4.

Dayan, A. D. (1990) Toxicological background. In O'Grady, J. & Kolar, O. (Eds) *Early Phase Drug Evaluation in Man*. Butterworth, London. 39–62

Dayan, A. D., Campbell, P. N. & Jukes, T. H. (1988) Hazards of biotechnology: real or imaginary? *J. Chem. Technol. Biotechnol.* **43** no. 4 Special Issue.

Food & Drug Administration (FDA) (1987) *Points-to-consider in the Manufacture and Testing of Monoclonal Antibody Products for Human Use*. FDA. Washington DC, USA.

Graham, C. E. (ed) (1987) Preclinical safety of biotechnology products intended for human use. *Progr. Clin. Biol. Res.* **235** 1–213.

Japanese Pharmaceutical Manufacturers' Association 1990. *Principles for the Safety Testing of Biotechnology Products*. JPMA, Tokyo (in Japanese).

Manning, G. J. & Deloria, L. B. (1986) The pharmacology & toxicology of the interferons: An overview. *Ann. Rev. Pharmacol. Toxicol.* **26** 455–515.

Marshak, D. & Liu, D. T. (Eds) (1988) *Therapeutic Peptides & Proteins*, Banbury Report 29, Cold Spring Harbor Laboratory, Jackson, Maine, USA.

Organisation for Economic Cooperation & Development (OECD) (1986) Recombinant DNA Safety Considerations. OECD, Paris.

Petricciani, J. & Hennessen, W. (Eds) (1987) Cells, Products, Safety. *Dev. Biol. Standard.* **68** 1–72.

Steering Committee on Future Health Scenarios (1988). Applications of the New Biotechnology. The Case of Vaccines. *Hlth. Care. Technol.* **6** 1–111, Kluwer, Dordrecht.

Trown, P. W., Wells, R. J. & Kamm, J. J. (1986) The preclinical development of Roferon. *Cancer* **57** 1648–1656.

UK CCCR (1986) Monoclonal Antibody Testing *Br. J. Cancer* **54** No. 3, Sept. 1986.

Weissinger, J. (1989) Non-clinical pharmacologic & toxicologic considerations for evaluating biologic products. *Regul. Toxicol. Pharmacol.* **10** 255–263.

9

Pyrogen testing of polypetide and protein drugs

Stephen Poole
National Institute for Biological Standards & Control,
Blanche Lane, South Mimms, Potters Bar, Hertfordshire EN6 3QG

9.1 PYROGENS

Polypeptide and protein drugs are usually administered parenterally to avoid their destruction in the gastrointestinal tract. Since pharmaceutical products intended for parenteral administration must be shown to be free from pyrogenic contamination, batches of polypeptide and protein drugs are tested for pyrogens.

While a pyrogen may in general be defined as any substance that causes fever, the pyrogens that are of particular concern to the pharmaceutical industry are bacterial endotoxins from gram negative bacteria. Endotoxins are large molecular weight complexes ($\sim 10^6$ Da) associated with and shed from the outer membranes of gram negative bacteria. Endotoxins consist of three distinct chemical regions: a lipid moiety — lipid A — which is linked to a polysaccharide core, which is itself linked to O-antigenic side-chains (Fig. 9.1).

Lipid A is made up of a disaccharide of glucosamine which is highly substituted with amide-linked and ester-linked long-chain fatty acids. The fatty acids are saturated and straight-chained. Lipid A is linked to core heteropolysaccharides by 2-keto-3-deoxyoctonic acid (KDo), an eight-carbon sugar acid. KDo together with the rest of the core polysaccharides serves to increase the solubility in aqueous phase of the hydrophobic lipid A in which resides most, if not all, of the biological (pathological) effects of bacterial endotoxin. Endotoxin or its lipid A moiety has more than 30 biological activities (Galanos 1979). These include the induction of fever and acute phase proteins, headache, changes in white blood cell counts, diarrhoea, severe hypotensive shock and disseminated intravascular coagulation.

9.2 RABBIT PYROGEN AND LAL TESTS

Pyrogen/endotoxin is detected in the rabbit pyrogen test. This test was first considered in 1912 (Hort & Penfold 1912) because of the fevers which sometimes

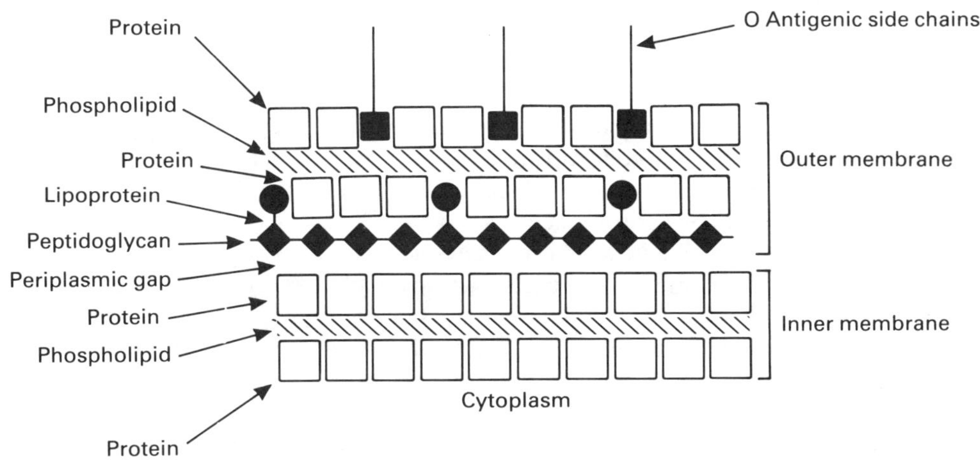

Fig. 9.1 — Schematic illustration of a gram-negative bacteria cell membrane.

accompanied injection. It was subsequently elaborated in 1925 (Seibert 1925) and included in the United States Pharmacopoeia in 1942 (Welch *et al.* 1943). The test involves measuring the rise in body temperature evoked in rabbits by the intravenous injection of a sterile solution of the substance to be examined (see European Pharmacopoeia 1971). The test is well-documented in the British, European and United States Pharmacopoeias and has proven its value over nearly fifty years for both quality control and quality assurance in the pharmaceutical industry. However, while the rabbit pyrogen test has proved reproducibly able to demonstrate the absence of pyrogens dependent on the dose administered (Cooper 1973) it has many disadvantages. These include cost, sensitivity, the need to spend between 1 and 4 days per test, and biological variability. Also the rabbit pyrogen test is poorly quantitative, not reliably 'standardized' (Bangham 1979) and uses experimental animals at a time when diminished use of animals is sought.

In 1964 Levin and Bang showed that endotoxin preparations from *Escherichia coli* (*E. coli*) and a marine Vibrio caused extracellular coagulation of the blood (haemolymph) of the horseshoe crab, *Limulus polyphemus* (Levin & Bang 1964a,b). This clotting reaction became the basis of a very sensitive and very specific assay of bacterial endotoxin. The clotting reagent, *Limulus* amoebocyte lysate (LAL) is prepared from amoebocytes isolated from the blood of the horseshoe crab and exposed to osmotic shock. Chloroform-extracted lysate in the presence of divalent cations will clot in the presence of as little as 2 pg/ml endotoxin and, when lyophilized, will remain stable for at least three years.

The simplest and most widely used variant of the LAL test is the LAL gelation test, which gives an 'all or none', i.e. pass or fail, result. Equal volumes of lysate and test solution (usually 100 μl of each) are mixed in (pyrogen-free) glass test tubes (usually 10×75 mm). The mixture is agitated gently and incubated at 37°C for 1 h. The endpoint is read by turning the tube 180°. If a solid clot is formed and remains solid on inversion, the test solution is positive for endotoxin. A number of two-fold serial dilutions of the test solution are tested and the clot endpoint determined. The endotoxin concentration is calculated by multiplying the reciprocal of the greatest dilution of the test solution giving a positive end-point by the sensitivity (to endotoxin) of the lysate preparation. A positive control consisting of a product sample spiked with a known concentration of endotoxin and a negative control using pyrogen-free water are included in the procedure. Other variants of the LAL test include turbidimetric and chromogenic assays and a Lowry Protein-Colorimetric method: these are described elsewhere (Pearson *et al.*, 1985).

Commercially available LAL reagents are available largely from manufacturers in the USA and are now well-standardized for sensitivity, which is given in terms of the US national standard for endotoxin, EC5, which contains 10 000 endotoxin units (EU) per vial. The requirement for a standard or reference endotoxin is one important difference between the LAL test and the rabbit pyrogen test. Monographs for the LAL test are to be found in the US (1980), European (1987) and British (1988) Pharmacopoeias. The test described in the US Pharmacopoeia specifies the use of a 'control standard endotoxin' calibrated against the US Reference Standard for endotoxin, currently EC5. The European Pharmacopoeia specifies the use of the European Biological Reference Preparation (BRP) of endotoxin and the British Pharmacopoeia specifies either the BRP or the international standard for endotoxin. The potencies of the latter two preparations of endotoxin are given in international units (IU) of endotoxin with 1 IU=1 EU for LAL gelation tests (Poole & Mussett 1989).

The rabbit pyrogen test and LAL test each have advantages and disadvantages. As stated above, the rabbit test cannot be made a quantitatively reliable bioassay: it is laborious, and animals and their housing are costly. It cannot be used for substances which interfere with the pharmacological responses (to pyrogen) of the rabbits or with radiopharmaceuticals. However, it does have the notable virtue of being an *in vivo* mammalian system in which the whole pathophysiological response to pyrogen is involved and is thus relevant to the response in man. The threshold pyrogenic dose of endotoxin is 7 IU/kg (~1 ng/kg) in rabbits (Poole & Mussett 1989) and 4–8 EU/kg (~0.4–0.8 ng/kg) in man (Hochstein & Fitzgerald 1990). Based upon an injection volume of 1 ml/kg in rabbits this gives a test sensitivity of 0.5–1.0 ng/ml compared with a sensitivity of about 3 pg/ml for the LAL test. In addition to greater sensitivity, the LALtest has the advantage that it may be used with radiopharmaceuticals. However, the LAL test detects bacterial endotoxin but not other pyrogens and gives false negative results with a number of products, e.g. enzyme inhibitors, certain antibiotics and viral vaccines.

9.3 OTHER PYROGEN TESTS

In view of the shortcomings of the rabbit pyrogen test and LAL test, other tests for pyrogen have been considered. Pyrogen tests based upon pyrogen-evoked depression of plasma zinc concentrations or elevation of the plasma acute phase protein Serum Amyloid P in mice have been described (Poole *et al.* 1986). Although such tests permitted quantitatively reliable bioassays *in vivo* in mammals, the requirement for test animals has meant that such tests have not found favour.

There is good evidence that the fever response to various exogenous pyrogens (e.g. endotoxin) is mediated by endogenous pyrogens generated internally by the host (Beeson 1948, Atkins & Wood 1955). The cells which produce endogeneous pyrogen are predominantly peripheral blood monocytes (Dinarello *et al.* 1974, Root *et al.* 1970) and it is believed that the endogenous pyrogen circulates and passes into the preoptic anterior hypothalamic area where it evokes synthesis and release of prostaglandins which have been shown to elevate body temperature when injected directly into this part of the brain (Milton & Wendlandt 1971). Thus a possible alternative test for endotoxin is the *in vitro* production of endogenous pyrogen (Duff & Atkins 1982). In 1984, Dinarello *et al.* reported that a batch of human recombinant growth hormone which had passed the rabbit pyrogen test and an LAL test was pyrogenic in man. Further, human peripheral blood monocytes incubated with the growth hormone produced endogenous pyrogen which evoked fever when injected intravenously in rabbits (Dinarello *et al.* 1984).

Endogenous pyrogen is believed to be a mixture of polypeptide hormones called cytokines, which have potent pyrogenic and inflammatory activities. The pyrogenic cytokines include interleukin-1α (IL-1α), interleukin-β (IL-1β), tumour necrosis factor α (TNFα) and interleukin-6 (IL-6). These molecules have many overlapping biological activities, can induce each other and self, and can act synergistically (Hamblin & Brannan 1989). The development of specific and sensitive bioassays and immunoassays for IL-1 (Gearing *et al.* 1987, Thorpe *et al.* 1988) and TNFα (Meager *et al.* 1986) permitted the development of a pyrogen test based on the direct assay of IL-1 and TNFα released *in vitro* by human peripheral blood monocytes. In this 'monocyte test' dilutions of medicinal products were added to cultures of 10^6 monocytes/ml and samples of supernatants taken from the cultures after 24 h for assay of IL-1 and TNFα. A standard curve was constructed for product spiked with the international standard for endotoxin (84/650, of which 1 ng= 7 IU). Detectable concentrations of IL-1 and TNFα were released in response to 2–10 pg/ml endotoxin. Using this monocyte-based test system, pyrogenic contamination was detected and quantified in batches of human serum albumin (HSA) which had passed the rabbit pyrogen test and the LAL test but which were reported to have caused adverse reactions in patients (Poole *et al.* 1988).

While useful in detecting pyrogenic contamination, a test system that requires monocytes freshly prepared from blood has limited applicability. Therefore a number of established cell lines with monocytic characteristics were evaluated for suitability to serve as substitutes for freshly prepared human monocytes. Also, besides measuring IL-1 production from these cell lines, endotoxin-evoked IL-6 production was measured since, unlike IL-1 and TNFα, IL-6 is a secreted protein rather than one which is retained predominantly within or attached to the stimulated cells; this makes its complete estimation easier. Of the cell lines evaluated, two appeared suitable for use in a 'monocyte test' for pyrogen: THP-1 cells (Tsuchiya *et al.* 1980) and MONO MAC 6 cells (Ziegler-Heitbrock *et al.* 1988). With endotoxin-evoked IL-6 secretion from MONO MAC 6 cells as the measured variable, the 'monocyte test' detected 5–10 pg/ml endotoxin (Taktak *et al.* 1990). The output of IL-6 (per 10^6 cells/ml) increased linearly to some 3 ng/ml immunoreactive

IL-6 in response to 500 pg/ml endotoxin; THP-1 cells responded to endotoxin only in the presence of interferon-γ (200 units/ml) and were less sensitive (Fig. 9.2). Similar

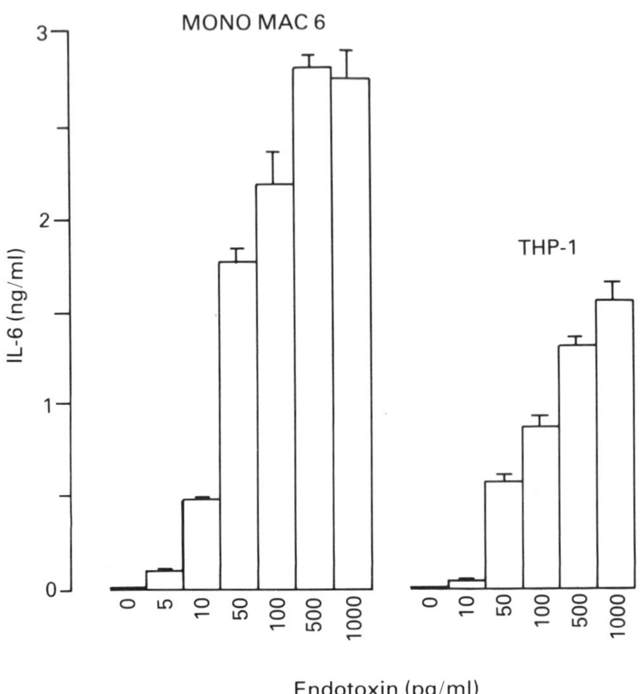

Fig. 9.2 — Dose–response relationships for endotoxin-evoked release of immunoreactive IL-6 from the monocytic cells MONO MAC 6 and THP-1 (10^6 cells/ml). The responses of the THP-1 cells were obtained in the presence of interferon-γ (200 units/ml).

results were obtained with immunoreactive intracelluar IL-1β as the measured variable. The 'monocyte test' using IL-6 release from MONO MAC 6 cells detected pyrogen in batches of HSA that was not detected in the LAL test or rabbit pyrogen test (Table 9.1) and thus represents a potential alternative to the existing tests for pyrogen. A comprehensive comparison of this novel test system with the rabbit pyrogen test and the LAL test using a wide variety of pyrogens and products will be required to determine its value.

9.4 SUMMARY AND CONCLUSIONS

In summary, polypeptide and protein drugs, in common with other pharmaceuticals that are administered by parenteral routes, will continue to be tested for pyrogen/endotoxin in the rabbit pyrogen test or in the LAL test. Where the LAL test is applicable to a product, that test will gradually replace the rabbit pyrogen test. Subsequently a 'monocyte test' for pyrogen may be adopted as a replacement for the

Table 9.1 — Comparative assay of four preparations of human serum albumin. 1 IU of endotoxin=1 EU=0.14 ng of preparation 84/650, the international standard for endotoxin. Preparations 1 and 2 caused adverse reactions in patients

Albumin	Endotoxin by IL-6 release from MONO MAC 6 cells (IU/ml)	Endotoxin by LAL test (IU/ml)	Rabbit pyrogen test
1	0.71	1.0–2.0	Pass
2	0.9	0.5–0.75	Pass
3	<0.06	<0.24	Pass
4	<0.06	0.72–0.96	Pass

rabbit test while the LAL test will become the end product test for the majority of parenteral pharmaceutical preparations.

REFERENCES

Atkins, E. & Wood Jr., W. B. (1955) Studies on the pathogenesis of fever and the presence of transferable pyrogen in the blood stream following the injection of typhoid vaccine. *J. Exp. Med.* **101** 519–528.

Bangham, D. R. (1979) Relevance and standardization in pyrogen tests. *J. Pharm. Belge.* **34** 134–136.

Beeson, P. B. (1948) Temperature-elevating effect of a substance obtained from polymorphonuclear leucocytes. *J. Clin. Invest.* **27** 524.

British Pharmacopoeia (1988), Addendum 1989, Appendix XIV A 312, HMSO.

Cooper, M. (1973) *Quality Control in the Pharmaceutical Industry*. Academic Press, New York, pp. 239.

Dinarello, C. A., Goldin, N. P. & Wolff, S. M. (1974) Demonstration and characterization of two distinct human leucocyte pyrogens. *J. Exp. Med.* **139** 1369–1381.

Dinarello, C. A., O'Connor, J. V. LoPreste, G. & Swift, R. L. (1984) Human leucocyte pyrogen test for detection of pyrogenic material in growth hormone produced by recombinant *Escherichia coli*. *J. Clin. Microbiol.* **20** 323–329.

Duff, G. W. & Atkins, E. (1982) The detection of endotoxin by *in vitro* production of endogenous pyrogen: comparison with amoebocyte lysate gelation. *J. Immunol. Methods* **52** 323–331.

European Pharmacopoeia, II, (1971) 58–60, Maisonneuve S.A.

European Pharmacopoeia, II, (1987). V.2.1.9 Maisonneuve S.A.

Galanos, C. (1979) Chemical, physical and biological properties of bacterial lipopolysaccharides. In: Cohen, E. (ed.) *Biomedical Applications of the Horseshoe Crab* (*Limulidae*). R. Liss, New York, pp. 319.

Gearing, A. J. H., Bird, C. R., Bristow, A., Poole, S. & Thorpe, R. (1987) A simple sensitive bioassay for interleukin-1 which is unresponsive to 10^3 U/ml of interleukin-2. *J. Immunol. Methods* **99** 7–11.

Hamblin, A. & Brannan, J. (1989) Cytokines. *TIPS/TIBS* July issues, centre page pull-out.

Hochstein, H. D. & Fitzgerald, E. A. (1990) Properties of US Standard Endotoxin (EC-5) in Human Volunteers. *First Congress of the International Endotoxin Society*, San Diego, USA, May 1990, p. 79.

Hort, E. & Penfold, W. (1912) Micro-organisms and their relation to fever. *J. Hyg.* **12** 361–390.

Levin, J. & Bang, F. B. (1964a) The role of endotoxin in the extracellular coagulation of *Limulus* blood. *Bull. Johns Hopkins Hosp.* **115** 265–274.

Levin, J. & Bang, F. B. (1964b) A description of cellular coagulation in the *Limulus*. *Bull. John Hopkins Hosp.* **115** 337–345.

Meager, A., Parti, S., Leung, H., Peil, E. & Mahon, B. (1986) Preparation and characterization of monoclonal antibodies directed against antigenic determinants of recombinant human tumour necrosis factor (rTNF). *Hybridoma.* **6** 305–311.

Milton, A. S. & Wendlandt, T. (1971) Effects on body temperature of prostaglandins of the A, E and F series on injection into the third ventricle of unanaesthetized cats and rabbits. *J. Physiol.* (*Lond.*) **218** 325–336.

Pearson, F. C., Weary, M. & Jorgensen, J. H. (1985) *Pyrogens: endotoxins, LAL testing and depyrogenation*. Marcel Dekker, New York.

Poole, S. & Mussett, M. A. (1989) The International Standard for Endotoxin: evaluation in an international collaborative study. *J. Biol. Std.* **17** 161–171.

Poole, S., Gaines Das, R. E., Baltz, M. & Pepys, J. (1986) Detection of endotoxin in mice by measurement of endotoxin-induced changes in plasma concentrations of zinc and of the acute-phase protein serum amyloid P-component. *J. Pharm. Pharmacol.* **38** 807–810.

Poole, S., Thorpe, R., Meager, A., Hubbard, A. & Gearing, A. J. H. (1988) Detection of pyrogen by cytokine release. *Lancet*, No. 8577, 130.

Root, R. K., Nordlund, J. J. & Wolff, S. M. (1970) Factors affecting the quantitative production and assay of human leucocyte pyrogen. *J. Lab. Clin. Med.* **75** 679–693.

Seibert, F. B. (1925) The cause of many febrile reactions following intravenous injections. *Am. J. Physiol.* **71** 621–651.

Taktak, Y. S., Bristow, A. F., Thavasu, P., Selkirk, S., Rafferty, B. & Poole, S. (1990) Detection of pyrogen by cytokine release using monocytic cell lines. *Proc. First International Congress on Cytokines: Basic Principles and Clinical Applications*. Florence, Italy, March 1990.

Thorpe, R., Wadhwa, M., Gearing, A., Mahon, B. & Poole, S. (1988) Sensitive and specific immunoradiometric assays for human interleukin-1α. *Lymphokine Research* **7** 119–127.

Tsuchiya, S., Yamabe, M., Yamaguchi, Y., Kobayashi, Y., Konno, T. & Tada, K. (1980) Establishment and characterization of a human acute monocytic leukemia cell line (THP-1). *Int. J. Cancer.* **26** 171–176.

US Pharmacopoeia (1980) 20th revision. Mack Publishing Co., Easton, Pennsylvania, 887.

Welch, H., Calvery, H. D., McClosky, W. T. & Price, C. W. (1943) Method of preparation and test for bacterial pyrogens. *J. Am. Pharm. Assoc.* **32** 65–69.

Ziegler-Heitbrock, H. W., Thiel, E., Fütterer, A., Herzog, V., Wiretz, A. & Rietmüller, G. (1988) Establishment of a human cell line (MONO MAC 6) with characteristics of mature monocytes. *Int. J. Cancer.* **41** 456–461.

10

Monoclonal antibodies

Robin Thorpe
National Institute for Biological Standards and Control, Blanche Lane, South Mimms, Potters Bar, HERTS EN6 3QG

10.1 INTRODUCTION

Monoclonal antibodies (mAbs) have very considerable potential for clinical use in humans. Although availability was originally limited to murine (almost always mouse or rat) antibodies produced by hybridoma technology (Kohler & Milstein 1975), advances in the past few years have enabled the production of human mAbs by various procedures (Steinitz *et al.* 1977), and also 'humanized' (i.e. human–mouse hybrid) antibodies by recombinant DNA methods (Neuberger *et al.* 1984). More recently the production of single immunoglobulin binding domains ('dAbs') from libraries of human V_H genes using the polymerase chain reaction has been described, which may revolutionize the clinical efficacy and usefulness of mAbs and their derivatives (Ward *et al.* 1989).

Despite the fairly obvious and widely publicized advantages of the use of mAbs for clinical application, there has been a rather slow acceptance of these products for therapeutic and even *in vitro* diagnostic purposes; to data there are very few mAbs licensed for *in vivo* clinical use. The problems with these products seem to be caused by difficulties in selecting 'ideal' antibodies with desirable characteristics, such as proven efficacy and safety, and difficulties with producing clinical grade material in pharmaceutical quantities. However, the scientific progress made in mAb technology, together with the availability of new production procedures, and increased knowledge concerning their characteristics and use, is leading to increased *in vivo* clinical use of these materials.

10.2 CLINICAL USES OF mAbs

mAbs are ideal reagents for many types of analytical procedures, and they are frequently employed in immunoassays. These assays are widely used for diagnostic purposes, ranging from the monitoring of physiological parameters to the detection of clinical abnormalities. There seems to be little difficulty in selecting mAbs with the desirable characteristics for such purposes, and relatively small amounts of antibody are needed for these applications. However, the use of mAbs for *in vivo* clinical purposes has been more problematic. Such uses can be summarized under three fairly broad headings:

(1) Direct therapy
(2) *In vivo* diagnosis ('imaging')
(3) Immunopurification by affinity chromatography of biological products to be used for therapy.

Direct therapy with mAbs has been used for anti-tumour treatment and to induce immunosuppression in organ graft recipients. Other applications include the treatment of septic shock, using either anti-lipid A mAbs, or antibodies specific for cytokines such as interleukin-1 (IL-1) or tumour necrosis factor (TNF). The prophylactic use of anti-Rh D mAbs is also considered an attractive possibility for the prevention of haemolytic disease of the newborn in cases of Rh incompatibility.

In vivo diagnostic use of mAbs has been used with some success for the location of tumours, e.g. melanoma, breast carcinoma and ovarian cancer. For such purposes, the antibodies are labelled with radioisotope, e.g. ^{111}In, and are detected using a gamma-camera. This approach has also been used for detection of myocardial infarction using mAbs specific for myosin heavy chain.

Immunoaffinity purification of biological products for therapeutic use has a potentially very wide application and has been used for the isolation of cytokines like interferons, and coagulation factors such as factors VIII and IX.

10.3 PRODUCTION OF mAbs

Production of murine mAbs using hybridoma technology is nowadays a fairly straightforward procedure which can be carried out almost routinely by expert staff in many academic and industrial laboratories. Antigen-primed lymphocytes are easily obtained as splenocytes from immunized mice or rats, and there are several mouse and rat myeloma cell lines available which are ideally suited for cell fusion to produce hybridomas. The required characteristics of these myeloma lines include:

(a) A high fusion efficiency with commonly used fusing agents (e.g. polyethylene glycol)
(b) Retention of stable characteristics to allow selection of hybridomas (e.g. remaining HAT- or azaserine-sensitive on continued passage)
(c) Do not secrete or (better) synthesize immunoglobulin chains themselves
(d) Are not contaminated with mycoplasmas or other possibly cytopathic/pathogenic microorganisms.

Resultant hybridomas can be easily cloned by limiting dilution or in soft agar to yield monoclonal lines which are usually stable in their ability to secrete antibody.

These lines can be grown in large scale fermenters in low concentrations of serum, or in defined 'synthetic medium' to produce large volumes of culture supernatant containing mAb. Alternatively, antibody can be produced *in vivo* as ascitic fluid by injecting hybridomas into the peritoneal cavity of appropriate mice or rats.

Production of human mAbs is considerably more problematic than the preparation of murine mAbs. This is due to the difficulty in obtaining suitable antigen-primed human lymphocytes and because, despite encouraging early reports, there is still no really satisfactory human myeloma cell line which is suitable for human hybridoma production. In view of the considerable advantages which human mAbs may have compared with their rodent equivalents (see 10.7) some effort has been made to provide alternative, more successful procedures for human mAb production. This has led to the development of the following approaches:

(a) Fusion of human lymphocytes with a murine or hybrid murine/human myeloma partner
(b) Transformation of human B lymphocytes with Epstein–Barr virus (EBV) to produce continuously growing antibody-secreting cell lines
(c) Fusion of an EBV transformed human line with a murine (usually mouse) cell line, to produce stable, rapidly growing antibody secreting cell lines
(d) Fusion of human B cells with a human lymphoblastoid cell line.

Fusion of human lymphocytes with murine myeloma cells may be regarded as a compromise owing to the lack of a suitable human myeloma fusion partner and the limited success of (d) above. This procedure is limited by the pronounced tendency of human/mouse hybridomas to preferentially lose human chromosomes, resulting in high instability as regards human antibody secretion. The use of human/mouse hybrid melanoma lines may overcome this in some cases.

Transformation of human B lymphocytes using EBV has been used for a considerable time to produce rapidly growing cell lines. The ability of these lines to secrete antibody is variable and such cells seem to undergo changes on continued culture, which often results in loss of their ability to secrete useful levels of mAb. The lines are also difficult to clone, as they strongly adhere to each other to form large polyclonal 'clumps'. Fusion of such cell lines with murine myeloma cells can confer stability and enables fairly easy cloning.

The difficulty of obtaining specific antigen-primed lymphocytes for production of cell lines secreting human mAbs is not easy to overcome. It is obviously potentially dangerous (or at least dubious on ethical grounds) to immunize humans with many antigens, and considerable effort has been invested in the evaluation and development of *in vitro* priming procedures and in the enrichment of antigen-primed lymphocytes using panning or rosetting. However, it is sometimes possible to immunize humans with non-pathogenic antigens such as erythrocytes (for production of anti-Rh D mAbs) or approved vaccines (for production of anti-bacterial or anti-viral mAbs). Antigen priming may also occur 'naturally' in some cases, e.g. with breast cancer patients. For practical and ethical reasons peripheral blood lymphocytes are the most commonly used source of cells for human mAb production, although splenocytes, or lymph node/tonsil-derived lymphocytes can occasionally be obtained as a 'spin-off' following surgery.

The difficulties in producing human mAbs with the appropriate specificities have led to the development of the alternative strategy of producing 'humanized' murine mAbs. For this, the genes coding for the variable or hypervariable parts of a mouse antibody are appropriately linked ('spliced') to the genes coding for the constant or variable parts of human immunoglobulin. These mouse/human hybrid monoclonal antibodies can then be expressed, usually in human myeloma cells. The secreted antibodies exhibit antigen specificity which closely mimics that of the murine monoclonal antibody, but most of the molecule is of human sequence.

Murine monoclonal antibodies can be produced *in vivo* as ascitic fluid or *in vitro* as culture supernatant. Production as ascitic fluid is unlikely to yield an mAb product suitable for licensing for *in vivo* clinical use in humans, although such preparations may be used for pre-clinical applications and experimental trials. *In vitro* production systems have been devised using airlift fermenters and hollow fibre cell carriers which enable relatively large scale production of clinical grade monoclonal antibodies. Batch and continuous culture procedures have been established; the former are usually easier to control, whereas the latter may be more efficient and cost-effective, and provide a relatively high concentration of starting material most suited for the available purification procedures.

10.4 PURIFICATION OF mAbs FOR *IN VIVO* CLINICAL USE

mAbs intended for clinical use should be as pure as possible, but the purification procedure should not impair the desired biological properties of the antibody. The antigen binding capacity of antibodies is robust and will be largely unaffected by most biochemical and physical procedures used for purification, but some immunobiological functions mediated by the Fc region are less hardy. Examples of procedures appropriate for purification of mAbs are salt or other precipitation fractionation, ion exchange chromatography, gel filtration and affinity chromatography. These are applicable to both hybridoma culture supernatant and ascitic fluid. It is difficult to produce mAbs of the quality required for *in vivo* use from ascitic fluid, and the simple isolation of murine IgG from such fluid may produce variable batches of product consisting of different proportions of the monoclonal (hybridoma) immunoglobulins, and those derived from the animal used for ascitic fluid production.

Affinity chromatography using immobilized bacteria-derived immunoglobulin binding proteins such as proteins A, G or C is a particularly useful and efficient procedure for isolation of monoclonal antibodies, especially if precautions are taken to prevent any 'leakage' of the immunoglobulin binding protein which may contaminate the product. (Ion exchange chromatography is ideal for this purpose.)

10.5 MONOCLONAL ANTIBODY CONJUGATES AND FRAGMENTS

The use of mAbs for imaging requires that they are conjugated to a radioisotope which allows their detection when bound to the target. This is usually achieved by introducing a chemical group into the antibody which firmly chelates an appropriate radiolabel. Several effective chelating compounds have been devised for this purpose which do not seem to affect the antigen binding properties of the mAb. The quality of the radioisotope used for preparation of such conjugates must be sufficiently high to generate an effective compound which produces efficient imaging.

mAbs intended for therapeutic uses may also be much more potent if conjugated to radioisotopes or toxins. In such cases the radioactive emission of the isotope or the cytotoxic effects of the toxin are used as effectors of cell death for tumour therapy or immunosuppressive purposes. Radioisotopes used for production of such conjugates are usually 'harder' than those used for imaging (e.g. ^{131}I) but as for imaging they are usually conjugated using chelation. This allows for the production of relatively large quantities of stable mAb–chelator complexes which can be rapidly conjugated to the relatively rapidly decaying radioisotope ($t_{\frac{1}{2}}$ values for such radioisotopes are usually fairly short) just prior to administration to the patient.

Toxin–mAb conjugates are usually fairly stable and numerous covalent chemical procedures have been devised for their construction. The ribosome-inactivating proteins derived from plant sources such as ricin and gelonin are particularly potent cytotoxins which form effective therapeutic agents when coupled to mAbs. Ricin has been particularly favoured for this purpose as it can be relatively easily split into the 'effector' A chain and the cell-binding B chain; separation of the A chain from the B chain is not very difficult, and so allows the production of mAb–Ricin A chain conjugates containing very few A–B chain dimers. Such conjugates have very low cytotoxicity for cells other than those targeted by the mAb.

An alternative very appealing approach is to produce ricin A chain by recombinant DNA technology. This entirely eliminates any possibility of ricin A–B chain species contaminating the mAb–conjugates. Bacterially derived toxins (such as diptheria toxin) can also be used to produce effective mAb-toxin conjugates (Thorpe *et al.* 1978).

For some purposes it may be advantageous to use fragments of mAb such as Fab or $F(ab')_2$ rather than the intact molecules. This avoids any possible adverse reactions in recipients due to the immunobiological functions of the Fc region, reduces immunogenicity and possibly (and most importantly) produces antigen binding molecules with a much shorter $t_{\frac{1}{2}}$ than intact IgG. This latter factor may be particularly valuable for providing a high target/background for *in vivo* diagnosis by imaging.

10.6 QUALITY CONTROL OF mAbs INTENDED FOR CLINICAL USE IN HUMANS

Numerous documents of the 'guidelines' and 'points to consider' nature have been produced by control authorities dealing with the production and control of mAbs and recombinant DNA-derived proteins intended for clinical use in humans; the reader is referred to these for detailed information (see end of reference section). In general, information should be available concerning all stages in the production of cell lines which secrete such antibodies. In particular, selection and cloning procedures should be shown to have produced a line which is stable in respect to both growth and secretion of monoclonal antibody and is actually monoclonal. The production procedure must be capable of producing consistent batches of product of a quality and purity suitable for use in humans. Production must be based on a carefully characterized and constructed cell bank ('seed lot') system which contains sufficient numbers of vials of cryopreserved cells to last for the expected life of the mAb product. This would normally comprise both 'master' and 'working' cell banks. The purified antibody should be shown to be efficacious and safe for the desired clinical indication(s). A reference preparation must be established for comparative purposes and the control of all future batches; this should consist of a large number of vials of product, if possible from a batch which was used in clinical trials.

Purification to apparent homogeneity of mAbs is not usually difficult (see Fig. 10.1), but particular attention should be given to low levels of potentially hazardous molecules which may contaminate mAbs if they are to be used *in vivo* in humans. Hybridomas and transformed human B cells can secrete cytokines which have very potent biological activity (Gearing *et al.* 1989a, b). Even low level contamination with such substances could cause adverse reactions in recipients. Ascitic fluid can contain fairly high levels of 'inflammatory' cytokines such as IL-1, IL-6 and TNF and these must be eliminated by purification procedures.

Two particular problems with clinical *in vivo* use of mAb which have been the subject of considerable discussion are possible undesirable cross-reaction of the mAb and the transmission of pathogenic viruses and nucleic acid which may contaminate the product. These are dealt with separately below.

10.6.1 Undesired cross-reactivity

Cross-reactivity of mAbs with tissues or cells other than the intended target is obviously undesirable, as it can limit the usefulness and interpretation of imaging, and may cause serious adverse reaction in recipients of therapeutic doses of such mAbs due to immunobiological consequences after binding to cross-reactive species. Cross-reactivity can arise either because the antigen occurs on or in more than one cell or tissue type, or because the antigenic determinant (epitope) recognized by the mAb is present on more than one molecular species. The former type of cross-reactivity is usually unrelated to the choice of mAb and generally reflects an inappropriate choice of antigen. An extreme example of this would be mAbs reacting with 'framework' determinants on HLA antigens, which although specific for antigen will bind to most human cells and tissues.

Cross-reactivity due to epitopes present on several molecules, most probably indicates an inappropriate choice of mAb and is very difficult to predict. It is now well-established that mAbs can be cross-reactive sometimes with molecules which appear to be unrelated (especially IgM mAbs and some human antibodies, Thorpe 1989, 1990) and so careful immunochemical characterization of the antigen binding properties of mAbs intended for use in humans is essential. Immunohistochemical and immunocytochemical studies using a wide range of tissues and cells probably provide the most valuable information pertinent to this issue (see Fig. 10.2), but often procedures such as immunoblotting and radio-immunoprecipitation analysis can also be very useful for demonstrating the molecular basis of cross-reactivity.

10.6.2 Transmission of viruses

Considering the manner of production of mAbs it would seem possible that viruses could derive from one or more sources, e.g.

(1) The myeloma cells used as fusion partners for hybridoma production

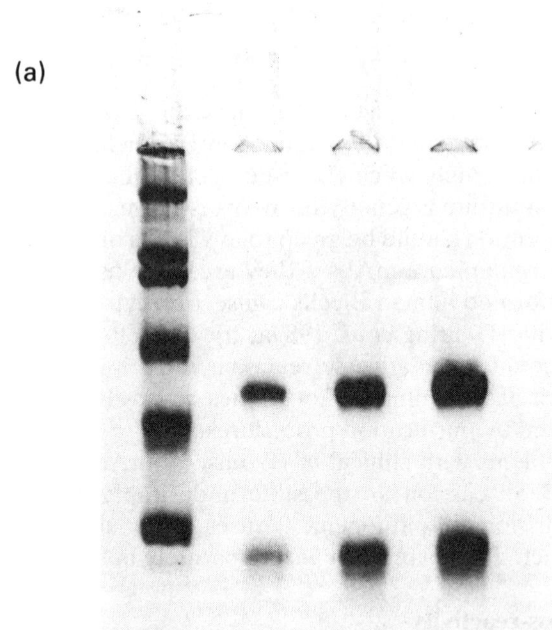

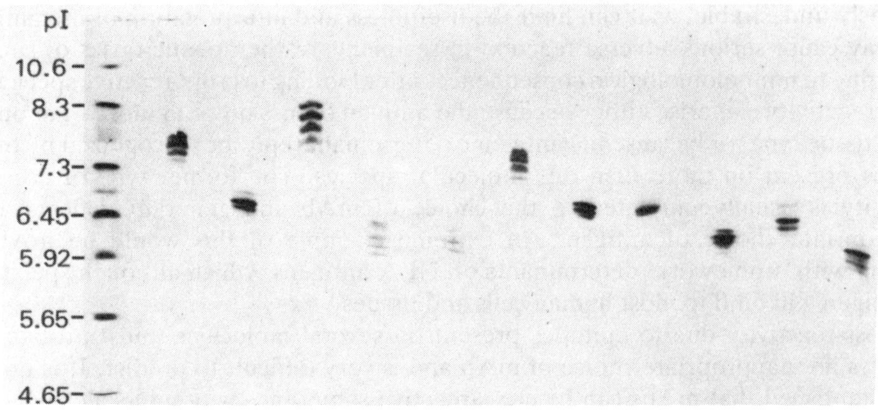

Fig. 10.1 — (a) SDS–polyacrylamide gel electrophoresis of a purified mouse IgG monoclonal antibody. The left-hand track shows molecular weight markers (205 kDa, 116 kDa, 97.4 kDa, 66 kDa, 45 kDa, 29 kDa from top to bottom). The other three tracks show different loadings of a purified mAb stained with Coomassie blue. Only heavy (γ) and light (κ) immunoglobulin chains are present (at 50 kDa and 23 kDa) even on 'overloaded' tracks. Courtesy of Chris Bird. (b) Isoelectric focusing on polyacrylamide gel of different mouse IgG mAbs (stained with Coomassie blue). The profiles show the preparations to be pure IgG and the closely spaced 'ladder' of bands in some preparations is due to post-translational changes to the immunoglobulin molecule resulting in change-dependent heterogeneity (differences in sialic acid content are particularly important for this effect). The p*I* value of the different mAbs can differ appreciably for mouse mAbs. The individual profile revealed by isolelectric focusing can be used as an identity test and is sometimes called the 'spectrotype' of the mAb. Courtesy of Maryvonne Brasher.

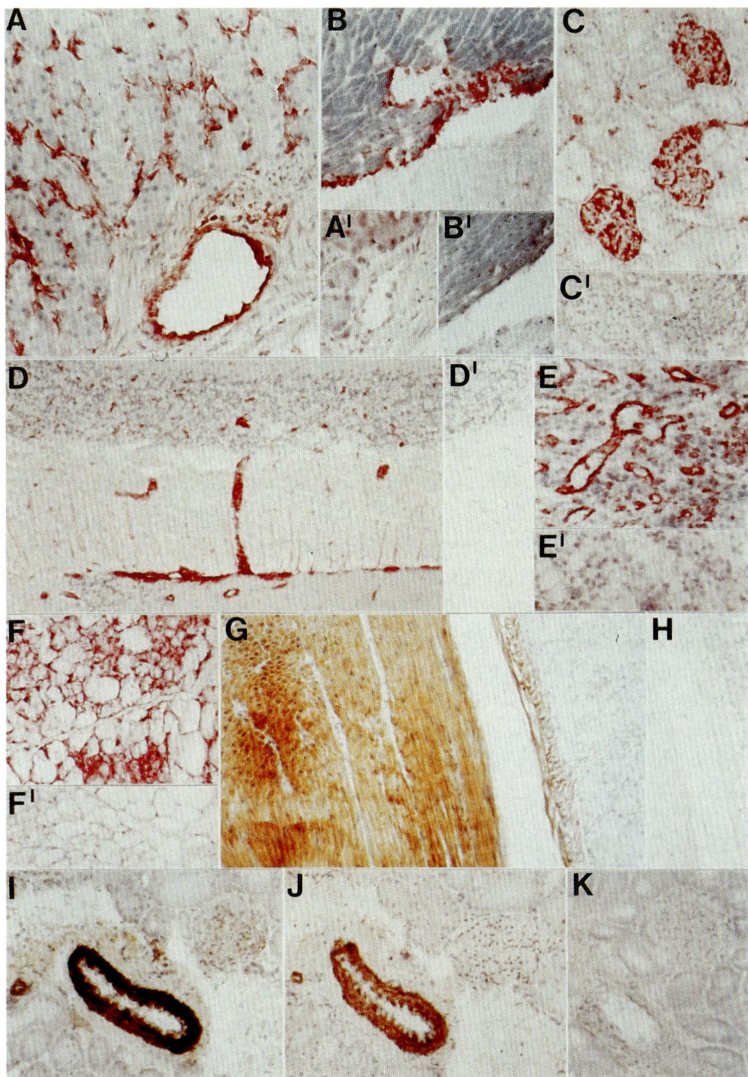

Fig. 10.2 — Cross-reactivity of human anti-Rh D mAbs (MAD-2, UCHD4, ESD1 and FOG-1) with various tissues as shown by immunohistochemistry. Panels A–F: alkaline phosphatase immunostaining using MAD-2. The red immunostaining is localized to fibroblasts in the lamina propria of stomach (A), endothelial cells lining blood vessels (A) and heart cavities (B), the glomuruli of kidney (C), vessels and cells having the appearance of Bergman glia in brain (D), the walls of the venous sinuses of spleen (E) and the alveolar walls of lung (F). A 'control' mAb (GB8) which was specific for a different red cell surface antigen showed no tissue cross-reactivity (panels A′–F′). Panels G and H: immunoperoxidase staining to show reactivity of UCHD4 (G) but not ESDI (H) with the smooth muscle of stomach. Panels I–K: immunoperoxidase staining of kidney sections to show staining of smooth muscle of vessels with UCHD4 (I) and FOG-1 (J), but not ESDI (K). UCHD4 also weakly stains glomeruli (I); sections were counterstained with haematoxylin.

This study shows that some anti-D mAbs recognize antigen(s) other than the Rh D antigen and that these are widely distributed in a range of tissues (Thorpe, S. J. 1989). Courtesy of Dr Susan Thorpe.

(2) Lymphocytes used as immune cells from mice/rats/humans, etc
(3) Mice or rats used for ascitic fluid production
(4) Tissue culture additives used during culture, e.g. foetal calf serum, insulin, transferrin
(5) EBV used to transform human lymphocytes
(6) Individuals involved in production procedures.

The range of viruses which could, at least theoretically, contaminate mAb preparations is very broad. Human mAb products are potentially more hazardous than murine preparations as any contaminating viruses are likely to be of human origin, and also cells secreting human mAbs have most probably been produced by transformation with EBV (a known human pathogen). However, some murine viruses are known to be pathogenic in humans and others may be undesirable, particularly when used in immunocompromised recipients.

One approach to viral detection involves investigations which attempt to detect specific viruses which are particularly hazardous or are considered possibly to be present or introduced at particular stages of mAb production. For murine antibodies (or murine/human heterohybridoma-derived human antibodies) mouse or rat antibody production (MAP or RAP tests) can be used to screen for specific mouse or rat viruses. Other, often immunochemical tests, are available for the detection of murine viruses. The methodology for detection of specific human viruses which may contaminate human mAb products is available, but the antibody production approach is obviously inappropriate.

Another strategy, designed to detect a broad range of viruses rather than pinpointing a particular virus, is to observe for cytopathic/cytostatic effects or altered growth characteristics using a panel of cell lines. The choice of lines depends upon the mAb and the particulars of the mode of production. Human diploid cells (such as MRC5) are particularly valuable for this purpose and transformed mouse, human and bovine lines as well as fibroblasts and mouse embryo primary cultures can provide useful information on this issue, Observing for disease, death or abnormal development of animals inoculated with mAb test material can also be used to test for viruses or possibly other hazardous contaminants of mAbs; suckling and adult mice as well as guinea pigs have been used for this purpose.

Detection of murine viruses which are pathogenic in humans, such as lymphocytic choriomeningitis virus and hanta virus or human pathogenic viruses would obviously suggest that the mAb which contains these viruses is unsuitable for use in humans. If master cell banks (or animal colonies used for ascitic fluid production) are contaminated with such viruses then they are unlikely to be of use for the production of safe products. Detection of other viruses in cell banks may not preclude the *in vivo* use of mAb derived from these cells as long as it can be demonstrated that the purification process removes or inactivates infectious virus. Most (if not all) murine myeloma cell lines used for hybridoma production contain retroviruses, and hybridomas derived using such lines are very likely to contain these viruses (see Fig. 10.3). Purification procedures should therefore be shown to remove these particles and pilot scale 'spiking' studies are probably best for this purpose. EBV is not usually secreted by transformed human B-cell lines which secrete mAb, but such lines do contain complex EBV gene sequences incorporated into their genome. Such DNA

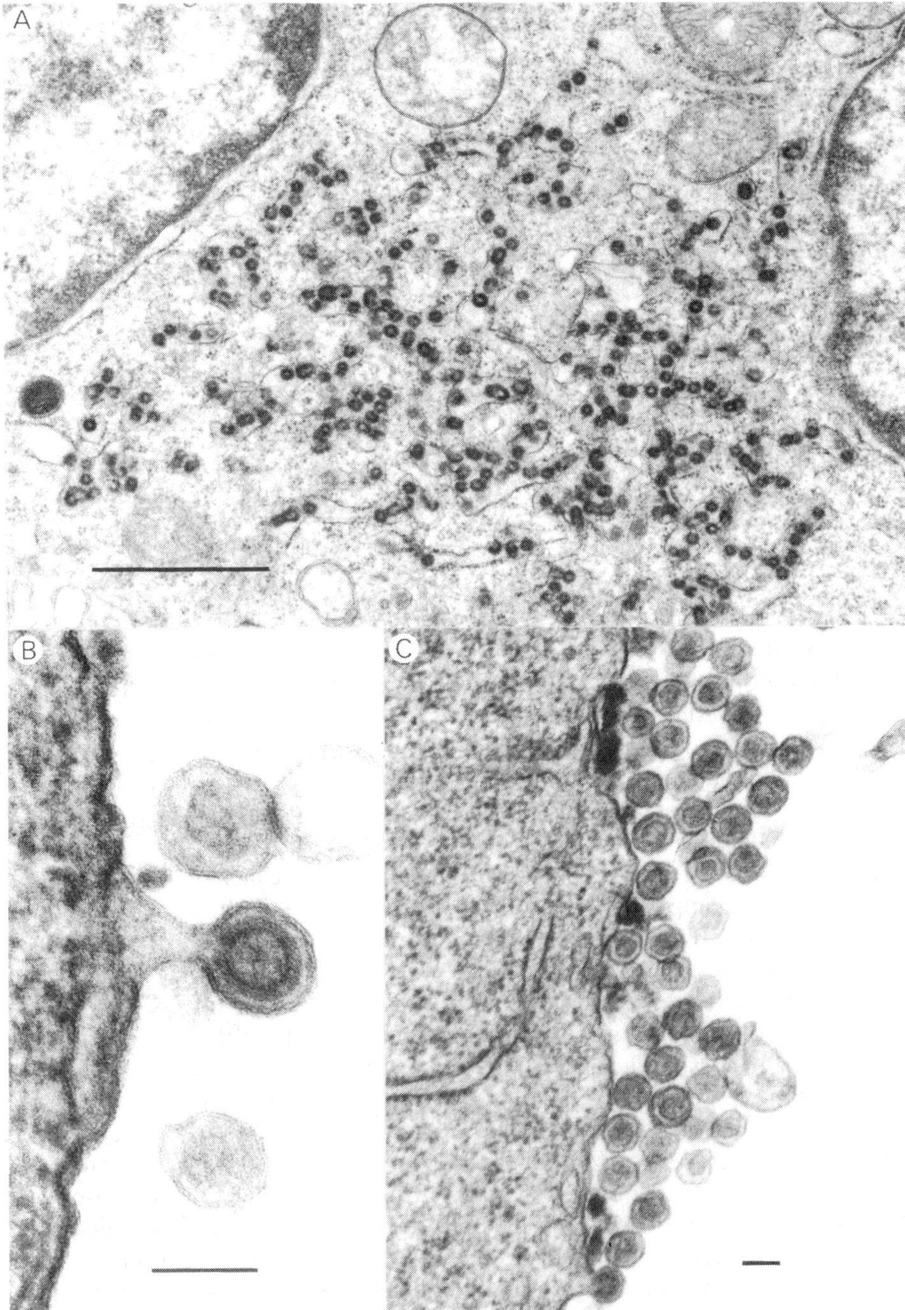

Fig. 10. 3 — Type C oncoviruses (Retroviridae) in hybridoma cells as revealed by transmission electron microscopy. (a) Viral cores budding into endoplasmic reticulum in the cytoplasm of a mouse hybridoma cell. Original magnification ×30 000; scale bar=1 μM. (b) Budding type C oncovirus at the surface of a human/mouse herterohybridoma. Original magnification ×180 000; scale bar=0.1 μM. (c) Mature type C oncoviruses released onto the surface of a human/mouse heterohybridoma. Original magnification ×66 000; scale bar=0.1 μM. Courtesy of Dr David Hockley.

sequences should not contaminate human mAb products intended for *in vivo* use in humans and subgenomic fragments of other human viruses are also potentially hazardous.

Tissue culture additives such as foetal calf serum should be from a source unlikely to be infected with virus or other pathogens and some testing for potentially hazardous viruses is advisable.

Avoiding contamination of mAb products with infectious agents derived from production personnel should be avoided by strict adherence to good manufacturing practice.

10.6.3 Transmission of oncogenic or pathogenic DNA

Contamination of mAb products with DNA should be minimized. Strict upper limits should be set for nucleic acid levels and DNA hybridization assays established using probes capable of detecting the host DNA from the mAb-secreting cell line. 'Spiking' studies using pilot-scale production are ideally suited for calculation of the reduction factor of the purification procedure for DNA.

10.7 PROBLEMS WITH THE *IN VIVO* USE OF mAbs

In vivo clinical use of mAbs has revealed some problems. Imaging and particularly therapy for solid tumours can be inefficient owing to poor penetration of the mAb. This can be improved in some cases by the use of fragments of IgG rather than intact molecules; dAbs or even their sub-fragments may improve efficiency of this application. Imaging for secondary tumours is often made difficult by the small target size and high background signal; this may be overcome by the use of computing and subtraction analysis using suitable controls.

Therapeutic use of murine mAbs is often limited by the lack of immunobiological functions such as complement fixation and opsonizing activity of such antibodies in human recipients. This is especially a problem for mouse antibodies and the approaches adopted to resolve this difficulty include: the use of toxin or radioisotope conjugates; the selection of specific murine IgG subclasses (rat IgG_{2b} seems most compatible with human immunobiological functions); and the substitution of humanized or if possible human mAbs for their murine equivalents. In some less common examples inherent activity of mAbs has been associated with adverse reactions in recipients, for example, mitogenic anti-T cell mAbs may produce undesirable side-effects due to mediator release by stimulated lymphocytes.

A serious problem with the use of mAbs in humans has been the induction of host antibodies against the mAb. This response consists of antibodies which bind to mouse immunoglobulin, presumably epitopes which represent differences between the mouse and human immunoglobulin sequences, and also sometimes anti-idiotypic antibodies which recognize the antigen binding part of the mAb. Antibodies of the latter type inhibit antigen binding and so preclude further use of the mAb in the recipient. Human anti-mouse antibodies (often called HAMAs) do not always block antigen–mAb interactions, but may cause adverse reactions or interfere with the efficient functioning of the mAb. Production of host antibodies against mAbs is obviously most serious if repeat dosing with mAb is intended. Development of anti-idiotypic antibodies obviously precludes subsequent use of a particular mAb, but a different mAb (which has a different idiotype) will be unaffected by the host antibodies. Development of a potent HAMA response in the recipient may well compromise the use of all mouse mAbs in that individual.

The use of murine mAb fragments such as Fab or $F(ab')_2$ may reduce the HAMA response in humans as these fragments are usually less immunogenic than intact mAbs. However, HAMAs will still be produced particularly on repeat dosing and anti-idiotypic antibodies are still a problem. Use of recombinant DNA-derived 'humanized' murine mAbs (particularly those containing only the murine hypervariable regions), dAbs, or human antibodies will obviously avoid HAMAs but anti-idiotypic and allotypic antibodies may still be produced in recipients.

10.8 CONCLUSIONS

The therapeutic, prophylactic and *in vitro* diagnostic value of mAbs is now clearly established and scientific/technical and pharmacological advances with these materials (and their derivatives) suggest that even greater potential will be realized in the future.

ACKNOWLEDGEMENTS

I thank Chris Bird and Maryvonne Brasher for Fig. 10.1, Dr Susan Thorpe for Fig. 10.2, Dr David Hockley for Fig. 10.3 and Deborah Kirk and Lisa Hudson for preparation of the manuscript. I am grateful to my colleagues at NIBSC, the MCA, and the Working Party on Biotechnology/Pharmacy of the CPMP for discussion of various issues relevant to this manuscript.

REFERENCES

Gearing, A. J. H., Leung, H., Bird, C. R. & Thorpe, R. (1989a) Presence of the inflammatory cytokines IL-1, TNF and IL-6 in preparations of mAbs. *Hybridoma* **8** 261–267.

Gearing, A. J. H., Bird, C. R., Priest, R., Cartwright, J. E., Wadhwa, M. & Thorpe, R. (1989b) Demonstration of cytokines in biological medicines produced in mammalian cell lines. *Lancet* (ii) 1011–1012.

Kohler, G. & Milstein, C. (1975) Continuous cultures of fused cells secreting antibody of predefined specificity. *Nature* **256** 495.

Neuberger, M. S., Williams, G. T. & Fox, R. O. (1984) Recombinant antibodies possessing novel effector functions. *Nature* **312** 604–608.

Steinitz, M., Klein, G., Koskimies, S. & Makel, O. (1977) EB virus-induced B lymphocyte cell lines producing specific antibody. *Nature* **269** 420.

Thorpe, P. E., Ross, W. C. J., Cumber, A. J., Hinson, C. A., Edwards, D. C. & Davies, A. J. S. (1978) Toxicity of diphtheria toxin for lymphoblastoid cells is increased by conjugation to antilymphocytic globulin. *Nature* **217** 752.

Thorpe, S. J. (1989) Detection of Rh D-associated epitopes in human and animal tissues using human monoclonal anti-D antibodies. *British Journal of Haematology* **73** 527–536.

Thorpe, S. J. (1990) Reactivity of a human monoclonal antibody against RhD with the intermediate filament protein vimentin. *British Journal of Haematology* **76** 116–120.

Ward, E. S., Gussow, D., Griffith, A. D., Jones, P. T. & Winter, G. (1989) Binding activities of a repetoire of single immunoglobulin variable domains secreted from *E. Coli. Nature* **341** 544–546.

Committee for proprietary medicinal products — Ad hoc working party on biotechnology/pharmacy. Guidelines on the production and quality control of monoclonal antibodies of murine origin intended for use in man. *Tibtech* **6** G5–G8.

Committee for proprietary medicinal products — Ad hoc working party on biotechnology/pharmacy. Guidelines on the production and quality control of medicinal products derived by recombinant DNA technology. *Tibtech* **5** G1–G4.

11

Development and quality assessment of vaccines

J. H. Peetermans
SmithKline Biologicals, Rixensart, Belgium

11.1 INTRODUCTION

Vaccines have been used for about 200 years and their development has always been linked to available techniques for obtaining the adequate antigen or immunogen. In bacterial vaccines, the evolution of technology has provided the world with vaccines containing whole cell microorganisms, detoxified toxins, purified chemically defined cell-wall polysaccharides, and isolated purified protein antigens. For viral vaccines, the development has been directed by the availability of isolation and multiplication technology and the presently available vaccines are produced in a variety of substrates including living animals, embryonated eggs and cell cultures. (Tables 11.1–11.5 illustrate the historical overview). Recombinant DNA technology has produced one worldwide licensed vaccine, namely Hepatitis B, extracted from transformed yeast cells. Other candidate vaccines developed by recombinant DNA methods are under evaluation. Examples are malaria, herpes, pertussis, foot and mouth disease, and poliomyelitis. Synthetic peptides representing the protective

Table 11.1 — Vaccine development — from idea to a final product

Definition: A vaccine
1. Is a product immunologically active against a given pathogen.
2. It does not cause short-term or long-term undesirable side effects (clinical acceptability).
3. It alters the immune system of the vaccinee in such a way that it confers protection against infection, invasion or clinical symptoms by the pathogen (efficacy).

Table 11.2 — Idea — feasability

Ideal (theoretical approach): in principle against any infectious organism or agent

Feasability (practical approach): conditioned by the present 'State of the Art' in

— Science: microbiology, parasitology, immunology, pathology
— Technique: isolation, culture or multiplication, inactivation, attenuation, expression

Remark: Economical considerations are also important

Table 11.3 — Historical evolution of bacterial vaccines

1. Killed, whole cells: pertussis
2. Detoxified exotoxins: diphtheria, tetanus
3. Purified subunits:
   — Polysaccharides
   — Proteins
   — Cell wall components — pili

Table 11.4 — Historical evolution of classical viral vaccines

Vaccine development is a function of the substrate:

1. Living animals
   Examples: Vaccinia — Cow
   Rabies — Rabbit, goat, baby mice
   Japanese B encephalitis — Baby mice
2. Embryonated eggs:
   Examples: Yellow fever
   Influenza
3. Cell cultures
   Examples: Poliomyelitis — Measles — Mumps — Rubella — Rota — Varicella — Hepatitis A

Table 11.5 — New generation vaccine production by genetic engineering, for example, hepatitis-B vaccine

1. Idea: HBsAg derived from the plasma of carriers induces protective immunity in chimpanzees and men
2. Feasibility: Techniques available:
   1. to clone the coding region for HBsAg in *E. coli, S. cerivisiae* and mammalian cells
   2. to bring the cloned genetic information to expression
   3. to purify the HBsAg from the crude cell extract

epitopes of the pathogen are a new approach for possible further applications. The vaccines obtained by the different technologies differ in the requirements for their quality assessment.

11.2 OVERVIEW OF VACCINES PRODUCED BY 'CLASSICAL' METHODS

Vaccines manufactured following 'classical' methods contain inactivated or infectious but attenuated viruses or bacteria. In some instances, the inactivated agents are purified. In other cases, the vaccines contain subunits or practically pure polysaccharides. It is generally accepted that the final product cannot be characterized completely and that extensive 'in-process' controls are therefore necessary to prove the consistent quality of such vaccines.

The quality of vaccines is generally determined by their safety, purity and immunogenicity. Clinical acceptability (Table 11.6) and efficacy (Table 11.7) are the

Table 11.6 — Clinical acceptability

1. Absence of clinical manifestations due to the active ingredient
   — Inactivation: non-infectious vaccine
     physical
     chemical
     combination of both
   — Attenuation: infectious vaccine
     — heterologous host's origin
     — artificial manipulations resulting in acceptable clinical reactions: passages, temperature sensitivity (ts-mutants) plaque purification

requirements for these three criteria.

For viral vaccines where the multiplication of the active agent takes place in mammalian cell cultures in the presence of media containing components of biological origin, such as trypsin and animal serum, the emphasis of safety and purity testing is put on the proof of absence of extraneous agents. Extraneous agents, such

Table 11.7 — Efficacy

Induction of the desired immune response
(1) Local immunity
(2) Humoral antibodies
(3) Cell-mediated immunity

Factors:
— Optimum quantity of antigen — potency or activity (inactivated or live)
— Antigen bearing the protective epitopes
— Presentation of the antigen:
  Three-dimensional structure
  Quality of the adjuvant
— Optimum vaccination schedule
— Satisfactory stability during storage

as bacteria, fungi, mycoplasma and viruses, may be introduced by the cell culture itself, by the media ingredients of biological origin and by the same infectious microorganism originating from the environment. The detection of extraneous agents takes place at different production steps: raw materials, cell cultures before production, bulk vaccine pool before clarification, and vaccine in the final container.

The methods for detection use a panel of sensitive cell cultures, embryonated eggs, laboratory animals and adequate growth media. For some live vaccines marker tests are required, as for example in the neurovirulence test for oral polio vaccine.

Most live viral vaccines are produced in primary cells or in validated diploid cell strains grown in chemically defined media, and the resulting viral suspension is very pure after the elimination of the cell debris. The final product generally consists of a dilution of the resulting bulks. For these reasons, no additional purification steps are required.

Some inactivated viral vaccines may be produced in heteroploid cells lines or may require significant concentration of the resulting bulk suspension. For such vaccines, careful monitoring of the purification efficacy for host-cell components and host-cell DNA is required.

The immunogenicity of live vaccines is based on titration for the minimum infective dose of each lot. The determination of the minimum required titre is based on dose-range studies carried out during clincal trials of the vaccine. The immunogenicity of inactivated vaccines is in most cases determined by the immune response in laboratory animals, where the new vaccine lot is tested in comparison with a validated reference preparation. The release of the new lot is based on an accepted limit for relative potency determined in a parallel dose–response assay.

For classical bacterial vaccines, which consist of inactivated whole cells or inactivated toxins, complete 'killing' and absence of reversion is the most important safety factor. Purity is important for all injectable preparations and adequate techniques for the elimination of contaminants from the host cell and the production medium have to be validated. The immunogenicity of bacterial vaccines is carried out in laboratory animals. A relative potency test, using international reference preparations, is the tool for evaluation.

11.3 OVERVIEW OF VACCINES OBTAINED BY RECOMBINANT DNA TECHNOLOGY

Recombinant DNA technology involves systematically arranging and manipulating specific segments of nucleic acid to produce a composite molecule which, when placed into an appropriate host environment, will yield a desired product.

There are three general methods for obtaining a specific coding segment:

(a) reverse transcription of mRNA to complementary DNA;
(b) isolation of genomic DNA or RNA;
(c) chemical synthesis as dictated by a peptide sequence.

The manufacturer should provide:

— a description of the method used to prepare the segment coding for the desired product, including both the cell type and origin of the source material.
— A detailed nucleotide sequence analysis. Normally, the gene of interest and flanking DNA are sequenced after cloning and the sequence of the encoded protein determined. If the particular gene in question was obtained from a natural source then the protein sequence it encodes should be similar to that of the natural protein.

The construction of the vector used for the expression of the cloned nucleotide segment should also be thoroughly described. This description should include a detailed explanation of the source and function of the component parts of the vector, e.g. origins of replication, markers such as antibiotic resistance genes, promoters and enhancers and whether or not the product is being synthesized as a fusion protein. The restriction enzyme digestion map of the entire constructed vector should be provided.

The host cell system which will generate the product is coordinated to fit the expression vector. It is therefore important that a complete description of the source, relevant phenotype and genotype of the host be provided.

The development of a hepatitis B vaccine from transformed yeast cells is detailed below as an example (Harford *et al.* 1987).

Serum from a human chronic carrier with hepatitis B virus (HBV) infection was used as the DNA source for cloning. The nucleotide sequence of the hepatitis B surface antigen (HBsAg) coding region was determined and identified. Fig. 11.1 shows the different forms of infectious virus and antigens found in human plasma. The aim of the transformation of recombinant yeast (*Saccharomyces cerevisiae*) was the induction of the coding sequence responsible for the expression of HBsAg, which was the antigenic component in the plasma vaccine (Valenzuela *et al.* 1982). The expression level of the 24 kDa HBsAg in the transformed yeast strain was in the 1% to 2% range of total protein, as judged by scanning of an autoradiograph displaying total cell proteins labelled by ^{35}S-methionine (Fig. 11.2). Lanes 1, 3 and 5 represent

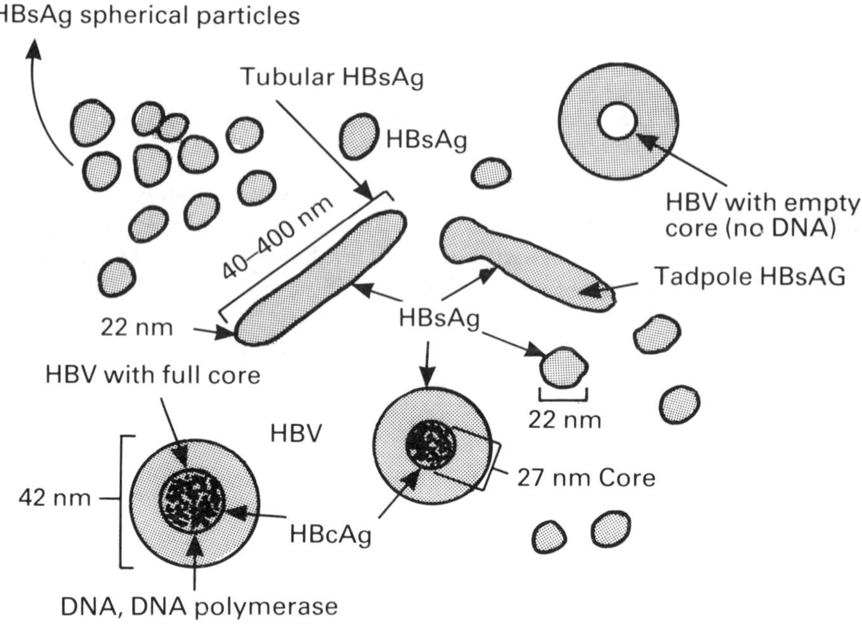

Fig. 11.1 — Schematic presentation of the different forms of HBV virus and antigens in the plasma of a chronic carrier.

the extract of the non-transformed host cell. Lanes 2 and 4 show the extract of the HBsAg producing cells. The bands at 24 kDa represent the HBsAg monomer. Intensive purification was necessary to obtain ~99% pure antigen in the form of particles of an average diameter of 20 nm (Fig. 11.3) very similar to those found in the plasma of chronic carriers (De Wilde *et al.* 1985).

11.4 PRODUCTION OF RECOMBINANT DNA VACCINES

The production of biologicals follows the seed lot principle. Once a suitable strain has been constructed, it is essential to hold it as the master seed lot. Starting from this master seed lot, a working seed lot has to be prepared under well-defined standard conditions.

The working seed should consist of several hundreds or thousands of containers stored under conditions that will ensure their stability (i.e. at −70°C or below).

During the fermentation cycles the working seed will multiply in order to provide the biomass necessary for industrial production of the vaccine. The multiplication cycles have to be very carefully determined in such a way that the end-product is obtained after a constant number of passages, irrespective of the fermentation volume used.

The stability of the seed lot has to be determined during storage and during the full fermentation run. The following criteria may be used to determine stability:

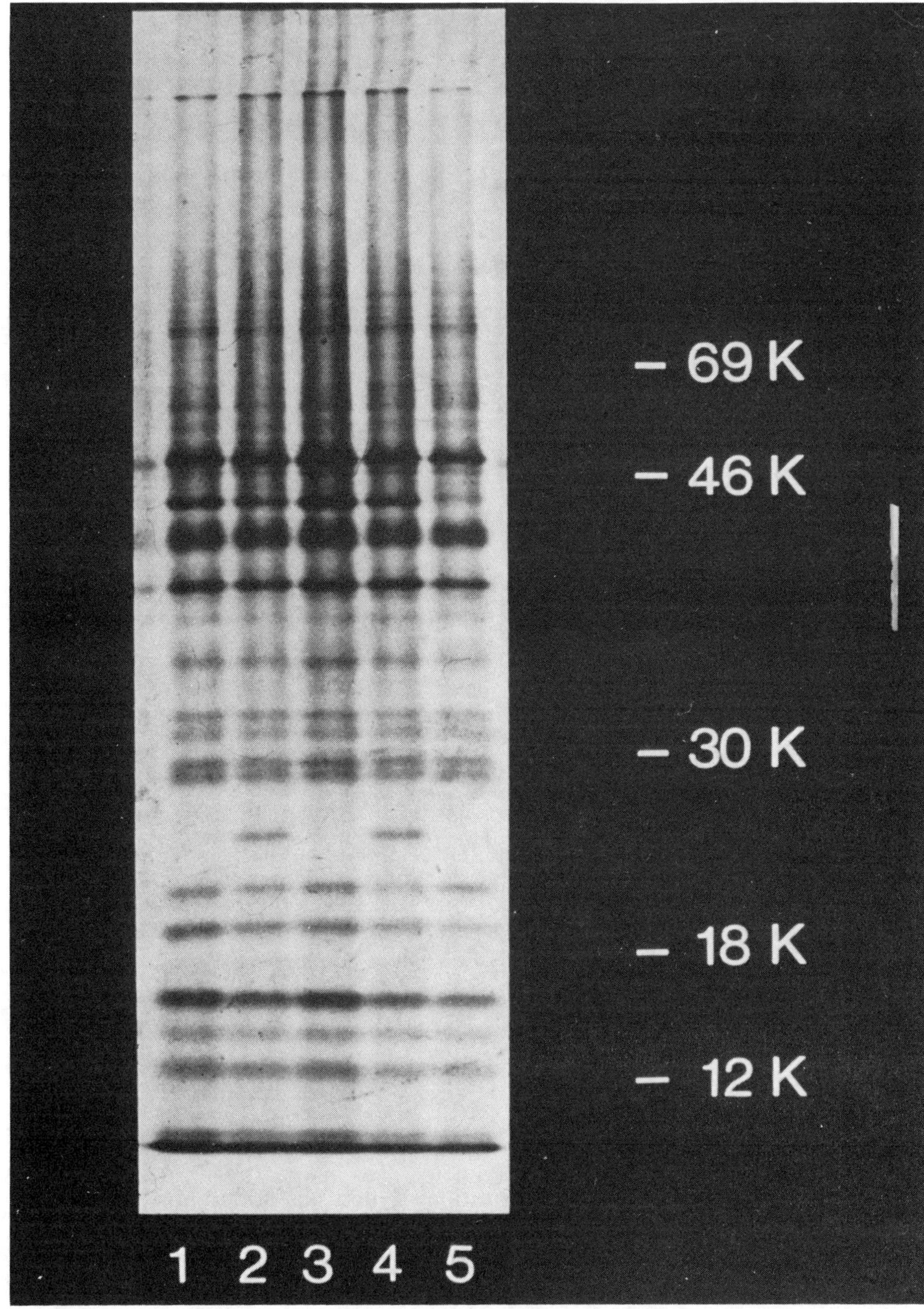

Fig. 11.2 — Fluorography of ^{35}S-methionine-labelled total yeast proteins. The extract of the non-transformed host cells is shown in lanes 1, 3 and 5. Lanes 2 and 4 show the yeast proteins and the 24 kDa HBsAg monomer of the producing strain.

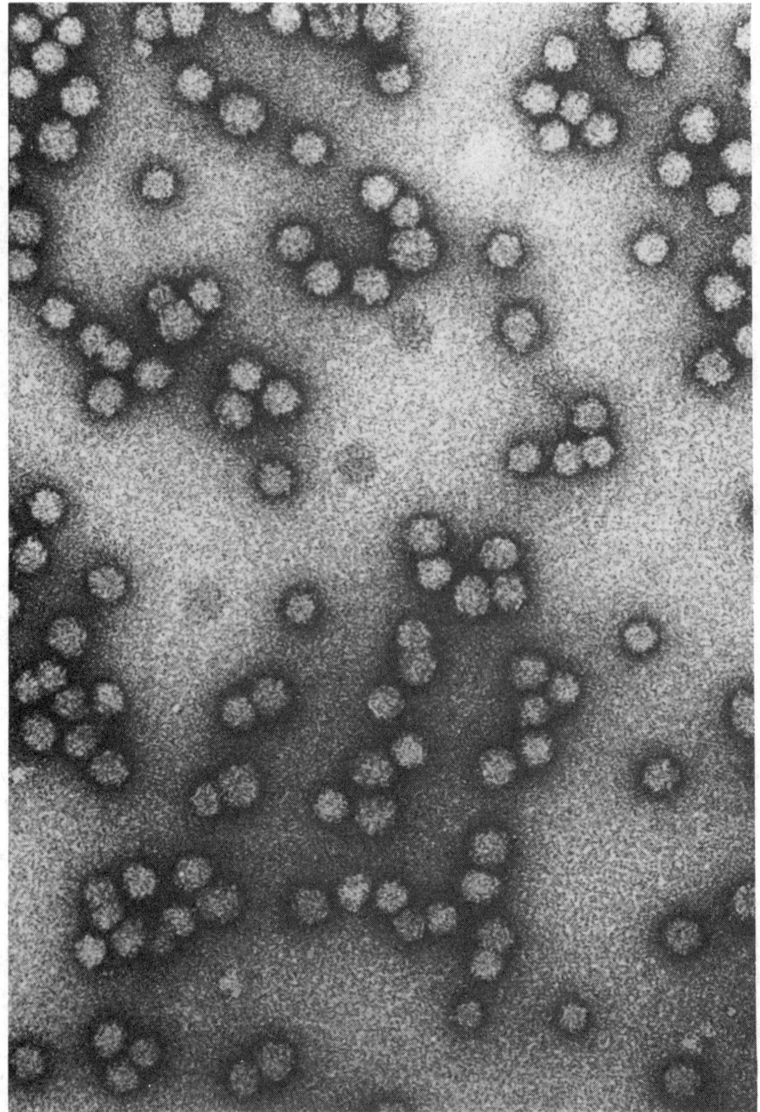

Fig. 11.3 — Electron microscopy of purified recombinant HBsAg particles.

- Genetic and biochemical characterization of the transformed host cell and the expressed gene product must show identity at all passage levels.
- The number of viable cells of the working seed lot must remain between acceptable limits during storage.
- Fermentations carried out after different time periods must give constant results meeting the specifications of the 'in-process' tests and the final product analysis.

The genetic stability should be proved by nucleotide sequencing of the coding region of the construct, as well as of the gene insert and the adjacent segments. This must be demonstrated at the seed lot level and at the end of a full fermentation cycle. Restriction endonuclease mapping of the entire recombinant vector at the end of representative fermentation cycles should give constant results.

The production of a product by biotechnology comprises the following steps:

— fermentation to provide the necessary bulk product;
— disruption of the fermentation product to free the antigen from the cell mass;
— purification of the antigen to eliminate all undesirable contaminants;
— formulation of the purified liquid vaccine by adsorption to an adjuvant;
— filling and packaging.

At each of these production steps, adequate tests and controls must be carried out according to specifications accepted by the national control authorities.

11.5 QUALITY ASSESSMENT OF DNA VACCINES USING A YEAST-DERIVED HEPTATITIS B VACCINE AS EXAMPLE

11.5.1 Overview

Over the last decade there has been rapid progress in the research and development of recombinant DNA biologicals. This has opened the way to numerous applications, but at the same time has created a need for adequate regulations covering the new product developments. Accordingly, the authorities have responded by issuing 'guidelines', 'points-to-consider', requirements and directives (Office of Biologics Research and Review, USA 1988; National Institute for Biological Standards and Control, UK 1987; World Health Organization 1987; Peetermans 1987).

The aim of quality assessment during production, also called 'in-process control', is to guarantee that the final products prepared in different production runs have identical quality. Detailed standard operating procedures for each of the steps are the keys to reproducibility, consistency and good manufacturing practices. An important first step in biotechnical production is fermentation. Besides the use of media with a standardized composition, the careful monitoring of temperature during the cycle and the strict application of the seed lot principle, certain tests have to be carried out during the cycle. Such controls cover monitoring of the pH and the dissolved oxygen, optical density, aspect of the colonies at the seed level and at the end of fermentation, sterility testing of the fermentation medium, microbiological purity of the harvest, and plasmid retention of the host cell at the end of each fermentation run. The limits for these parameters have to be determined during the development phases of the product and must be strictly followed during each fermentation cycle.

The most important and most difficult step seems to be purification. This is generally a multistep procedure combining centrifugation, precipitation, ultrafiltration, chromatography and other separation methods. Each step has to be validated during development.

During routine production, tests have to be carried out at each stage in order to determine the efficacy of each step for the elimination of contaminants and, even more important, for the optimal recovery of the active ingredient. Efficient purification will result in a purified concentrated product, which has to be assessed in the first instance by a full characterization of the HBsAg obtained. Secondly, routine control tests have to be carried out on each lot of the purified concentrated bulk vaccine.

11.5.2 Characterization of the gene product

For a vaccine, this characterization can be achieved by the immunological study of the product, the physico-chemical analysis of the expressed antigen and the evaluation of the absence of impurities and sensitizing material originating from the host cell.

11.5.2.1 Immunoglogical characterization (Hauser et al. 1987)

The immological characteristics of yeast- and plasma-derived antigen can be shown by the reaction of the yeast HBsAg with polyclonal and monoclonal anti-plasma HGsAg antibodies. The results should show that the polyclonal antibodies recognize perfectly the yeast-derived HBsAg, and that the epitopes corresponding to the plasma-derived product are also present in the yeast-derived vaccine. Table 11.8 shows the analysis of specific plasma monoclonal antibodies in vaccinees with recombinant and plasma vaccines.

Table 11.8 — Analysis of the antibody response of specific plasma monoclonal antibodies in vaccinees with recombinant and plasma vaccines

Monoclonal antibody	Percentage responders in vaccinees with	
	Recombinant Vaccine	Plasma vaccine
RF1	94	92
RF6	26	11
RF7	89	77
RF13	65	47

Inhibition experiments between the yeast- and plasma-derived antigens against anti-plasma antibodies obtained in animals and in human vaccinees should indicate that similar amounts of yeast HBsAg and plasma HBsAg are needed to obtain an equivalent inhibition, and that the plasma-derived vaccine induces anitibodies reacting similarly with both yeast and plasma HbsAg.

Monoclonal antibodies induced by the yeast-derived HBsAg should show that *a* and *d* determinants are present on the yeast antigen. Western blot analysis of the yeast HBsAg with polyclonal or monoclonal anti-plasma antigen should show that the yeast-derived vaccine contains epitopes which are also present on the plasma-derived vaccine. The yeast-derived vaccine should induce specific antibodies in animals; mice, guinea-pigs, Ceropithecus monkeys and chimpanzees are readily immunized by the vaccine.

Antibodies induced in human vaccinees with the recombinant vaccine should be characterized by determination of the percentage of antibodies raised against the common *a* determinant, which is considered to induce protective immunity, by monoclonal antibodies of defined specificity against the HBV and by competition studies using immunoglobulins from convalescent subjects and from vaccinees. Results should show that all seroconverters produce antibodies predominantly against the *a* determinant, that the nature of the immune response of human subjects to the yeast-derived vaccine corresponds closely to the response after the plasma-derived vaccine, and that the antibody response after yeast-derived vaccine is similar to the immune response in convalescent subjects.

Although the vaccine, as mentioned in the immunological charcterization, should induce antibodies similar to those induced by the plasma-derived vaccine, the study in chimpanzees should clearly demonstrate protection conferred by the vaccine. Chimpanzees vaccinated following the basic immunization scheme with vaccine from a representative lot are challenged with a dose of about 3000 chimpanzee infectious units of a virulent HBV, preferably of the non-homologous serotype, i.e. if the vaccine contains the *ad* type the challenge should be carried out with the *ay* type. Non-vaccinated control animals are challenged with the same preparation. During the six-month observation period, the control animals should show evidence of hepatitis B infection, while the vaccinated animals remain negative for markers of the infection.

11.5.2.2 Physico-chemical characterization (Pêtre et al. 1987)

Several methods useful for obtaining information on the chemical structure of the antigen are listed in the general guidelines. The following methods are to be preferred: SDS–PAGE analysis followed by Coomassie Blue and Silver staining or Western blot with polyclonal and monoclonal antibodies. The determination of amino acid composition should give values comparable with the data obtained for the natural product or comparable with the amino acid sequence predicted from the nucleotide sequence. Finally, N/C terminal sequence determination will complete the information required on the correct expression of the antigen. Peptide mapping after cleavage with enzymes of different specificity will enable reconstruction of the predicted amino acid sequence. The UV spectrum of the vaccine should show the characteristic UV features of the plasma product, with maximum absorption at 280 nm and a typical plateau at about 290 nm. The electron microscope pictures should show particles with a diameter of about 20 nm. Such particles are composed of the antigen protein and lipids. The proportions of both components should be determined and the acceptance limits fixed.

The purity of the vaccine has to be determined by standardized methods for each production lot. The presence or absence of sensitizing materials in the bulk product can be determined by sensitive tests such as the guinea-pig sensitization assay, Western blot analysis of impurities using an antiserum against crude whole yeast cell extract, and by the determination of anti-yeast antibodies in the sera of vaccinees. Most if not all vaccinees do have natural anti-yeast antibodies in their sera. Sera after the full vaccination course should not show a significant increase of these antibodies.

The comparison of the effect of vaccination schedule, the vaccine dose used in dose-range studies and the purity of experimental preparations on the antibody titre could show interesting correlations.

11.5.3 Routine quality control on production lots

The routine quality control of vaccine lots is carried out at five different consecutive production steps: raw and starting materials, 'in-process' monitoring, at the stage of the purified concentrated vaccine pool, on the final bulk vaccine and on the final containers. The most important control on the starting materials is the test to be carried out on control cell cultures for the absence of contaminants and viruses. This test is carried out on part of the substrate or host cells. During incubation of the control cells, microscopic monitoring takes place and specific tests in media, cell cultures and laboratory animals on samples taken during and at the end of the observation period should prove the purity of the substrate.

The fermentation of 'engineered' microorganisms and the cell culture of antigen-producing transformed cells is monitored by standard parameters of the multiplication cycles: purity of the inoculum, identity of the host cells during the whole incubation period, the consistency of the expressing system, quantity of the crude antigen, and limits of microbiological contaminants. Physical parameters such as pH, oxygen concentration, and cell density should be kept within predetermined close limits. In addition, if the anitigens are harvested by disruption of the producing cells, there should be careful monitoring of the different parameters and conditions of this process.

After a series of purification steps, the concentrated purified antigens are submitted to a series of tests in order to show freedom of extraneous agents, microbiological sterility, and (if applicable) effective inactivation. The protein content can be determined by several methods such as UV spectrophotometry, Kjeldahl nitrogen determination, Lowry assay or amino acid analysis. Amino acid analysis, however, only yields valuable information for low molecular weight peptides (not larger than 16 kDa). Each of these methods need accurate validation by well-defined standards.

The antigen titre is determined by titration methods for live viruses or micro-organisms. The antigenic activity of polypeptides can be quantified by adequate immunological tests such as RIA or ELISA. The ratio antigen to protein is a valuable indicator of consistency and immunogenicity of the product. The identity testing of the purified antigen can be completed by gel electrophoresis (SDS–PAGE) followed by staining of the separated bands with Coomassie Blue or Silver staining. The bands after electrophoresis can also be identified by Western blot techniques using specific poly- or monoclonal antibodies.

The quantitative determination of trace impurities is in most cases very difficult. Chromatographic profile methods useful for the separation of recombinant proteins and impurities are reverse-phase (RP), ion-exchange (IE) and size-exclusion (SE) HPLC. Other methods involve SDS–PAGE, Western blots using polyclonal antibodies against host-cell proteins, and immunoassays such as ELISA. Special problems may arise for antigens which aggregate in particles wherein impurities are most likely to be entrapped also, and where for purity determination, the particles have to be disrupted while keeping the proteins in solution. In some cases, the impurities may be separated or eluted in the profiles of the antigen. In these difficult cases, a combination of methods has to be used. Correct validation by incorporating small quantities of several pure proteins in the antigen prepartion to be tested or by running them in parallel is necessary.

DNA from the host cell can be determined by dot-blot hybridzation while total DNA may be estimated using the recently developed biosensor technology. If DNA is present in the product in significant quantities, it is important to evaluate the molecular size. Tests for components of the growth media and purification processes have to be developed for each case. The endotoxin content is determined by the classical LAL test (see Chapter 9).

Most of the routine testing on the final bulk vaccine and/or on the final containers is identical to that carried out on classical vaccines: sterility according to compendial requirements; endotoxin content using the routine LAL method, which can also be carried out on vaccines containing $Al(OH)_3$ as adjuvant; pH; limits of the filled volumes; identity and concentration of antiseptics; adjuvants and inactivating substances. Absence of abnormal toxicity or general safety is carried out in mice and guinea pigs.

The identity of the final product can be tested by SDS–PAGE on the antigen eluted from the adjuvant, followed by staining or Western blotting using specific antibodies. In some cases, immunological identity tests can be carried out directly on the absorbed antigen.

The potency testing of live vaccine preparations involves the determination of the infectious titre for which the minimum dose has been determined in clinical trails. For inactivated vaccines or for protein antigens, the minimum immunizing dose is also determined in clinical trails. This minimum effective quantity of antigen is used as the basis for the calculation of the quantities to be used in the formulation of the final bulk vaccine. In general the immunogenicity or potency is tested in laboratory animals. Several approaches are possible, but the method generally recommended is the determination of the fifty percent effective dose (ED_{50}) from the results of the dose-range test. The potency can be expressed as ED_{50} and the specifications drawn from the results obtained on the consistency lots and clinical trial lots. The method which is mostly used is the expression of the immunogenicity of the vaccine to test by relative potency calculated by Probit analysis. The reference vaccine in the relative potency test could be a vaccine lot tested in clinical trials and shown to have given satisfactory results in man. The potency test in animals is in most cases only a marker for consistency and immunogenicity of each new vaccine lot. The test is not adequate for the comparison of vaccines from different manufacturers because the response in animals is determined by factors such as the kind of adjuvant for which there is not correlation in man.

Good immunogenicity of routine lots is finally the result of identity, consistency and content of the antigen confirmed in a valid animal test, preferably combined with accelerated stability testing on samples of each lot.

11.6 CONCLUSION

'Classical' vaccines have been successfully produced and tested for several decades. The experience obtained in their quality assessment has been used as a basis and completed by 'newer' analytical techniques developed for biomolecules obtained by recombinant DNA processes. It has been shown that the different guidelines and points-to-consider issued by national and supranational licensing and marketing authorities form a valid basis for the quality control of the 'newer' vaccines.

REFERENCES

De Wilde, M., Cabezon, T., Harford, N., Rutgers, T., Simoen, E. & Van Wijnendaele, F. (1985) Production in yeast of hepatitis B surface antigen by R-DNA technology. *Develop. Biol. Stand.* **59** 99–107.

Harford, N., Cabezon, T., Colau, B., Delisse, A-M., Rutgers, T. & De Wilde, M. (1987) Construction and characterization of a *Saccharomyces cerevisiae* strain (RIT4376) expressing hepatitis B surface antigen. *Postgrad. Med. J.* **63** (Suppl. 2) 65–70.

Hauser, P., Voet, P., Simoen, E., Thomas, H. C., Pêtre, J., De Wilde, M. & Stephenne, J. (1987) Immunological properties of recombinant HBsAg produced in yeast. *Postgrad. Med. J.* **63** (Suppl. 2) 83–91.

National Institute for Biological Stardards and Control, Holly Hill, Hampstead, London NW3 6RB. Standardization and control of the biological medicinal products of recombinant DNA and hybridoma technologies and of products derived from aneuploid cells. (1987) *TIB Tech.* **5** (Suppl) G1–G4.

Office of Biologics Research and Review, Center for Drugs and Biologics, USA (1988) Points to consider in the production and testing of new drugs and biologicals produced by recombinant DNA technology, Draft (April 10, 1988).

Peetermans, J. H. (1987) Specifications and quality control of a yeast-derived hepatitis B vaccine. *Postgrad. Med. J.* **63** (Suppl. 2) 97–100.

Pêtre, J., Van Wijnendaele, F., De Neys, B., Conrath, K., Van Opstal, O., Hauser, P., Rutgers, T., Cabezon, T., Capiau, C., Harford, N., De Wilde, M., Stephenne, J., Carr, S., Hemling, H. & Swadesh, J. (1987) Development of a hepatitis B vaccine from tranformed yeast cells. *Postgrad. Med. J.* **63** (Suppl. 2) 73–81.

Valenzuela, P., Medina, A., Rutter, W. J., Ammerer, G. & Hall, B. D. (1982) Synthesis and assembly of hepatitis B virus surface antigen particles in yeast, *Nature* **298** 347–350.

World Health Organization (1987) Requirements for hepatitis B vaccines made by recombinant DNA techniques in yeast. *Technical Report Series* **760**.

12

Assuring the quality of plasma products

T. J. Snape
Blood Products Laboratory, Elstree, UK

12.1 INTRODUCTION — HUMAN PLASMA, A VALUABLE RESOURCE

Human plasma, the non-cellular component of human blood, is a valuable source of therapeutic proteins, exploited by national and commercial fractionators throughout the developed world. A thought-provoking argument for the effective use of human blood and plasma, and for the management (in the UK context) of a national blood transfusion service, was presented by Professor John Cash in his editorial in the British Medical Journal (Cash 1987), and extended in a more recent editorial by the same author (Cash 1990).

Notwithstanding the labour-intensive nature of large-scale plasma fractionation, and the high capital and revenue costs of a modern fractionation facility, plasma products are essentially low-value-added therapeutic materials, source plasma representing more than 50% of operational costs (in the UK at least). For this reason, and also because availability of source plasma is currently capacity-limiting in the UK, it is desirable that the processes for the manufacture of plasma fractions be mutually compatible, allowing recovery of most fractions, at optimized yield, from every pool of plasma fractionated.

National (UK) and international guidelines are available setting out good manufacturing practice and manufacturing specifications for plasma products: UK Health Departments (1983 & 1989a), WHO (1989). The monograph *Blood Products — Function and Use*, Bright (1989) provides a useful overview of the subject, with practical reference to the BPL operation. Minimum finished-product specifications for the more important plasma fractions are to be found in the relevant monographs in the British and European Pharmacopoeias.

12.2 THERAPEUTIC PLASMA FRACTIONS

The fractionated products of human plasma fall into three categories:

(i) Coagulation factor concentrates, factor VIII and factor IX (and other less frequently required concentrates — factor VII, factor XI and factor XIII) and the coagulation inhibitor antithrombin II, are used for replacement therapy in congenital or acquired deficiencies of these factors. These concentrates are freeze-dried preparations, reconstituted in water for injections for administration by intravenous injection.

(ii) Human albumin solutions are used to restore/maintain circulatory volume in the treatment of shock following accidental or surgical trauma and blood loss or severe burns, for the treatment of hypoalbuminaemia and for total plasma replacement in disorders such as Guillan–Barre syndrome. Preparations containing 4.5% and 20 w/v albumin (5% and 25% outside the UK) are available for administration by intravenous infusion.

(iii) Normal and antibody-specific immunoglobulins are used to provide passive immunity in individuals especially at risk of infection. Rh D immunoglobulin is used in the prevention of haemolytic disease of the newborn. Preparations are available for both intramuscular and intravenous injection, the intravenous route facilitating the administration of much larger doses with more immediate therapeutic effect.

The text *Handbook of Transfusion Medicine* (UK Health Departments, 1989b) provides useful, if limited, information on blood and plasma products and their applications.

12.3 ASSURING QUALITY — SOURCE MATERIALS

As with any manufactured product, the quality of plasma products is affected by the quality characteristics of the source materials used in their manufacture. The quality of the starting plasma is of particular importance. Plasma recovered from individual donations of blood, or obtained by plasmapheresis of individuals donors, has the biological individuality of its donor. This variability is further compounded by possible collection and storage lesions resulting in the loss/activation of the more labile components.

These problems are minimized, and economies of scale achieved, by the fractionation of 'pools' of several thousand plasma donations to produce a single batch of any product. At the Blood Products Laboratory (BPL) 10–12 000 donations are pooled to give a starting batch of 3 to 5 t frozen plasma.

Of particular current significance, if donor selection, plasma testing and fractionation procedures are not well-chosen and securely implemented, is that blood-borne microorganisms (specifically viruses) may be transmitted from an infective donor to the recipient of the plasma product. The viral contaminants of most significance in this context are hepatitis, B, hepatitis C (previously designated hepatitis non-A, non-B), HIV-1 and HIV-2.

Screening tests for hepatitis B surface antigen and for antibody to HIV-1 and HIV-2 are applied to all donations of blood or plasma collected from donors in the UK. Routine testing for markers for hepatitis C virus is not currently performed in the UK, although a test for antibody to the virus is under evaluation. In North America and mainland Europe donors are tested for elevated liver enzymes as a surrogate marker for hepatitis C infection.

Although tests for viral markers remain important, both in respect of tests on individual donations and of tests on each plasma pool and its derived products, the introduction of viral inactivation procedures in the manufacture of plasma products has greatly reduced the risk of viral transmission by such products. These procedures are elaborated in section 12.5.

12.4 ASSURING QUALITY — PROCESS DESIGN

The design of the processes for the recovery of plasma fractions is currently dominated by three considerations:

(1) factor VIII, the product which usually determines the scale of fractionation cannot currently be recovered by direct capture from plasma — it is obtained by cryoprecipitation during controlled thawing of plasma;
(2) economy of plasma utilization demands that as many products as possible be recovered from a single batch of plasma, by mutually compatible processes;
(3) with the exception of the manufacture of immunoglobulin fractions, the manufacture of all other plasma products requires the inclusion of some kind of viral inactivation procedure — increasingly it becomes difficult to exclude even immunoglobulins from such considerations.

Two fractionation schematics (Fig. 12.1 and 12.2) set out in simplified form the comprehensive fractionation scheme used by BPL for the maximal recovery of plasma fractions from a single plasma pool.

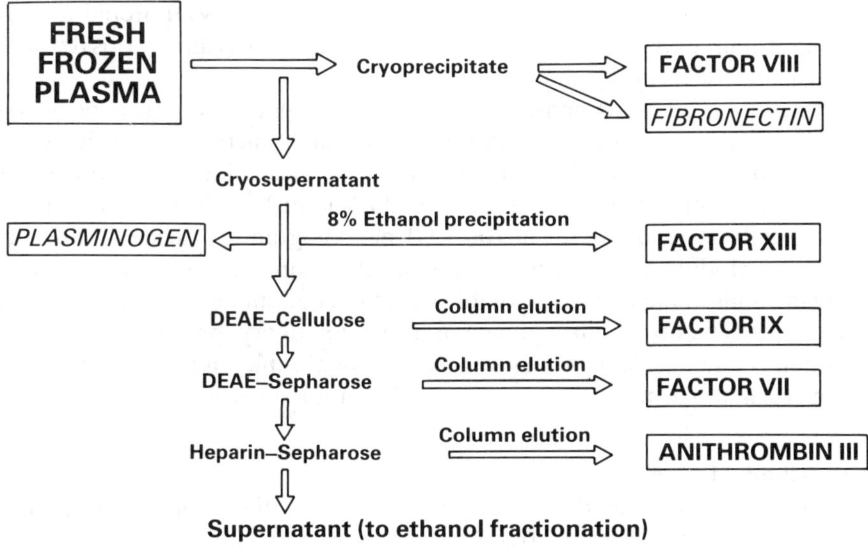

Fig. 12.1 — Plasma fractionation of coagulation factors.

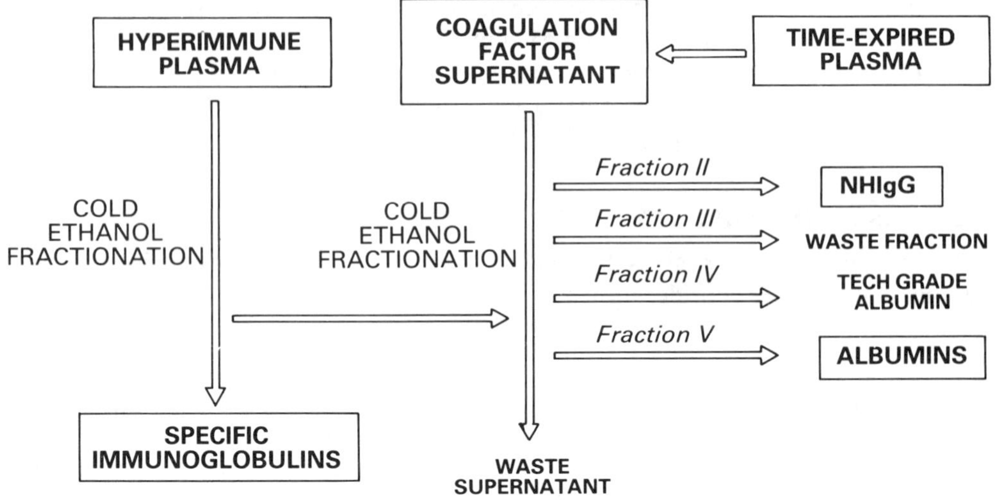

Fig. 12.2 — Plasma fractionation of ethanol fractions.

12.4.1 Coagulation factors — and an inhibitor

In Fig. 12.1 the steps necesssary for the recovery of the most important coagulation factors, antithrombin III and two potential products — fibronectin and plasminogen — are set out. In principle, all of these products could be recovered from every plasma pool; in practice the particular fractions recovered from any given pool would be determined by the requirement to satisfy current clinical demand. For most fractionators, the total amount of plasma required for the manufacture of factor VIII results in potential over-capacity of other products. As factor VIII preparations from recombinant sources become available on a commerical basis during this decade, this picture will almost certainly change.

Cryoprecipitation, the common starting point for all therapeutic concentrates of factor VIII, imposes considerable limitations on manufacturers, and in particular limits yield of finished product. Typically, the cryoprecipitation process allows recovery of little more than half of the available factor VIII from the frozen source plasma (usually around 400 IU factor VIII per kilogram of plasma). Subsequent processing, including some form of viral inactivation step, reduces this further, resulting in yields of factor VIII available net to the patient of 200 IU per kilogram of plasma or less. The other consequence of utilization of cryoprecipitation as the first process step is that a homogeneous sample of pool plasma is almost never available, since the crushed plasma cannot be effectively sampled before thawing.

14.4.2 Ethanol fractions

Fig. 12.2 sets out in simplified form a typical ethanol fractionation scheme, providing for the recovery of all usable fractions. Normally the only products recovered would be normal immunoglobulin (for intramuscular or intravenous injection) and albumin solutions. The supernatant from coagulation factor production may be supplemented by the addition of time-expired or, in some countries through not the UK, pooled liquid plasma.

Perhaps the single most significant factor in the control of the plasma pool for ethanol fractionation is the requirement to minimize contamination of the pool by viable organisms. Despite the sub-zero fractionation temperatures, the utilization of ethanol at up to 25% v/v in the fractionation process and terminal sterilizing filtration of the therapeutic products, bacterial growth during the several days of processing may contribute significantly to the endotoxin levels of the products. For solutions infused intravenously in litre quantities, at rates of up to 40 ml per minute, the risk of serious febrile reaction cannot be over-stated.

No longer a common problem with therapeutic albumin preparations, the inducement of life-threatening hypotension (usually seen in patients undergoing surgery involving cardio-pulmonary bypass, and therefore lacking the normal mechanism for inactivation of kinins and kinin precursors) was seen in preparations containing high levels of prekallikrein activator, formed as a result of contact activation of the coagulation–kinin pathways during long processing times (Snape *et al.* 1979).

12.5 ASSURING QUALITY — ELIMINATING VIRAL CONTAMINANTS

The viral safety of blood products is determined by many contributing factors — donor selection, donor screening for viral markers, donor follow-up (with plasma exclusion as necessary), tests for markers on the plasma pool, on in-process samples and on the finished product, as well as by the viral virucidal nature of some process steps.

Viral inactivation procedures for coagulation factor concentrates in current use are:

(i) in-process treatment with solvent detergent mixtures capable of disrupting lipid-enveloped viruses;
(ii) in-process pasteurization, the labile coagulation factor protected by added stabilizer;
(iii) in-process exposure of the freeze-dried product to controlled heating (either by low-pressure steam, or by heating as a slurry in a non-aqueous solvent);
(iv) end-process heating (in the final container) under validated conditions of time and temperature.

The method of choice for UK fractionators has been end-process heating, the freeze-dried concentrate being subjected to heating at 80°C for 72 hours (validated conditions for inactivation of hepatitis viruses and HIV). The advantages of this approach are that inactivation is performed in the final container, with no risk of reinfection. Against this, control of dried product characteristics, in particular moisture content is critical to effective virus inactivation. Factor VIII, factor IX and factor VII concentrates may be treated in this way. Antithrombin III concentrate is pasteurized at 60°C for 10 hours earlier in processing (Fig. 12.3).

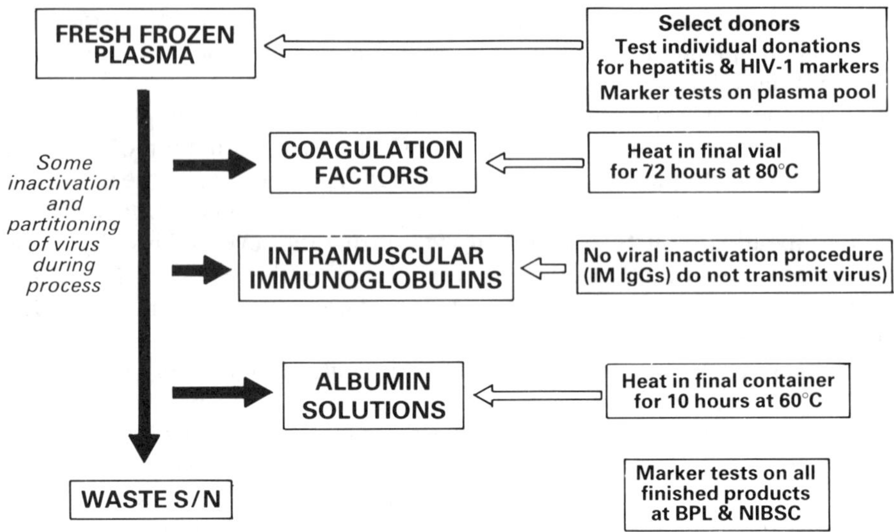

Fig. 12.3 — Elimination of viral contaminants.

The approach with greatest acceptance world-wide is the use of solvent detergent mixtures (TNBP + Tween-80 or TNBP + Triton X-100) on an early process stage. Advantages are the rate and security of inactivation of lipid-enveloped viruses. The major disadvantages are yield impairment, the need to effectively remove the viral inactivation components and the very real risk of downstream recontamination by cross-infection from parallel processes.

Albumin products are rendered virus-safe by pasteurization for 10 hours at 60C in the final container. Immunoglobulin preparations have not normally been subjected to a viral inactivation step, but many years of safe and effective use would suggest that intramuscular immunoglobulin preparations made by traditional cold ethanol fractionation techniques do not transmit blood-borne virus infections. Increasingly however, and for intravenous immunoglobulin preparations in particular, manufacturers are reviewing their position and considering the benefits of introducing a defined viral inactivation step in immunoglobulin manufacturing processes.

12.6 ASSURING QUALITY — QUALITY SYSTEMS

The manufacture of plasma fractions requires the application of sound quality assurance principles to the total process beginning with product and process design, and with the application of good (pharmaceutical) manufacturing practice at all stages of processing, from plasma collection through to despatch of the finished product to users. Of special significance is the requirement for full traceability from individual plasma donations to finished product manufactured from pools of several thousand such donations.

The implementation of effective quality systems ensures that compliance with quality assurance principles is both effective and demonstrable. In the context of

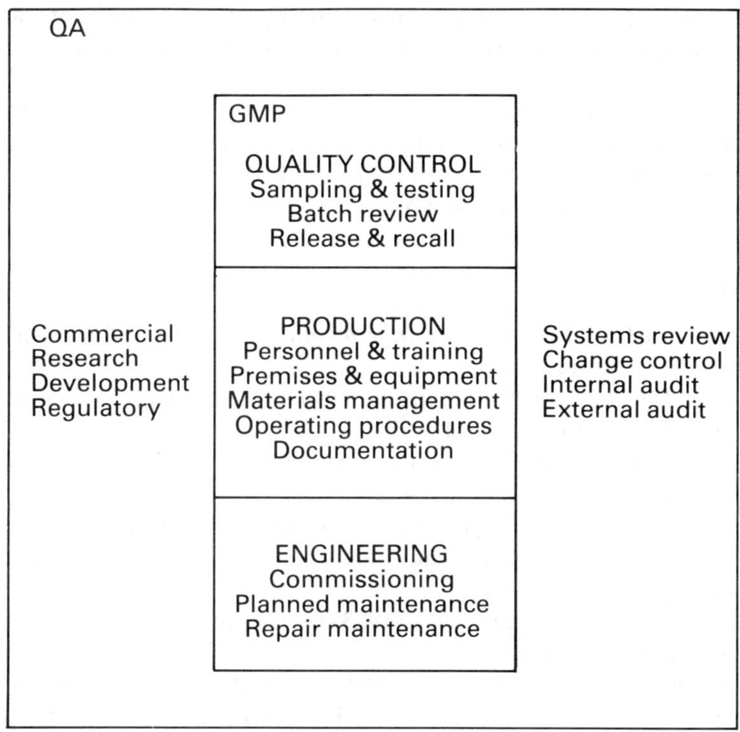

Fig. 12.4 — Operational interactions.

plasma products manufacture, and probably most other manufacturing operations, quality systems may be oriented according to the variables they are designed to control, specifically:

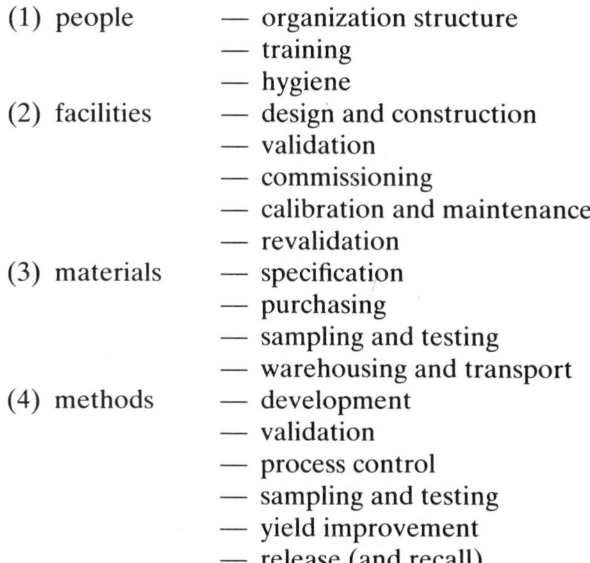

(1) people — organization structure
— training
— hygiene

(2) facilities — design and construction
— validation
— commissioning
— calibration and maintenance
— revalidation

(3) materials — specification
— purchasing
— sampling and testing
— warehousing and transport

(4) methods — development
— validation
— process control
— sampling and testing
— yield improvement
— release (and recall)

With attention to quality systems covering these variables, and with appropriate overlays of documentation and audit, maximum assurance of product quality may be obtained.

The operational interactions of the component parts of an operation for the manufacture of plasma products (in this case BPL) may be set out as in Fig. 12.4. Production is shown as central, with an overlay of quality control activities and underpinned by a sound engineering base.

REFERENCES

Bright, S. (1989) *Blood Products — Function and Use*, Higher Education Enterprise Link Series, Hobsons.

Cash, J. D. (1987) The Blood Transfusion Service and the National Health Service, *British Medical Journal*, **295** 617–619.

Cash, J. D. (1990) Blood Transfusion Services and the European Community, *British Medical Journal*, **300** 481–482.

Snape, T. J., Griffith, D., Vallet, L. & Wesley, E. D. (1979) The assay of prekallikrein activator in human blood products, *Developments in Biological Standardisation*, **44**, 115–120.

United Kingdom Health Departments (1983) *Guide to Good Pharmaceutical Manufacturing Practice*, HMSO.

United Kingdom Health Departments (1989a) *Guidelines for the Blood Transfusion Services in the United Kingdom*. Vol II — Guidelines for the Preparation of Plasma Fractions.

United Kingdom Health Departments (1989b) *Handbook of Transfusion Medicine*, HMSO.

World Health Organization (1989) *Requirements for the Collection, Processing and Quality Control of Blood, Blood Components and Plasma Derivatives*, WHO Publications.

13

Interleukin-2: a case history

P. A. Nessi
Glaxo Institute for Molecular Biology S.A.,
Route des Acacias 46, 1211, Geneva 24,
Switzerland

13.1 INTRODUCTION

Interleukin-2 (IL-2) was the first lymphokine to be recognized and completely characterized (Smith 1988, Tsudo *et al.* 1986, Teshigawara *et al.* 1987, Robb *et al.* 1987, Dukovich *et al.* 1987). It was first identified by its ability to promote the long-term proliferation of T-cells (Morgan *et al.* 1976), and was originally called T-cell growth factor. Subsequent research has shown that its functions are more diverse, and that, in addition to stimulating the proliferation of IL-2-dependent T-cells, it also modulates the immunological effects of cytotoxic T-cells, natural killer cells and activated B-cells.

The various functions of the hormone are mediated by binding to a specific IL-2 receptor. This receptor consists of two separate polypeptides (P75 and P55), and has two independent but co-operative IL-2 binding sites (one per polypeptide) (Smith 1990).

IL-2 itself is a 133 amino acid protein with a molecular weight of about 15kDa; it has a disulphide bridge linking residues Cys 58 and Cys 105, and a free cysteine at position 125. X-ray crystallography (Brandhuber *et al.* 1987) has shown that the tertiary structure of the molecule is formed from six α-helices (A–F). Four of these helices (B–D, F) constitute the structural scaffold of the molecule, and the remaining two (A, E) provide the IL-2 receptor binding site. The integrity of the tertiary structure is essential for high affinity receptor binding.

From studies on IL-2 and related lymphokines it has become apparent that the immune system uses hormonal mechanisms to direct all aspects of the response to antigens, including the induction of specific gene expression, proliferation of clonal expansion, and lymphocyte differentiation. These observations have produced widespread interest in the clinical applications of IL-2, as a means of stimulating immune responses. In the research carried out by Rosenberg, for example, IL-2 is demonstrated to have potential as a new form of immunotherapy for cancer (Rosenberg *et al.* 1985).

13.2 IL-2 PRODUCTION PROCESS (see Fig. 13.1)

The microorganism, *Escherichia Coli*, used for production of IL-2 has been described elsewhere (Liang *et al.* 1988). During fermentation, IL-2 is secreted as an insoluble aggregate in the host cell. Once harvested, the cells are broken and the aggregated protein is separated from soluble constituents by differential centrifugation. In the resulting pellet, which contains aggregated IL-2 and cell debris, the IL-2 is about 50–70% pure.

The next stage of purification involves selective extraction of the *Escherichia coli* (*E. coli*) phospholipids, by washing the precipitate with phosphate buffer. The aggregated IL-2 is solubilized in 5% acetic acid and loaded on a gel filtration column, where monomeric IL-2 is separated from polymerized IL-2, peptidoglycans and phospholipids.

Monomeric IL-2 is then air oxidized (using a tangential flow filtration system) and the folded product is separated from disulphide-mismatched, reduced and residual polymeric IL-2, on a reverse-phase C-8 silica column using 30% acetonitrile and 1% TFA (Browning *et al.* 1986).

A final gel filtration in acetic acid buffer gives the IL-2 bulk drug substance. This is formulated in lactic acid and D-mannitol, and is freeze-dried to give the injectable drug.

13.3 QUALITY CONTROL OF IL-2

13.3.1 'In-process' quality control

The various tests and determinations involved in the 'in-process' quality control (QC) of IL-2 are summarized in Fig. 13.1. The first tests performed are carried out on the seed stock, the purity of which must be rigorously controlled. Then at all stages throughout the subsequent fermentation, the numbers of microorganisms per unit volume are monitored, along with the purity of the cell culture. After fermentation, the pelleted, and aggregated IL-2 is checked on SDS gels, and following air oxidation, the monitoring is performed with analytical HPLC. Fig. 13.2 shows HPLC traces obtained in monitoring the air oxidation, and Fig. 13.3 shows preparative HPLC profiles. In addition to the analytical HPLC carried out on the final bulk drug, the product is also checked for protein content and pyrogens.

13.3.2 Quality control of the bulk drug

For the bulk drug, there are two different types of QC employed:

- the routine QC procedure, which involves quick and easily performed tests applied for the release of each drug lot;
- the control QC procedure which employs a more sophisticated type of analysis and which is applied only twice a year.

13.2.2.1 Routine quality control

The routine QC procedure involves a series of observations and simple determinations, including appearance, pH, osmolarity, protein concentration and endotoxin content. In addition it includes:

Protein status	Production step	Batch size	Purity	In process QC
Frozen GMP seed stock (TET)				Culture purity Host identification Plasmid characterization
Inoculum				
Fermentation		1000 l		Optical density
	Centrifugation			Culture purity
Whole cell pellet		50–70 kg	10–15%	SDS-PAGE
	Cell rupture			Expression level
Washed pellet		3 kg	50–80%	SDS-PAGE Weight
	Washing	30–40 g		
Aggregated IL-2				
	Acid extraction (5%)	12–15 g		
Soluble IL-2			60–70%	OD_{280}
	G-100 gel filtration	3–4 g		SDS-page
Monomeric IL-2 (reduced)				HPLC
	Diafiltration #1	1 g		Lowry-OD_{280}
Monooomeric IL-2 (oxidized)			80–85%	HPLC
	RP HPLC			OD_{280}
Single conformer IL-2		300 mg		HPLC
	Diafiltration #2			Lowry-OD_{280}
Bulk drug			97–99%	Lowry-OD_{280}
	G-75 gel filtration			
New drug substance		250 mg	99%	Final QC

Fig. 13.1 — Interleukin-2 phase I–II purification process and related in-process quality control.

UV spectral analysis The shape of the curve (as well as its first and second derivatives) provides a check on modifications arising as a result of product degradation.

Determination of amino acid composition To protect against any deletion or changes of amino acid residues caused by mutations of the strain or by the purification process.

Analytical HPLC Provides the only reliable means to detect the presence of reduced, dimeric and misfolded IL-2 (all of which must be excluded from the final product).

SDS–gel electrophesis To check for the presence of foreign proteins.

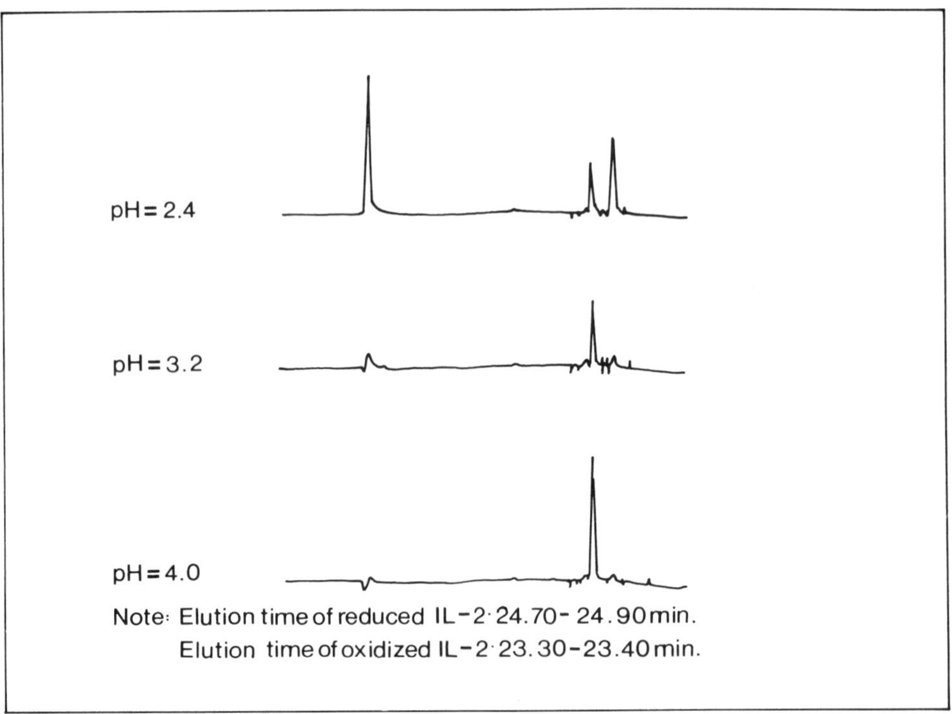

Fig. 13.2 — Monitoring of interleukin-2 oxidation by HPLC.

Determination of bioactivity This must be superior to 10^6 U/mg and is checked by the incorporation of either ^{3}H-thymidine (Gillis *et al.* 1978) or MTT* (Mosmann 1983) using a cell line whose growth is stimulated by IL-2.

If the final bulk drug passes all of these tests, it is released for further processing, i.e. formulation, filling and lyophilization. If not, it is rejected but may be used simply for *in vitro* work.

13.3.2.2 Control quality control

The control QC procedure involves more time-consuming determinations, including analysis of disulphide conformation, peptide mapping, sizing, circular dichroism, and N- and C-terminal analyses. (All of these analyses are performed according to standard procedures using an IL-2 in-house reference for comparison.) Finally, to protect IL-2 against unwanted contaminations, there are quantitative determinations made for DNA residues, *E. coli* proteins, tetracycline, dithiothreitol, acetonitrile, acrylonitrile and heavy metals.

13.3.3 Quality control and stability of the lyophilized product

The quality control of the final (lyophilized) product is essentially the same as the routine QC procedure applied to the IL-2 bulk drug. The analyses specific to an injectable drug that are added to the procedure include moisture and pyrogen

*3-(4,5-dimethylthiazol-2-yl)-2,5 dophenyl tetrazolium bromide.

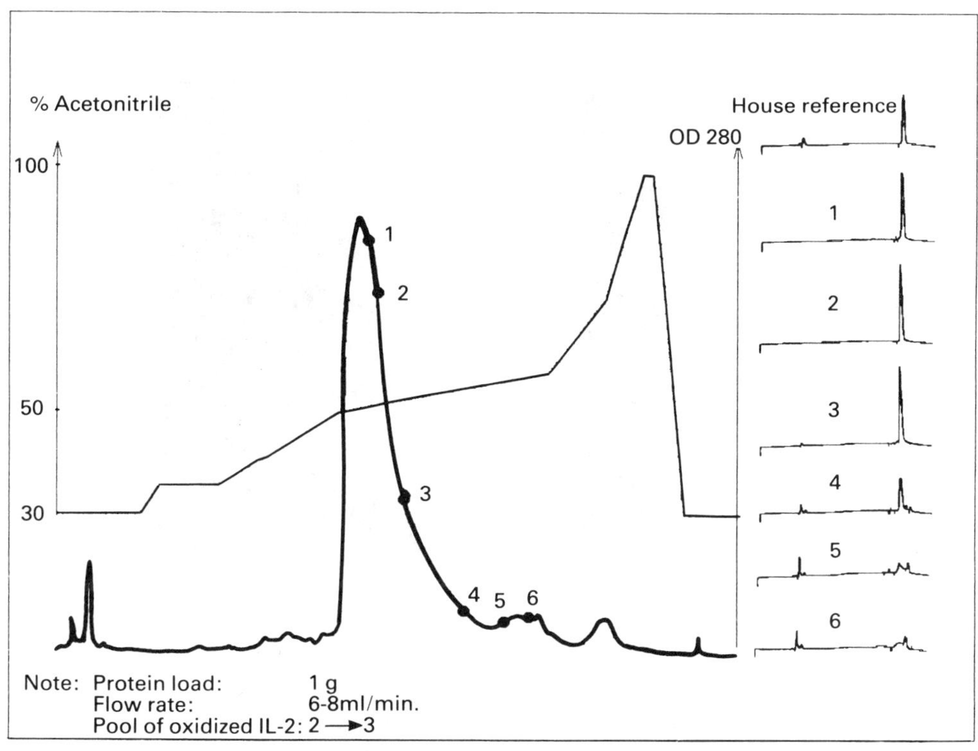

Fig. 13.3 — Interleukin-2 preparative HPLC elution profile.

content determinations, together with monitoring of the product's stability and murine toxicity.

The biological activity of the finished drug is again determined by means of incorporation of ^{3}H-thymidine or MTT. The two tests do not give exactly the same specific activity, and our criterion for the release of IL-2 is based on the thymidine assay, for which the lower limit is fixed as 1.0×10^{6} U/mg. The experimental values are calibrated against an in-house standard which has been compared with the WHO 1st International standard lot 86/504.

Once released by QC, each new batch of finished product is put on an accelerated stability study at five different levels of temperature (−80°C, 4°C, 20°C, 37°C and 55°C). The results at three months are used to calculate (by means of the Arrhenius equation) the shelf life to be assigned to each lot. The latter is generally 14 months but can be extended depending upon the results of the stability testing carried out over this period.

13.4 FROM LABORATORY TO PATIENT: TECHNICAL AND REGULATORY PROCESSES

13.4.1 Production

In scaling-up the IL-2 production from a laboratory process to a phase I–II process, there were several technical and regulatory issues which presented difficulties.

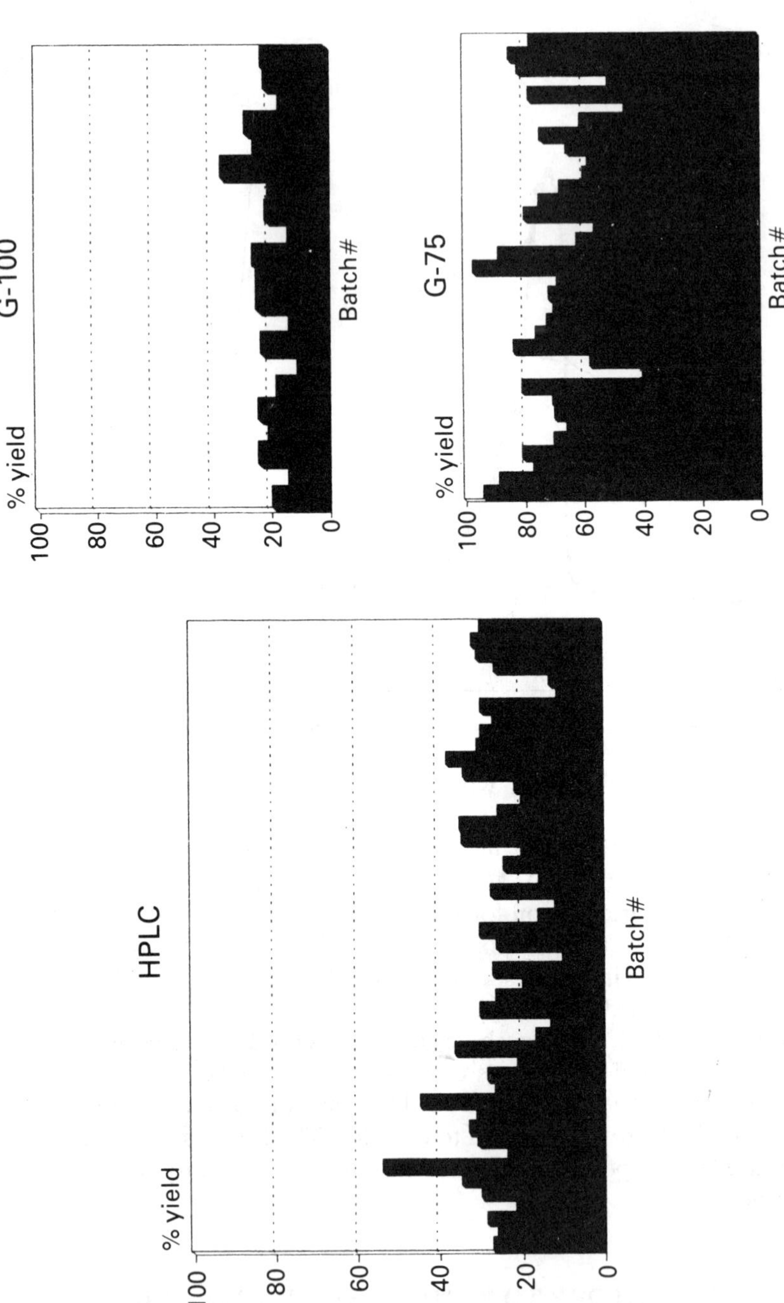

Fig. 13.4 — Yields of the three main purification steps of interleukin-2.

Table 13.1 — IL-2 House reference bioactivity

I. Intra-assay variation Date	Experimental values ($\times 10^{-6}$ U mg^{-1})				Mean±SD
20/07/89	4.92	4.87	5.40	4.91	5.02±0.25
25/07/89	2.55	2.65	2.94	2.87	2.75±0.18
31/07/89	4.29	4.38	4.23	4.33	4.31±0.06
07/08/89	6.93	5.93	5.28	4.98	5.78±0.86
08/08/89	7.63	7.73	6.36	5.30	6.75±1.15
09/08/89	6.37	6.46	7.53	5.07	6.35±1.00
14/08/89	4.43	4.91	5.35	4.87	4.89±0.38
15/08/89	5.56	5.36	5.72	5.62	5.56±0.15

II. Inter-assay variation Period	Mean±SD	(n)	Extreme values
07/89–11/89	5.19±1.40	(63)	2.48–8.78

These included the replacement of the ampicillin-resistant strain by a tetracycline-resistant strain; the establishment of a contained process for the part of the production dealing with living organisms; the replacement of dialysis by diafiltration (tangential flow filtration); and the development of the IL-2 formulation. The first three of these issues posed regulatory rather than technical difficulties, in the sense that each has demanded a complete revalidation of the IL-2 production and/or purification process. The problem of IL-2 formulation arises because of the isoelectric point of the protein, which inconveniently lies in the neutral region (pH 7.5–7.8).

Other, more serious difficulties which had to be resolved in the scaling-up process were a consequence of the batch-to-batch variability, not only in terms of IL-2 yield (Fig. 13.4) but also the drug bioactivity (Table 13.1). These difficulties also cause related problems in quality control (see below).

13.4.2 Quality control issues

In addition to the problems in production caused by the scaling-up process, there were also problems with the associated quality control. These included the setting up of assays to quantify the levels of DNA and virus contamination of IL-2 preparations; the determination of the stability of the intermediate product; and the problems arising from IL-2 self-aggregation (caused by the higher protein concentrations employed in the phase I–II process; Fleischmann *et al.* 1989).

A certain level of batch-to-batch variability, which made it difficult to set up suitable acceptance and rejection criteria for the drug. In this context it is important to note that when the release specifications for a new drug, such as IL-2, are established, it is clearly necessary to be severe enough to ensure the rejection of all unwanted lots, but, at the same time, to define sensible specifications appropriate for biological derived material which has a degree of intrinsic variability.

REFERENCES

Brandhuber, B. J., Boone, T., Kenney, W. C. & McKay, D. B. (1987) Three-dimensional structure of Interleukin-2. *Science* **238** 1707–1709.

Browning, J. L., Mattaliano, R. J., Chow, E. P., Liang, S. M., Allet, B., Rosa, J. & Smart, J. (1986) Disulphide scrambling of Interleukin-2: HPLC resolution of the three possible isomers. *Analytical Biochem* **155** 123–128.

Dukovich, M., Wano, Y., Thu, L. T. B., Katz, P., Cullens, B. R., Kehr, J. H. & Greene, W. C. (1987) A second Interleukin-2 binding protein that may be a component of the high affinity Interleukin-2 receptor. *Nature* **327** 518–522.

Fleischmann, J. D., Wentworth, D., Valencic, F., Imbembo, A. L. & Koehler, K. A. (1989) Interleukin-2 self-association. *Biochem. Biophys. Res. Comm.* **152** 879–885.

Gillis, S., Ferm, M. F., Ou, W. & Smith, K. A. (1978) T-cell growth factor parameters of production and a quantitative microassay for activity. *J. Immunology* **120** 2027–2032.

Liang, S. M., Allet, B., Rose, K., Hirschi, M., Liang, C. M. & Thatcher, D. R. (1988) Characterisation of human Interleukin-2 derived from *Escherichia col. Biochem. J.* **229** 429–439.

Morgan, D. A., Ruscetti, F. W. & Gallo, R. C. (1976) Selective *in vitro* growth of T-lymphocytes from normal human bone marrow. *Science* **193** 1007–1008.

Mosmann, T. (1983) Rapid colorimetric assay for cellular growth and survival — application to proliferation and cytotoxicity assays. *J. Immunological Methods* **65** 55–63.

Robb, R. J., Rusk, C. M., Yodoi, J. & Greene, W. C. (1987) Interleukin-2 binding molecule distinct from the *Tac* protein: analysis of its role in formation of high affinity receptors. *Proc. Natl. Acad. Sci. U.S.A.* **84** 2002–2006.

Rosenberg, S. A., Mule, J. J., Spiess, P. J., Reichest, C. M. & Schwarz, S. L. (1985) Regression of established pulmonary metastases and subcutaneous tumour by the systemic administration of high-dose recombinant IL-2. *J. Exp. Med.* **161** 1169–1188.

Smith, K. A. (1988) Interleukin-2: inception, impact and implications. *Science* **240** 1169–1175.

Smith, K. A. (1990) Interleukin-2. *Scientific American* **262** (No. 3) 26–33.

Teshigawara, K., Wang, H. M., Kato, K. & Smith, K. A. (1987) Interleukin-2 high affinity receptor expression requires two distinct binding proteins. *J. Exp. Med.* **165** 223–238.

Tsudo, M., Kozak, R. W., Goldman, C. K. & Waldermann, T. A. (1986) Demonstration of a non-*Tac* peptide that binds Interleukin-2: a potential participant in a multi-chain Interleukin-2 receptor complex. *Proc. Natl. Acad. Sci. U.S.A.* **83** 9694—9698.

14

Genotropin® — recombinant human growth hormone

A. Wielburski
Department of Protein Analysis, Kabi Pharmacia, Stockholm, Sweden

14.1 INTRODUCTION

14.1.1 Growth hormone

The recent use of recombinant DNA technologies has allowed the expression, isolation and characterization of a number of clinically important proteins, one of which is human growth hormone (hGH). In mammals, growth hormone is produced and secreted by the anterior pituitary (Kostyo and Reagan 1976), and the cells secreting the hormone are themselves under the control of factors released from the hypothalamus. The actions of the hormone are many and complex but are primarily concerned with stimulating growth. In all cases these actions are achieved indirectly, being mediated by secondary growth factors (somatomedins) which are produced in other tissues and organs in response to growth hormone.

14.1.2 History of growth hormone production

The involvement of KABI in growth research began in the late 1960s and pituitary-derived hGH was introduced onto the market in 1971. Later, in 1985, the pituitary-derived preparations were abandoned for therapeutic use owing to the reports from the USA, indicating contamination with infectious agents causing Creutzfeldt–Jacob Disease. Attention was then focused on producing hGH by recombinant DNA technology, expressing a DNA sequence coding for hGH in bacteria.

After a period of six years of research and development a methionyl-hGH, Somatonorm®, was introduced onto the market. The availability of the recombinant hormone effectively eliminated the risks of transmitting infectious agents, and also opened up the possibility of producing sufficient hGH to treat all children diagnosed with growth hormone deficiency.

In 1987, KABI introduced a recombinant human growth hormone, Genotropin®, which was structurally and functionally identical with the endogeneous hGH.

14.2 BIOSYNTHESIS OF GROWTH HORMONE

The production of recombinant hGH (rhGH) is achieved using *Escherichia coli* that are transformed with a plasmid that is genetically manipulated to carry the gene for growth hormone (Floodh 1986, Fryklund 1987). The hGH gene is derived partly from chemical synthesis (amino acids 1–23), and partly from a reverse transcript of RNA obtained from human pituitary. Expression of the rhGH during fermentation is regulated by an alkaline phosphatase promoter. Between the promoter and hGH gene is inserted a DNA sequence coding for a 23-amino-acid bacterial signal peptide (ST II). This construction provides a system that ensures efficient secretion of the hGH molecule from a cytosol, through the inner cell membrane out in the periplasmic space. During the secretory process a bacterial signal peptidase removes the signal peptide from the hGH molecule, leaving authentic hGH in the periplasmic space for purification (Fig. 14.1).

14.3 IDENTITY OF RECOMBINANT HUMAN GROWTH HORMONE

After purification of the final rhGH, it is essential to demonstrate that the product obtained is identical to the pituitary-derived hormone. Table 14.1 summarizes the techniques used to demonstrate this identity and further details are presented below:

Amino acid analysis
The amino acid composition of the rhGH is determined and compared with the known composition of pituitary-derived hGH.

N-terminal analysis
The N-terminal residue of the rhGH is identified by reacting the N-terminal α-amino group with phenylisothiocyanate (PTH). The resulting PTH-amino acid is cleaved from the peptide by the action of trifluoroacetic acid, and the identity of this residue is then determined by comparing against standard PTH-amino acids using thin layer chromatography.

C-terminal analysis
The C-terminal residue of the rhGH is identified either by sequencing a short C-terminal fragment generated by tryptic digestion of the peptide, or by enzymatic cleavage with carboxypeptidase A.

Protein sequence
The complete amino acid sequence of the rhGH is determined by classical Edman degradation. The analysis is performed on a fully automated pulse-liquid amino acid sequencer and the identity of each amino acid is confirmed using reverse-phase HPLC.

Tryptic fingerprint analysis
The identity and disulphide pattern of the rhGH are also confirmed by tryptic fingerprint analysis, the results obtained for the recombinant peptide being compared against those for the pituitary-derived hGH. The treatment with trypsin (which cleaves at the C-terminal side of arginine and lysine residues) generates 19 peptide fragments and a free lysine (Lys 168). (Note that there is no cleavage at Lys 172.) The 19 tryptic peptides generated are separated by reverse-phase HPLC, giving the characteristic fingerprint shown in Fig. 14.2.

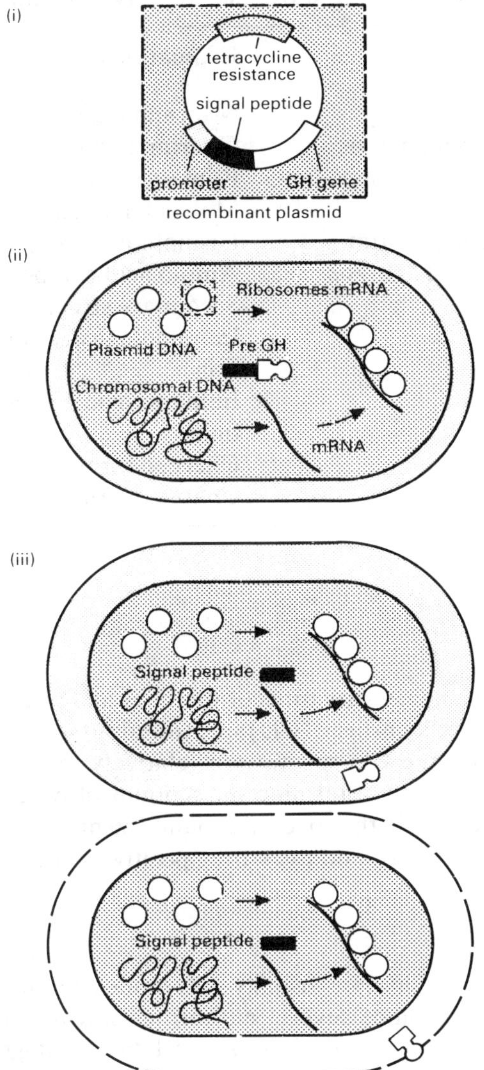

Fig. 14.1 — Production of hGH in *E. coli* (i) Plasmid construction for synthesis of hGH. (ii) Pre-GH synthesis in the *E. coli* cell. (iii) Processing and secretion of GH into periplasmic space. Release of the hormone is achieved by selective disruption of the outer cell wall. The inner cell wall is left intact to facilitate the GH purification.

Disulphide analysis

The locations of the disulphide links in the rhGH are confirmed by the characterization of the tryptic fragments T6–16 and T18–19. These two fragments are separated and isolated under non-denaturing conditions using reverse-phase HPLC (see Fig. 14.2). They are then treated with 2-β-mercaptoethanol, and the amino acid compositions are determined for the reduced fragments T6, T16, T18 and T19.

Table 14.1 — Drug substance identity of Genotropin®

Technique	Evaluation of
N-terminal analysis	N-terminal amino acid
C-terminal analysis	C-terminal amino acid
Sequence analysis	Primary structure
SDS-PAGE	Molecular weight
Tryptic fingerprint	Disulphide arrangement
Amino acid analysis	Amino acid composition
CD	Secondary structure
ELISA	Immunological identification
RIA	Immunological identification
PAGE	Charged forms
IEF	Charge distribution

Circular dichroism
The identity of the secondary structures of the rGH and pituitary derived hGH are compared using circular diochroism (CD) spectroscopy (see Fig. 14.3, Urry 1985). In the region 190–250 nm, the CD spectra are relatively sensitive to conformational changes, and can be used to determine the amount of α-helix and β-sheet in the protein. The fine structures in the spectra, around 260 nm and 267 nm originate from the phenylalanine side-chains, and the strong positive signal at 292 nm is caused by tryptophan transitions.

Potency
The biological potency of hGH is usually determined by the weight gain assay in hypophysectomized rats, since several authorities still recommend this assay for batch analysis. Bioassays, however, have several disadvantages: they are generally time-consuming, expensive and offer poor precision when compared with chemical, receptor-based or immunological assays. For ethical, as well as for economic reasons, it would be advantageous, therefore, to replace the *in vivo* bioassay for growth hormone with a simpler *in vitro* assay. Receptor assays for hGH have been described using rat liver membranes. (Nederman and Sjödin 1987), pregnant rabbit liver membranes (Kelly *et al.* 1974), IM-9 lymphocytes (Fagin *et al.* 1980) and human fibroblasts (Murphy *et al.* 1983). These assays have advantages over bioassays, since they are faster, employ smaller amounts of material and can be used to process large numbers of samples. However, they do have the disadvantage that they do not provide a proper test of the peptide's biological efficacy. It is therefore very important to demonstrate that the results obtained from the receptor assay and the bioassay are similar in terms of potency of the same material (batch, etc.).

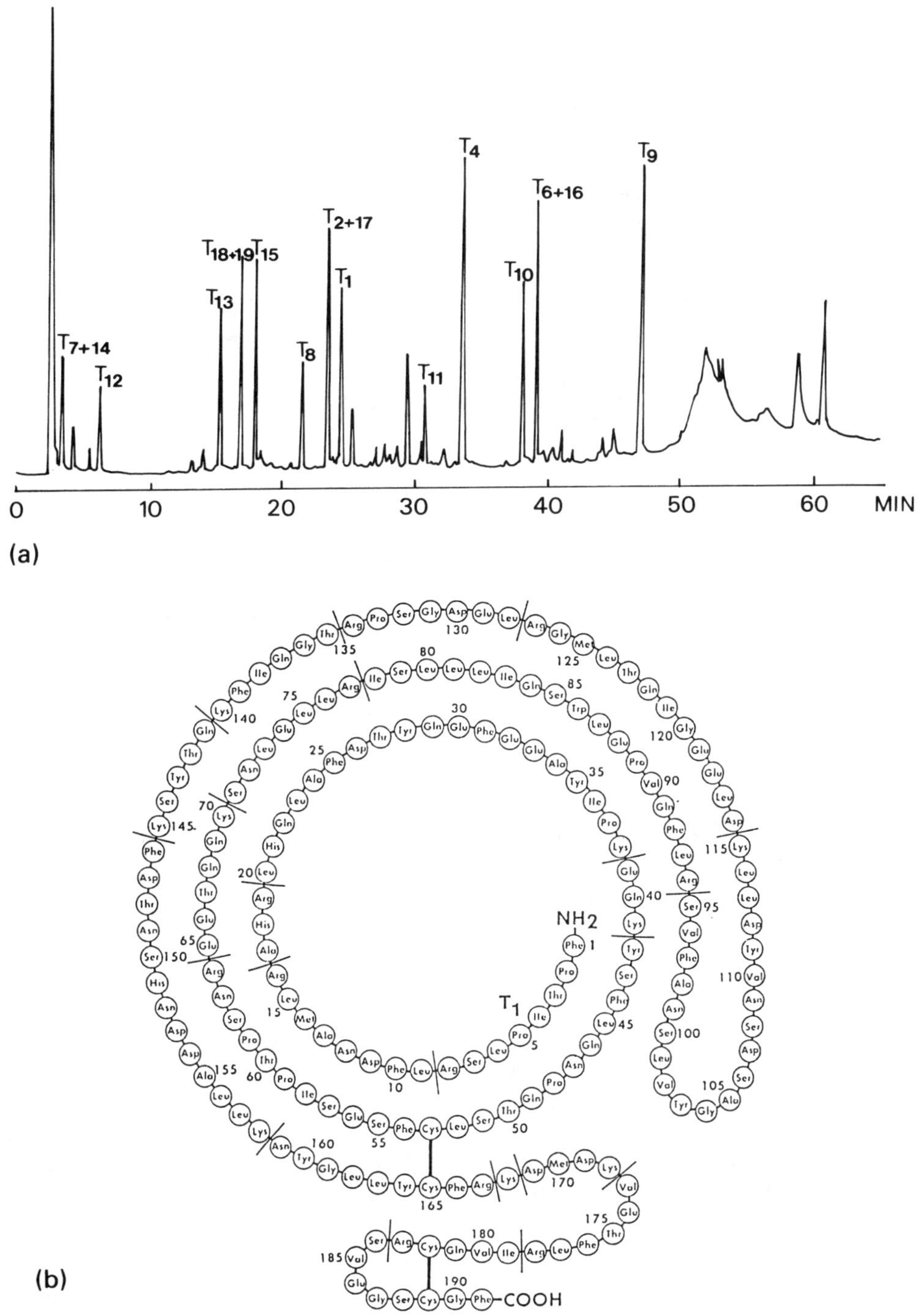

Fig. 14.2—Tryptic fingerprint of rhGH. (A) Chromatogram of trypsin-digested hGH analysed on reverse-phase HPLC. (B) The amino acid sequence of rhGH with trypsin cleavage sites indicated.

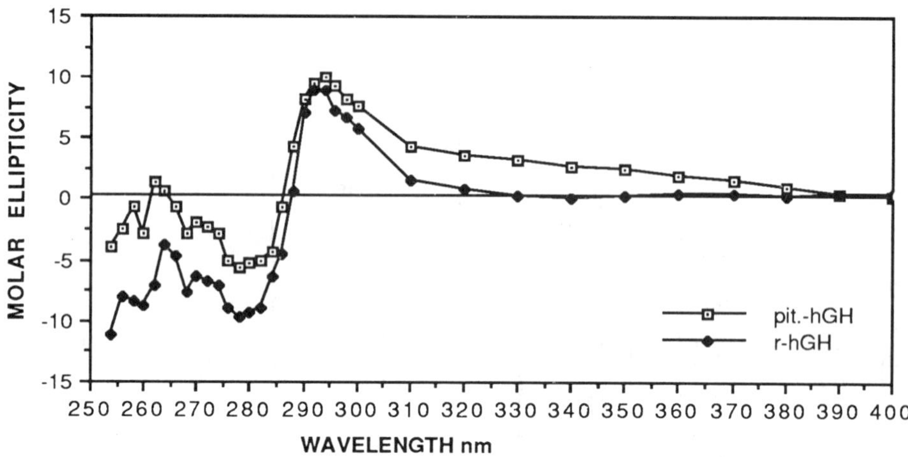

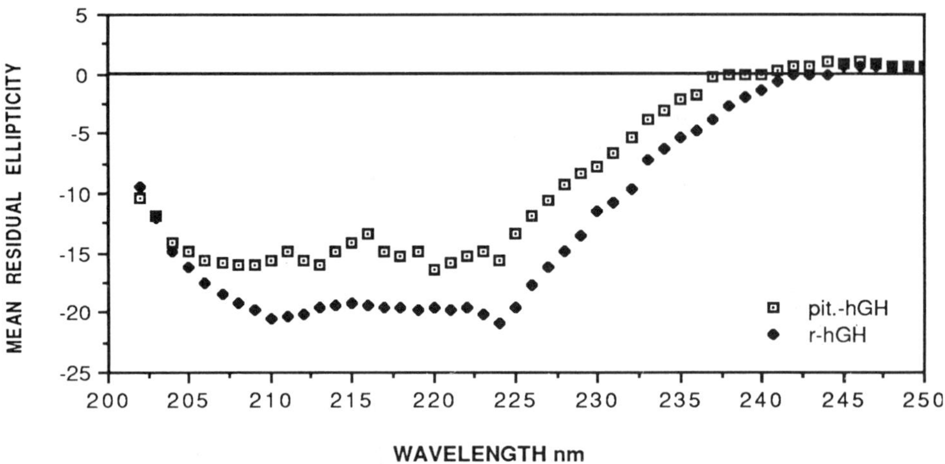

Fig. 14.3 — CD spectra of pituitary-derived hGH and recombinant hGH. Upper panel: side-chain absorption region. Lower panel: amide-bond-dominated absorption region.

14.4 ANALYTICAL TECHNIQUES UTILIZED FOR DETERMINATION OF PURITY OF RECOMBINANT HUMAN GROWTH HORMONE

14.4.1 Isoelectric focusing

The technique of isoelectric focusing (IEF) is used to determine the isoelectric point of the rhGH samples, and also to detect modifications such as deamidations (see section 14.5). The technique is essentially an electrophoretic one, but instead of the separation being carried out at constant pH, it is carried out in a pH gradient which is generated by low molecular weight amphoteric reagents. The proteins in the sample applied to the gel migrate in the pH gradient until they reach the pH corresponding to their isoelectric point. At this point, the protein then possesses zero net charge and migration ceases. IEF is thus an equilibrium technique in which the effects of diffusion are overcome.

Fig. 14.4 shows a typical IEF run for growth hormone samples stored at 30°C for different time periods; the increasing amounts of an acidic component present suggest a degradation of the peptide involving deamidation (see section 14.5).

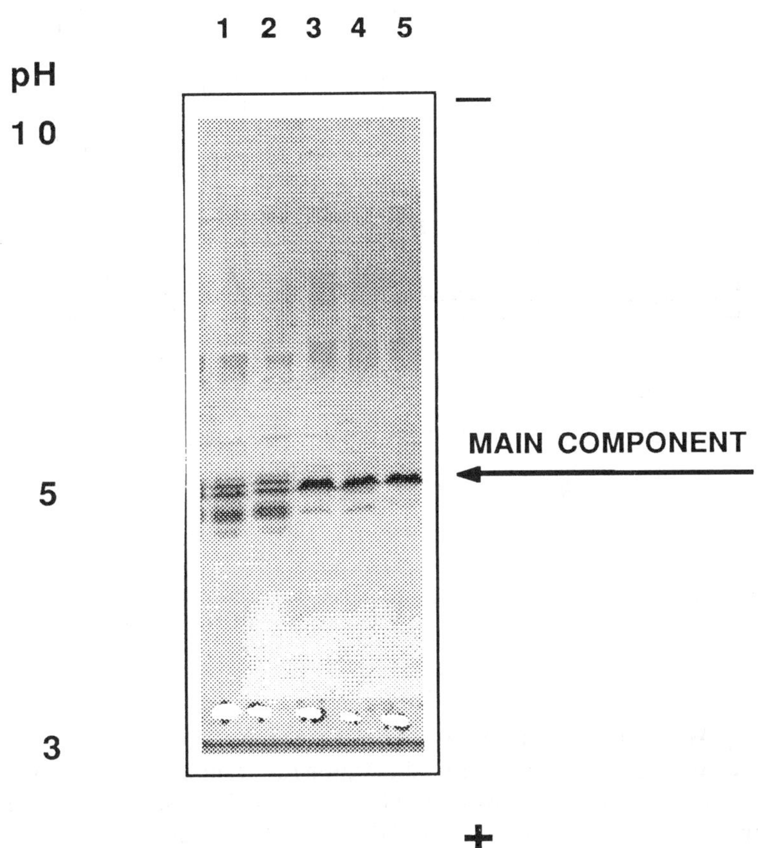

Fig. 14.4 — IEF of rhGH samples. Lanes 1, 2: samples stored at 30°C for one month; lanes 3, 4: samples stored at 30°C for one week; lane 5: reference rhGH sample.

14.4.2 SDS-PAGE

In SDS–PAGE the rGH samples are treated with the negatively charged detergent sodium dodecyl sulfate. The samples are also reduced with 2-β-mercaptoethanol, which leads to reduction of the two disulfide bridges. The resulting unfolded, negatively charged sample is then applied at the cathodic side of the vertical plate, and is allowed to migrate in an electric field towards the anode. The separation of the peptide that is achieved is based mainly on molecular weight. The bands obtained are visualized by silver staining, using two loads of hGH for densitometric evaluation (0.25 μg and 10.0 μg), in order to ensure reliable quantitation (Fig. 14.5).

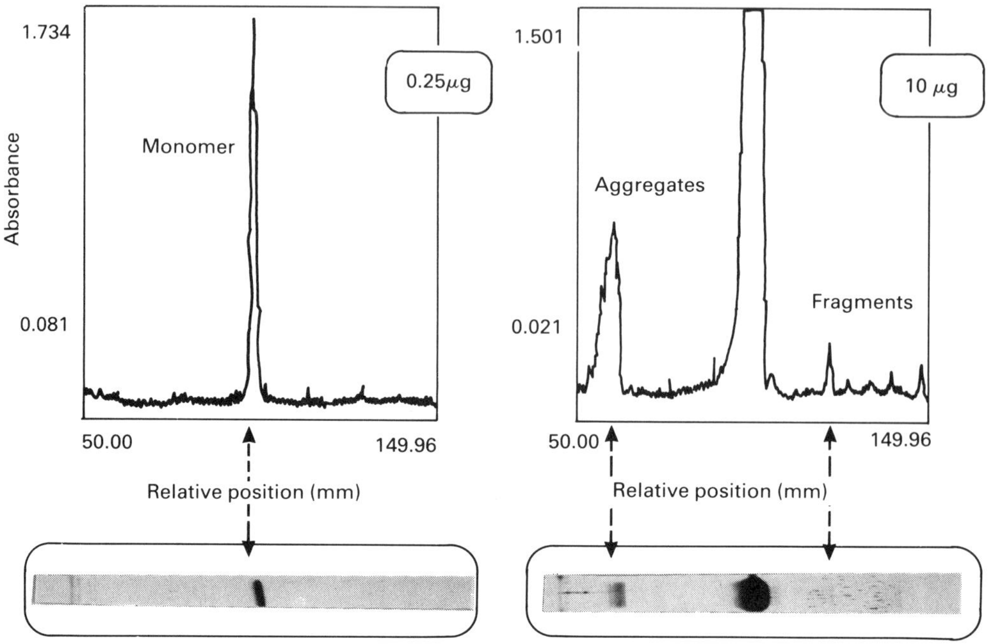

Fig. 14.5 — Densitometric scanning (upper panel) of a silver-stained and dried SDS-PAGE gel (lower plane) when 0.25 μg and 10 μg of rhGH sample were applied.

14.4.3 Hydrophobic interaction chromatography

This technique was recently introduced for rapid separation of proteins under mild adsorption and elution conditions. The system is highly selective and has been shown to separate methionyl-hGH from hGH, as well as oxidized, clipped and unmodified authentic hGH.

14.5 DEGRADATION PRODUCTS OF GENOTROPIN®

As with any recombinant DNA product it is important to carry out a full range of stability tests in order to detect degradation products (see Table 14.2). In the case of rhGH, the degradation products include: denatured, aggregated, disulphide-mismatched, and oligomeric species; methionyl sulphoxide derivatives; and deamidated forms in which asparagine and glutamine residues are converted to aspartate and

Table 14.2 — Possible modifications in recombinant proteins

Deamidation
Aggregation
Proteolytic
Oxidation
Glycosylation (with eukaryotic hosts)
Carbamylation (caused by urea employed in chromatographic separation)

glutamate. In addition, there are various forms of clipped protein which can be produced through intramolecular proteolytic cleavages. One such clipped form which has been isolated and characterized by protein sequence analysis is shown to be caused by cleavage at the site Thr142–Tyr143 (Gellerfors *et al.* 1989). Fig. 14.6

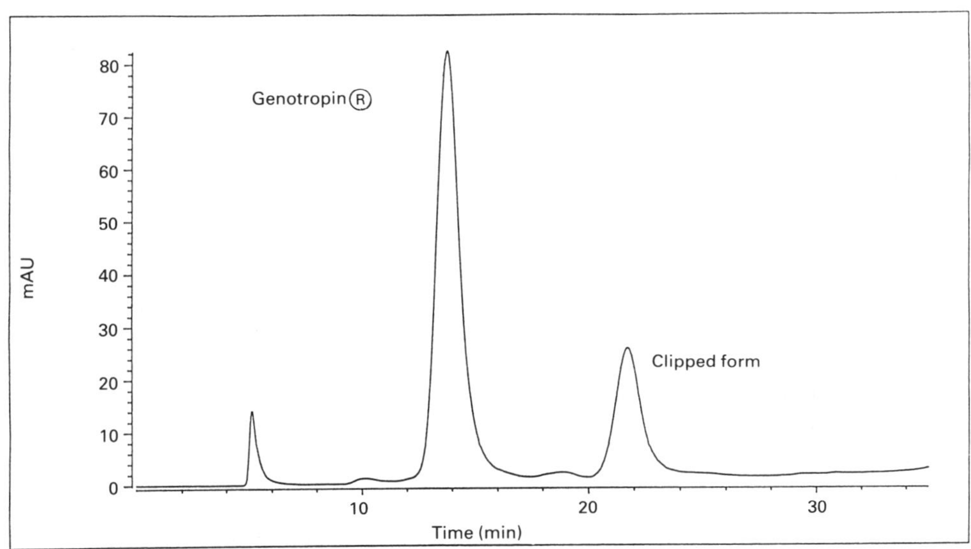

Fig. 14.6 — Hydrophobic interaction chromatography of rhGH.

shows a chromatogram in which this clipped form of hGH is separated from the intact molecule; when analysed under reducing conditions on SDS–PAGE, the clipped species appear as 16-kDa and 6-kDa fragments (Gellefors *et al.* 1989).

Fig. 14.7 shows the separation of unmodified and oxidized rhGH which is achieved using reverse-phase HPLC.

Fig. 14.8 shows an anion-exchange chromatogram of rhGH samples stored at 30°C for six months. The two major deamidated species observed are the mono- and didesamido forms of rhGH. The monodesamido form has been isolated and characterized by tryptic digestion (see Fig. 14.9); a comparison of this tryptic fingerprint with that obtained for authentic hGH demonstrates that the modification is localized in fragment T15, suggesting a deamidation of either Asn 149 or Asn 152.

14.6 IMPURITIES

Further contamination of the final product can result from impurities that derive from the *Escherichia coli* (*E coli*) expression system and/or from the purification process. These contaminants are checked in the rhGH batch analysis (see Table 14.3) and the analytical methods employed are detailed below:

Periplasmic E. coli proteins (*PECP*)

These proteins represent a large number of potential impurities. In the rhGH batch analysis the aim is therefore to detect as many of these components as possible using a multiantigen ELISA test. The first major requirement for such a test is to obtain a pool of host-cell protein impurities that is representative of the manufacturing process and devoid of the final product. This preparation then serves as the immunogen for production of polyclonal antibodies. After demonstrating that the antibodies raised bind to most, if not all, of the observed host-cell impurities, the number of immunoreactive components in the rhGH batch can then be determined by 2-D SDS–PAGE and Western blotting.

As a further refinement, it is possible to use a double-antibody sandwich ELISA, in which horseradish peroxidase is coupled to form an enzyme–polyclonal antibody conjugate. Firstly, the rhGH product is added to microtitre plate coated with purified antibodies to the host cell protein impurities. The plate is then incubated, and is subsequently washed to remove any unbound material (including all hGH). Following a second incubation, in the presence of the peroxidase conjugate, an appropriate substrate is added for colour development. This is then analysed using a spectrophotometric plate reader and the concentration of the host-cell protein is determined from a standard curve.

Host-cell DNA

The technique of DNA hybridization, commonly referred to as dot blot analysis, is used to estimate the DNA content in every rhGH lot. This method is capable of quantitating DNA at the nanogram to picogram level and is the most sensitive routine DNA assay currently available. The dot blot assay relies on the hybridization of cellular DNA from the sample with specific ^{32}P-labelled DNA probes obtained from the DNA of the host cell. The hybridization is performed on a nitrocellulose filter that is subsequently placed between two X-ray films and exposed to produce an autoradiograph. The DNA of the sample is estimated by a visual comparison of the dot intensity of the sample against that of diluted DNA standards.

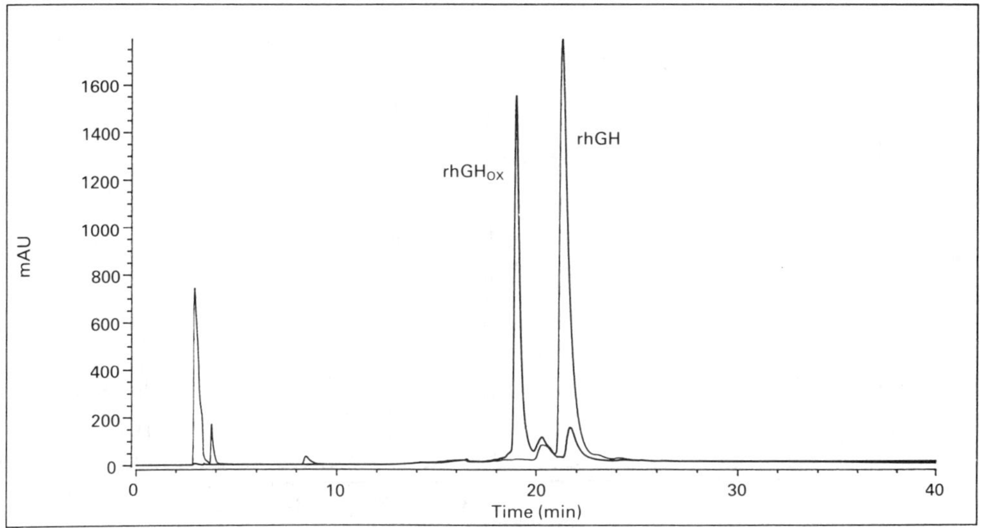

Fig. 14.7 — Reverse-phase HPLC trace for rhGH and oxidized rhGH.

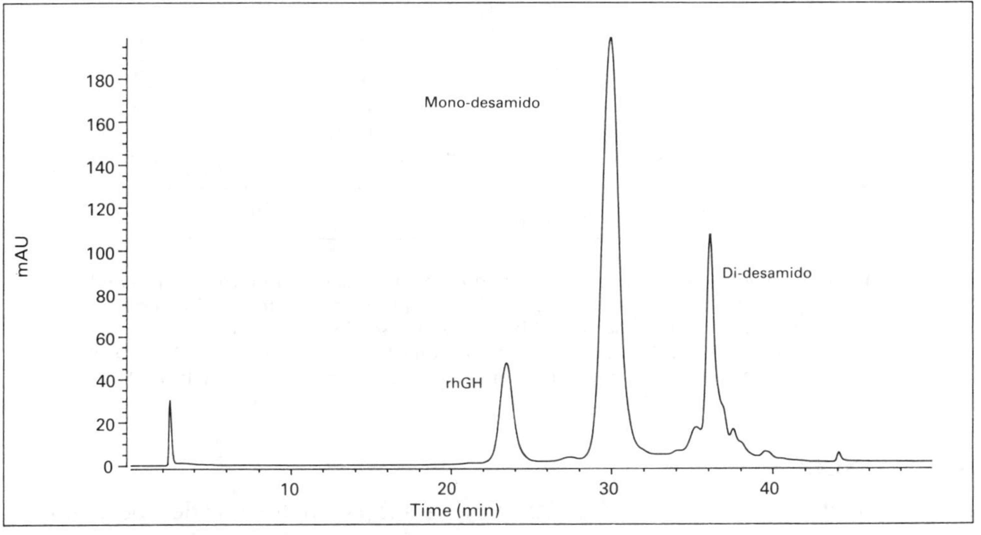

Fig. 14.8 — Anion-exchange chromatogram showing deamidated forms of rhGH.

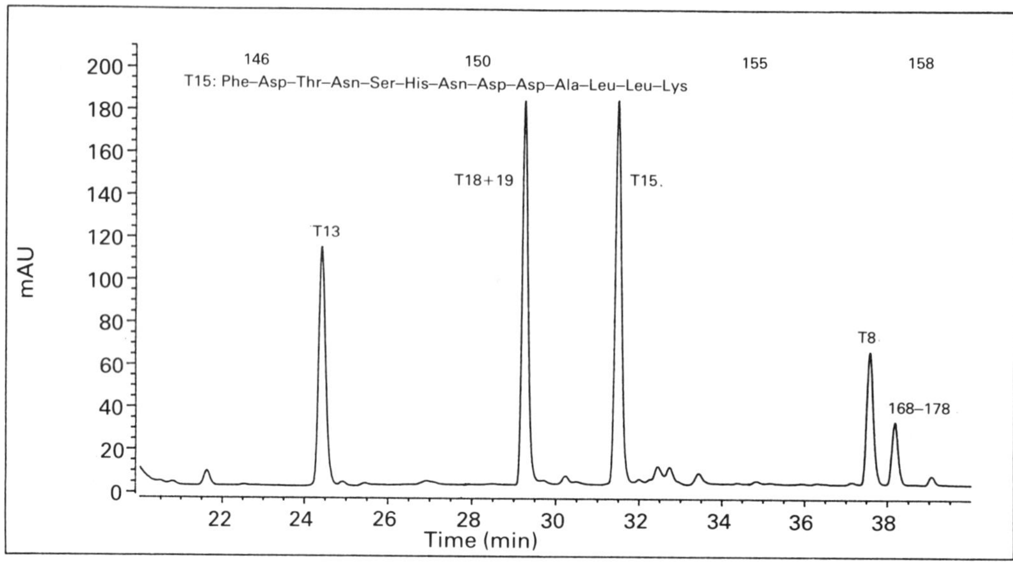

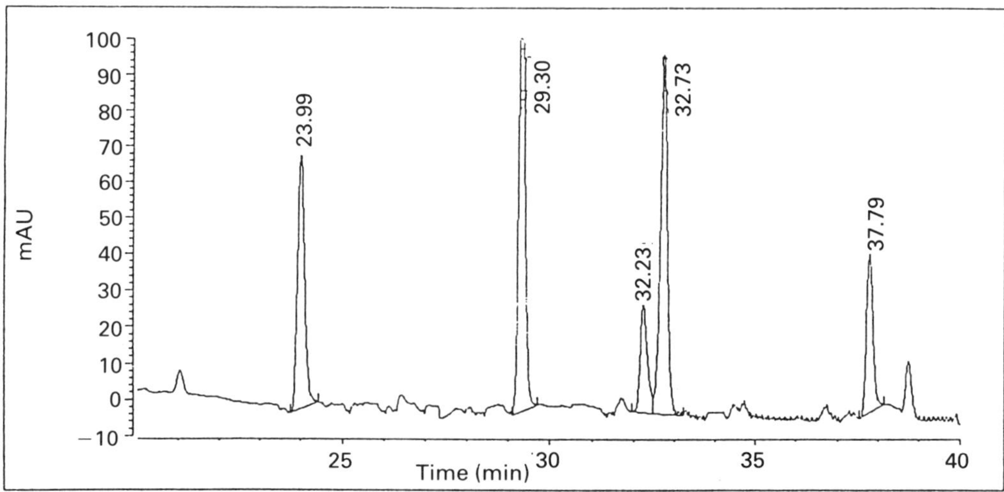

Fig. 14.9 — Tryptic fingerprint of monodesamido. The monodesamido form was isolated by anion-exchange chromatography (see Fig. 14.8; retention time 30 min), and digested with trypsin. Tryptic fragments were separated by reverse-phase HPLC. The figure shows only a part (22–30 min) of the total chromatogram encompassing tryptic fragments T13, T18–19, T15, T8 and residues 168–178. Upper panel; monodesamido form. Lower panel: rhGH.

14.7 CONCLUSION

Recombinant human growth, Genotropic®, is subjected to a wide spectrum of quality control analyses to ascertain safety and efficacy. These include HPLC, peptide mapping and SDS–PAGE using a sensitive silver staining detection system.

Table 14.3 — Batch analysis of Genotropin®

Purity	Molecular weight Isohormone Dimer content Polymer content
Indentity	Tryptic fingerprint
Potency	Immunoactivity Bioactivity
Impurities	DNA Dot blot analysis PECP SDS–PAGE Western blot ELISA
General tests	pH Colour Solubility

Immunoassays for quantitation of host cell protein impurities at the parts per million level as well as a sensitive assay to determine nanogram to picogram levels of residual DNA have been developed. Immunoassays and bioassays are used for potency determinations. The preparation has been proven to be clinically safe and effective, with no side-effects in the treatment of growth hormone deficient children.

ACKNOWLEDGEMENTS

I would to thank K. Axelsson, S. Lindqvist, C. Nyhlen and B. Pavlu for performing LC analysis, S. Johansson for performing SDS-PAGE analysis and P. Gellerfors for valuable discussions.

REFERENCES

Fagin, K. D., Lackey, S. L., Reagan, C. R. & DiGirolamo, M. (1980) Specific binding of GH to rat adipocytes. *Endiocrinology,* **197** 608–611.

Floodh, H. (1986) hGH produced with recombinant DNA technology; development and production. *Acta Paediatr. Scan. (suppl.)* **325** 1–9.

Fryklund, L. (1987) Production of authentic recombinant somatropin. *Acta Peadiatr. Scan. (suppl.)* **331** 5–8.

Gellerfors, P., Eketorp, G., Fhölenhag, K., Pavlu, B., Johansson, S. & Fryklund, L. (1989) Characterization of a secretal form of recombinant derived human GH Expressed in *E. coli* cells. *J. Pharm. and Biomed. Anal.* **7** 173–183.

Kelly, P. A., Posner, B. I., Tsushima, T. & Friesen, H. G. (1974) Studies on insulin, growth hormone and prolactin binding: ontogenesis, effects of sex and pregnancy. *Endicrinology* **95** 532–537.

Kostyo, J. L. & Reagan, C. R. (1976) The biology of growth hormone, *Pharmacol. Ther.* **2** 591–604.

Murphy, L. J., Vrhousek, E. & Lazarus, L. (1983) Identification and characterization of specific GH receptors in cultured human fibroblasts. *J. Clin. Endocrinol. Metab.* **57** 1117–1121.

Nederman, T. & Sjödin, L. (1987) The quantitation of hGH by a radioreceptor assay using an established human cell line. *J. Biol. Standards* **15** 199–202.

Urry, D. W. (1985) Absorption, circular dichroism and optical rotary dispersion of polypeptides, proteins, prosthetic groups and biomembranes. In: Neuberger, A. & Van Deenen, L. L. M. (eds) *Modern Physical Methods in Biochemistry*. Elsevier, Amsterdam part A, pp. 275–343.

15

The development of peptide and protein pharmaceuticals

D. Ganderton
Chelsea Department of Pharmacy, King's College, Manresa Road, London SW3 6LX

15.1 INTRODUCTION

The formulation of peptides as clinically effective, stable medicines presents a great challenge to the pharmaceutical technologist. An innovative component in meeting the challenge can have a dominant influence on commercial outcome, the dosage form causing the selection of a particular product even though other presentations of the same peptide or its analogue are available. In the extreme case, the entire project may only be viable if an acceptable dosage form can be divised. Companies engaged in this exciting area should, therefore, devote appropriate resources to drug delivery.

This situation arises from the complexity and lability of molecules which range in size from a few hundred to over 150 000 Da. Unless administered parenterally, these molecules must cross epithelia, which always possess a high barrier function and may have a high degradative enzymic activity. The effects on peptide absorption have been reviewed recently by Lee and Yamamota (1990). For these reasons, even the smallest peptide may demand new pharmaceutical principles for the development of a satisfactory product and this has important organizational implications.

15.2 ORGANIZATION OF PRODUCT DEVELOPMENT

Although the generation of a conventional pharmaceutical may be seen within the continuum from research through development to manufacture, these phases can be discrete. Product development can begin on the basis of a good performulation programme and the developed product and process can be handed on to manufacture by a disciplined scale-up and validation procedure. This is unlikely to be the case with peptides and proteins, and the contrasting development sequence is shown in Fig. 15.1. In these circumstances, the technologist must maintain very close association with all the other elements in a development programme.

The development chemistry will be hugely important. The way in which variations in bulk drug manufacture may influence quality and the complexity of the

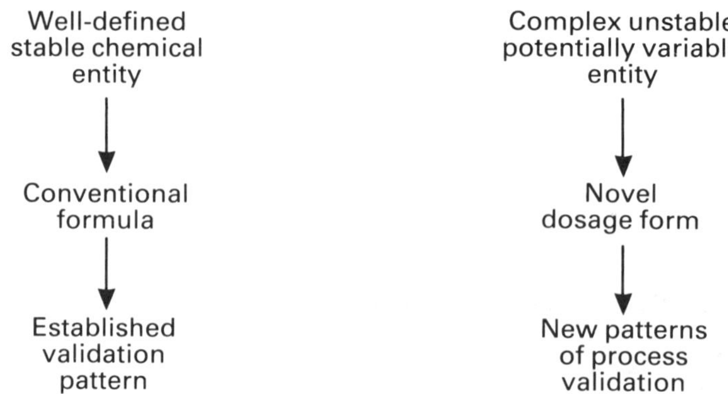

Fig. 15.1 — The development sequence for conventional (left) and peptide pharmaceuticals (right).

analytical procedures used to establish purity demand the closest integration of chemist, analyst and formulator. The formulator must often work with very small amounts of material and important decisions will be made when the bulk manufacturing process is still evolving, the stability-indicating powers of analysis are uncertain and no validated reference sample is available. In these circumstances, technical judgement should be made within the framework of extremely competent project management.

Within this framework, the formulator will adopt the normal criteria for assessing a product:

— efficiency (bioavailability)
— stability
— 'user' factors
— registration
— manufacture

15.3 ORAL ADMINISTRATION OF PEPTIDES

Although experimental interest in oral administration remain high, studies so far have shown that the membrane and enzyme barriers of the gastrointestinal tract catastrophically reduce bioavailability. Until a consistent bioavailability, which may be acceptable even if low, has been achieved, formulations devised for the oral route should not be included in a development programme. No further consideration will be given to it in this chapter.

15.4 PARENTERAL FORMULATIONS

15.4.1 Bioavailability

Physical and enzymic barriers are avoided by the direct systemic injection of an aqueous solution. An assured bioavailability will make this route the first development objective.

15.4.2 Stability

Development of a sales formulation will be an extension of the simple aqueous solutions used in clinical evaluation. A range of analytical procedures will have evolved to assay these presentations and assess their short-term stability. These early studies should define whether an aqueous solution with a shelf life adequate for marketing is possible, or whether the product should be presented as a freeze-dried powder for reconstitution.

Owing to complexity in thermal degradation, acceleration studies cannot be used with the same precision as in conventional development. However, lightly stressed conditions at, say, 40°C, are used to probe the stabilization of either aqueous or freeze-dried product by means of pH control or added stabilizers. This is exemplified in the work of Geigert *et al.* (1988) who studied a solution of recombinant (Serine-17) human interferon β stabilized by acetate, sodium dodecyl sulphate (SDS) and EDTA. Studies were conducted over the temperature range −70° to +37°C using three stability indicating assays: potency, SDS–PAGE and RP-HPLC. The results of the latter analysis are shown in Fig. 15.2. A highly tentative prediction of shelf-life was derived from an Arrhenius plot (Fig. 15.3). An estimate of 10% degradation in eight months by potency assay compared with seven years by SDS–PAGE and RP-HPLC illustrates the complexities of analytical judgement.

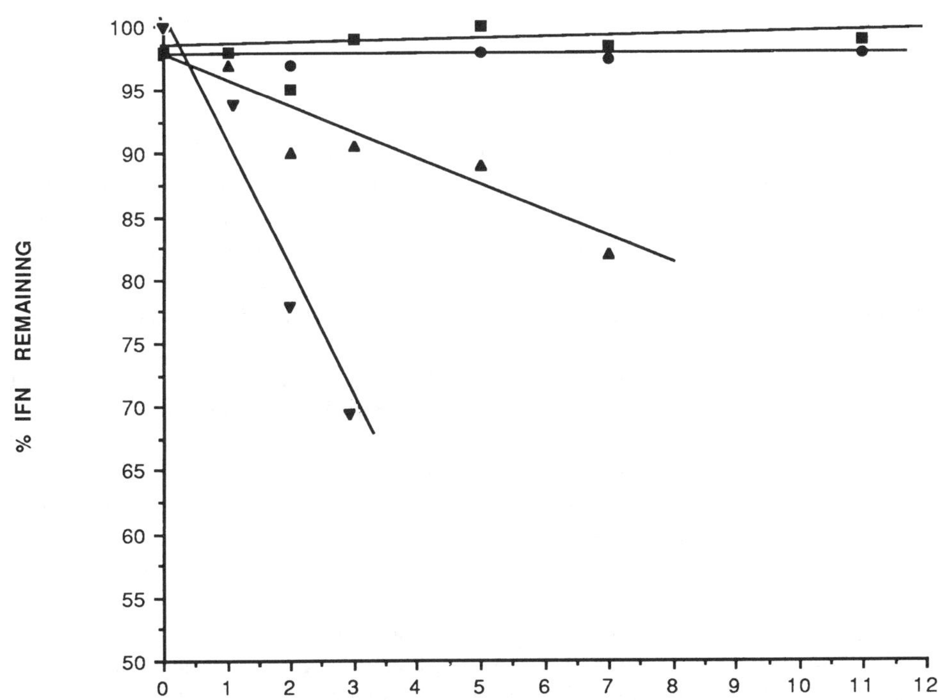

Fig. 15.2 — Protein purity of stabilized HuIFN-βSER assessed by SDS-PAGE at −70°C (■), 4°C (●), 25°C (▲) and 37°C (▼).

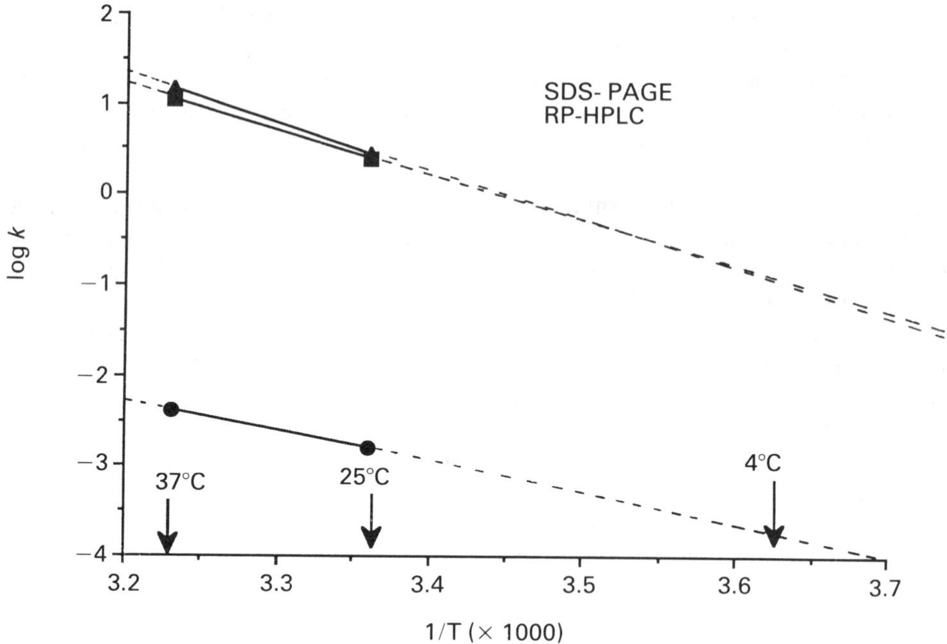

Fig. 15.3 — Arrhenius plot for HuIFN-βSER using potency (●), SDS–PAGE (■) and RP-HPLC (▲) to assess degradation.

Ideally, a conventional pH–stability profile showing a minimum value at some characteristic point will be obtained. Studies at a suitable pH will then be carried out using phosphate, citrate, acetate or other buffer systems and varying molar concentrations. The selection of stabilizers is less systematic. Examination of marketed products shows that mannitol, glycine, human serum albumin and polysorbates are widely used and their constitutional variety suggests that selection is somewhat empirical.

The development pharmaceutics consists of permuting these variables under lightly stressed conditions to establish a suitable formula. Other elements, such as container and closure selection, solution sterilization, etc., present no special problems. However, full confidence in the stability of the product will only come from real time exposure at the proposed storage temperature of material prepared by the approved method and distributed in the selected pack.

15.4.3 User factors

The medical or paramedical skill required for parenteral administration is a major constraint and thought should be given to facilitating the use of the product. For example, the packaging of Roche's form of α-interferon, ROFERON A (Anon. 1990) as a complete administration unit (Fig. 15.4) helps the increasing number of patients who administer their own interferon at home. There will be a proliferation of devices which facilitate subcutaneous or intramuscular injection and the formulator must be able to exploit their advantages whilst solving the problems which arise, for example, from interaction with the formulation.

An alternative strategy is to reduce the frequency of administration by converting the peptide to an insoluble form. As suspensions in water or oils, they may be effective for a period of up to several days. Insulin products which exploit variations of solubility and particle size to reduce frequency in injections are well-documented. However, these modifications are largely specific to the insulin molecule and a more general solution is to encapsulate the peptide with a barrier membrane or incorporate it as a molecular or particulate dispersion in a matrix. The latter is technically easier and the matrix may be fabricated as a solid implant or it may be subdivided and presented as an injectable suspension. Peptide solubility, particle size, drug loading and method of manufacture can be varied to produce a wide pattern of release. Fig. 15.5, which displays some of these factors, is drawn from the work of Brown *et al.* (1986) who incorporated insulin in a copolymer of ethylene and vinyl acetate (EVAc). *In vitro*, these matrices released their contents over a period of up to one month into a pH 7.4 buffer and the same authors (1988) later showed that a sintered matrix containing 30% sodium insulin of particle size less than 300 μm depressed the glucose level of diabetic rats to levels comparable to normal animals for up to 100 days.

The non-biodegradable nature of EVAc and many other polymers probably proscribes their use in modern development programmes, the need to recover the exhausted device constituting an unacceptable limitation. Degradable polymers avoid this problem but are more difficult to use. Ideally, erosion should be confined to a narrow, advancing layer by ensuring that the rate of erosion is not less than the rate of aqueous penetration. Most of the peptide will then be protected from the destabilizing process and the rate of release will be determined solely by the peptide loading, the geometry of the device and the rate of erosion. Polyorthoesters have been used in this way by Heller *et al.* (1981) for the delivery of steroids. The cleavage of these polymers is very pH-dependent and basic substances are added to the matrix to suppress degradation. Erosion occurs when these additives are leached from the surface. More recently, Leong *et al.* (1985) have devised polyanhydrides based on a combination of poly[*bis*(*p*-carboxyphenoxy)methane] with varying quantities of the more hydrophilic sebacic acid. No additives are necessary and, as shown in Fig. 15.6, the release of a marker compound (*p*-nitroaniline) is entirely controlled by surface erosion.

More commonly, matrices based on polymers of lactic acid and glycolic acid have been used because of established biocompatibility. However, peptide release occurs by a much more complex process in which channels capable of transporting large molecules are created by bulk degradation. Depolymerization is affected by molecular weight, glycolide–actide ratio and polydisperity, factors which can be used to control release rate. For example, Sanders *et al.* (1984) prepared microspheres of the copolymer containing the decapeptide, nafarelin acetate. Release after subcutaneous injection to rats was assessed by the suppression of oestrus and Fig. 15.7 shows the effects of microspheres of varying constitution. A 69:31 lactide–glycolide copolymer of relatively low moleculer weight (intrinsic viscosity 0.97 dl/g) gave diphasic release (Fig. 15.7a) whereas a more labile polymer of higher glycolide

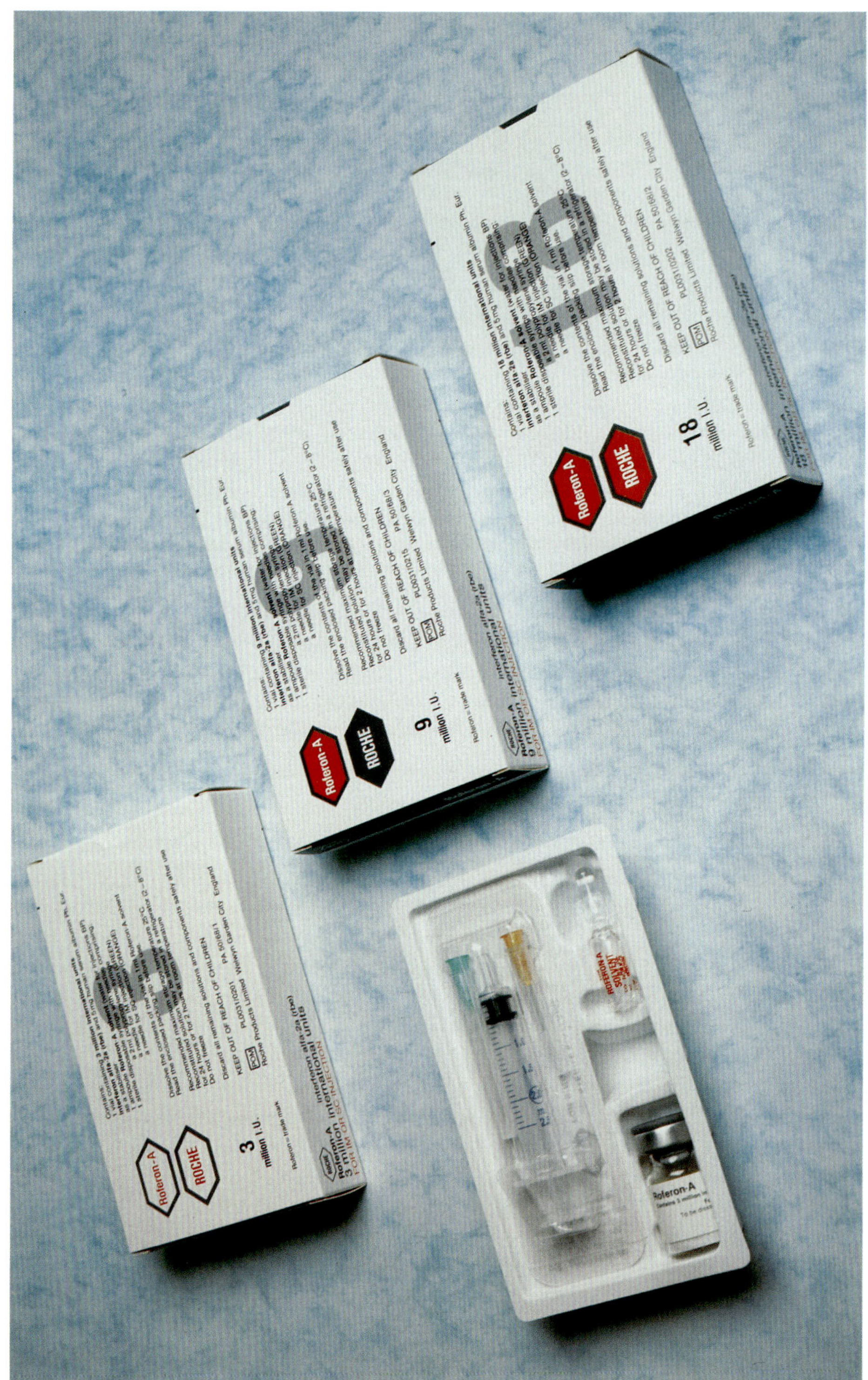

Fig. 15.4 — Administration set devised by Roche Products Limited for α-interferon.

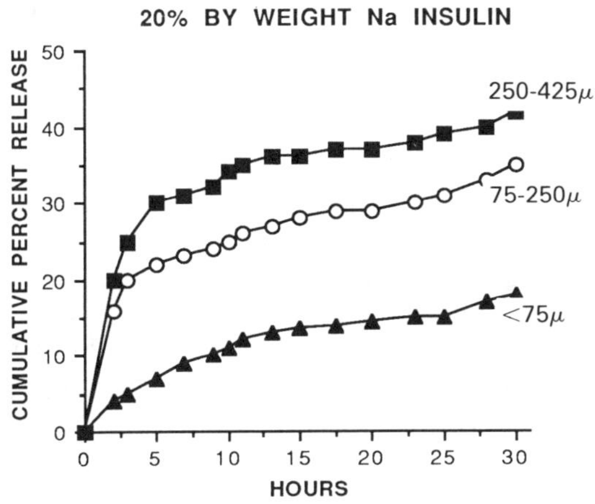

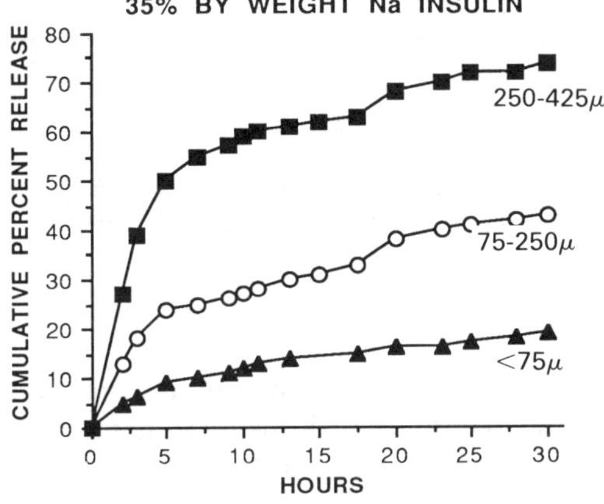

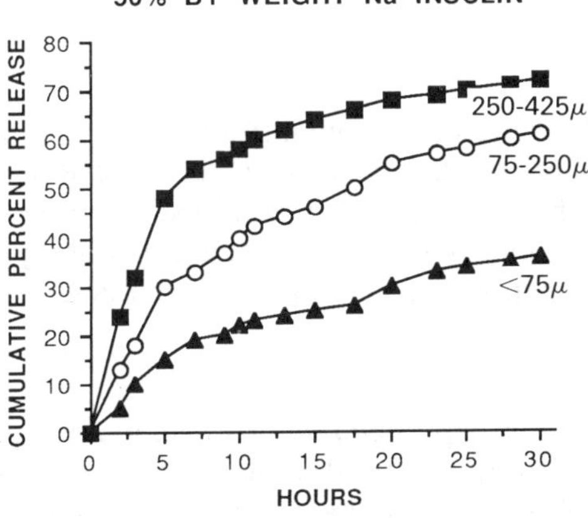

Fig. 15.5 — Release of insulin from EVAc matrices.

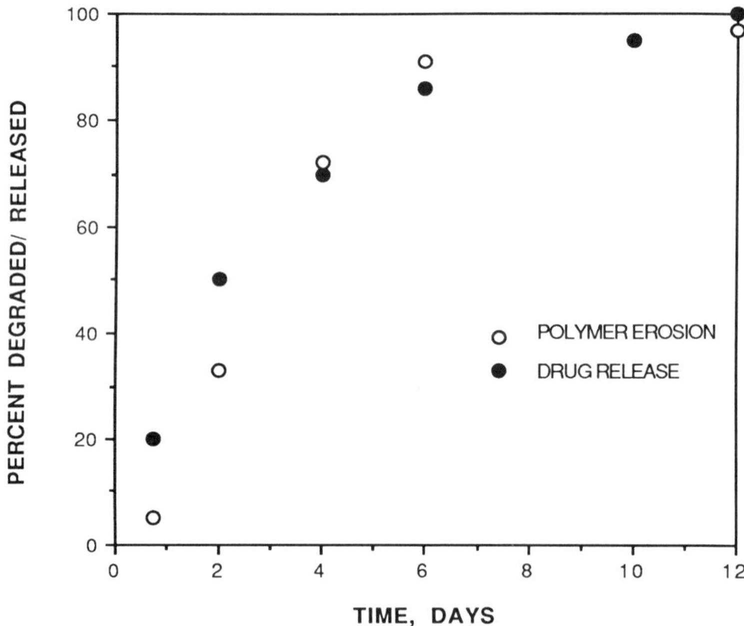

Fig. 15.6 — Release of *p*-nitroaniline (10% loading) from a poly[*bis*(*p*-carboxyphenoxy)methane]/sebacic acid (45:55) matrix into 0.1 M pH 7.4 phosphate buffer at 37°C.

content and molecular weight gave shorter, monophasic release (Fig. 15.7b). It was assumed that release occurred initially by diffusion from material close to the surface of the microsphere, but was only sustained by bulk hydration and hydrolysis. In one case, the mechanisms are separated in time whereas in the other they overlap. The resolution of these problems and the development of a long-acting implant is the subject of Chapter 16.

15.4.4 Registration

Any refined drug delivery system may well introduce new materials and methods of manufacture, examples of which are given in the preceding section. It may be difficult to associate such innovation with an already complex development plan unless drug delivery has been pursued as a separate strategy, which can then be harnessed to a particular peptide as a mature, well-characterized contribution.

15.4.5 Manufacture

Product conception and its expression in a development plan must give due consideration to manufacture. The product devised by Sanders *et al.* (1984) and described in section 15.4.3 was made by preparing a water in oil emulsion of an aqueous solution of nafarelin acetate in a solution of the copolymer in dichloromethane. The copolymer was precipitated around the droplets and the suspended

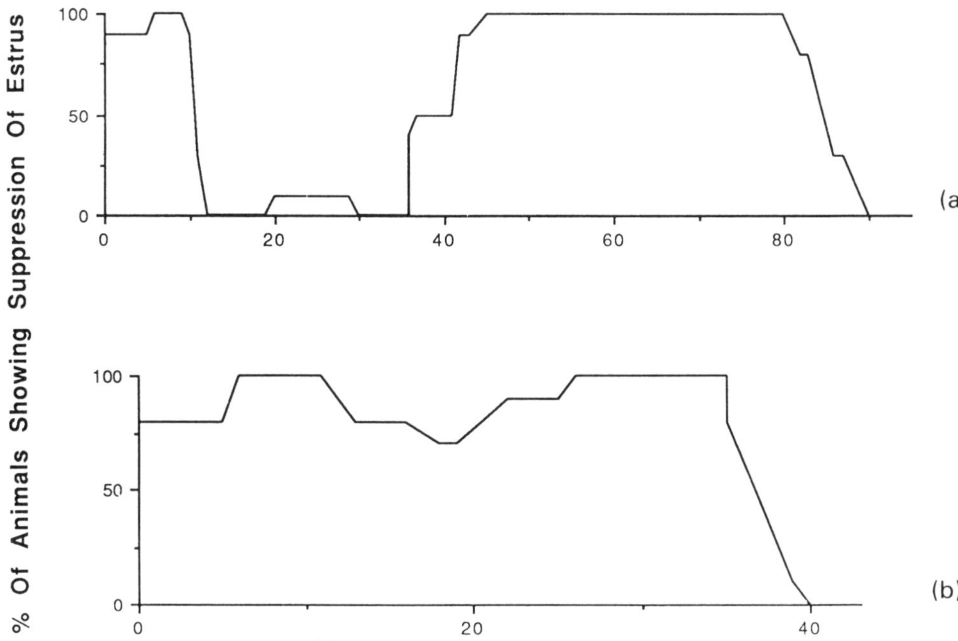

Fig. 15.7 — Oestrus suppression profiles provided by (a) 69:31 ($\eta = 0.97$ dl/g) and by (b) 50:50 ($\eta = 1.52$ dl/g) lactide–glycolide copolymer microspheres.

microspheres hardened, sieved, washed, dried and suspended in an aqueous medium. The final capsules contained about 1% of nafarelin. This procedure would present formidable difficulties on a large scale, which would be further complicated by the lack of a terminal sterilization procedure. Products must be capable of a far simpler translation to a manufacturing scale.

15.5 NASAL DELIVERY OF PEPTIDES

Although parenteral preparations will always command a major place in peptide delivery, administration demands skills found only in medical personnel or trained, well-motivated patients. Other, more convenient, routes are urgently needed. The nasal mucosa are quite permeable to some peptides and, although the bioavailability is usually low (below 5%), highly potent materials may be given via the nose. Oxytocin, desmopressin, buserelin and calcitonin are currently marketed in this form, as either drops or sprays.

15.5.1 Formulation of drops and sprays

Although the use of non-aqueous suspensions in a pressurized metered-dose inhaler is possible, peptides are usually presented as preserved, isotonic aqueous solutions. Their deposition and clearance, both of which affect the absorption of the peptide, depend on whether they are administered as drops or as a spray. Using gamma scintigraphy, Harris *et al.* (1986) showed that sprays of desmopressin solution were deposited mainly anteriorly, from which small portions were cleared slowly into the nasal pharynx. In contrast, drops, gave mainly posterior deposition and rapid clearance. Some of the solution was swallowed immediately. These effects are reflected in absorption of desmopressin shown in Fig. 15.8. Peak blood levels and

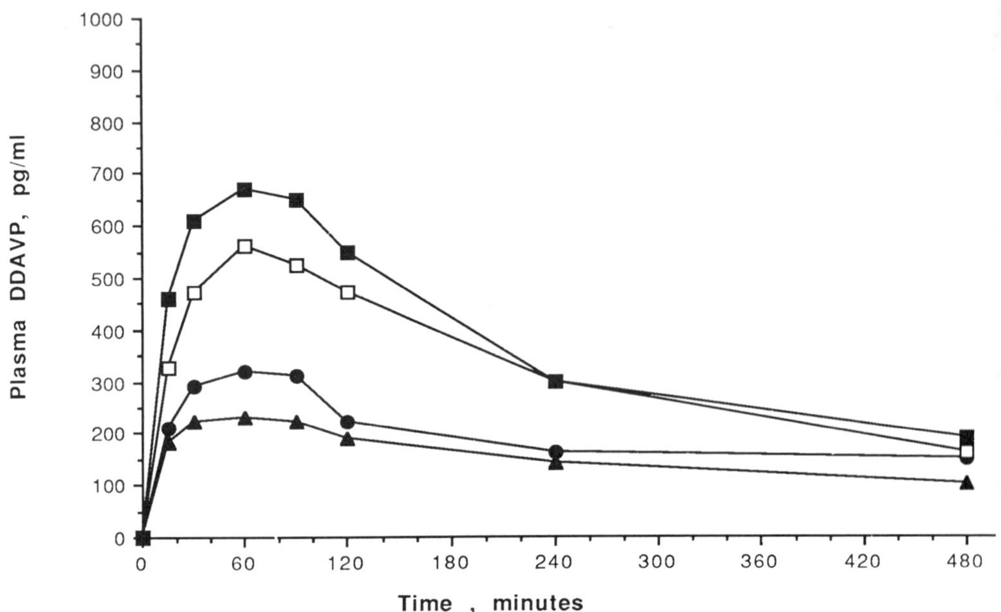

Fig. 15.8 — Plasma levels of desmopressin following nasal administration of 300 μg in 100 μl spray (■), 200 μl spray (□), drops by rhinal catheter (●) and pipette (▲).

AUC values for sprays are more than double those of drops. The work also indicated the influence of dose volume, 300 μg presented in 100 μl of spray giving significantly better absorption than the same amount of drug in 200 μl. The complexity of dose/volume relationships was also explored by Anik *et al.* (1984) who sprayed solutions of nafarelin acetate of various strengths into the noses of rhesus monkeys. The non-linear relation between dose and amount absorbed is shown in Fig. 15.9. Over the dose ranges 133–431 μg the bioavailability, compared with subcutaneous injection, increased from 0.2% to 2%, the lower doses probably being greatly affected by saturatable binding or metabolism. In this case, the potency and clinical profile of this LHRH analogue is such that the highest strength solution is acceptable for product development.

Although sprays appear to be more efficient than drops, this advantage must be weighed against a more complex development pattern. A fluent association with a manufacturer must be established to ensure the supply of spray devices which meet the needs of the programme.

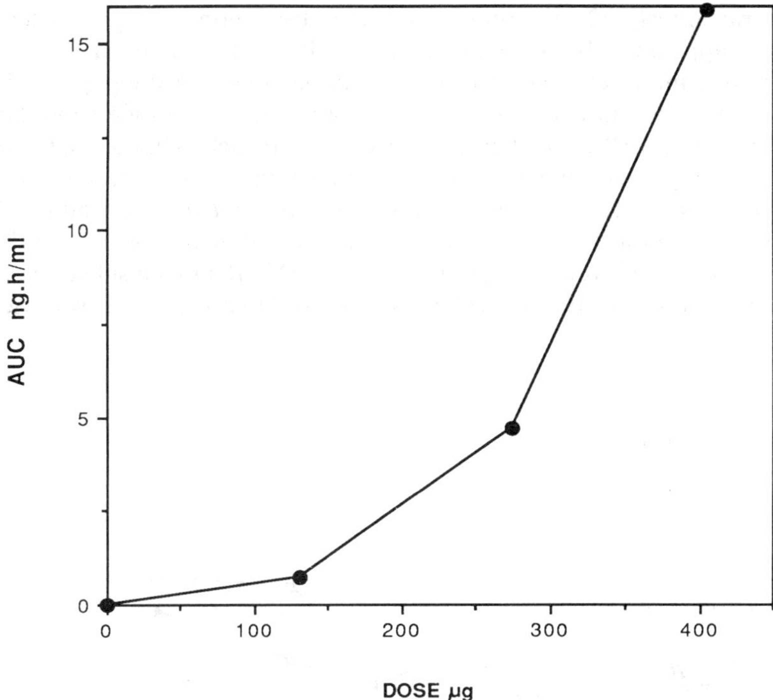

Fig. 15.9 — Influence of dose on the nasal absorption of nafarelin acetate in rhesus monkeys.

Deposition of the spray will depend upon droplet size distribution and the fraction of droplets less than 10 μm should be controlled. The collecting efficiency of the nose decreases rapidly below this value and finer particles will be deposited in the lower respiratory tract and lost. Dose uniformity must be assured by demonstrating that the volume delivered by the spray over the lifetime of the product lies within specified limits.

User factors probably favour the choice of spray. To maximize absorption from drops requires complex movements of the head after administration and many patients will not comply with directions so that the efficiency of the product will be compromised.

Registration and manufacture should not present special problems provided that reproducibility of absorption has been demonstrated and a good stability profile established.

15.5.2 Enhancement of nasal absorption

The permeability of the nasal mucosum to peptides varies widely: a 2% bioavailability of nafarelin acetate, a decapeptide of molecular weight 1322, is described in section 15.5.1. In contrast, Su *et al.* (1985), using surgically modified anaesthetized rats, showed that intranasal administration of the pentapeptide metkephamide (MW 661), gave serum levels which were not significantly different from those of the same dose given intravenously. On the other hand, the absorption or larger molecules, such as insulin, appears to be negligible and it is clear that molecular weight and constitution have a major effect on the intrinsic permeability of the peptide. For a wider application of intranasal delivery, the mucosum must be modified and there is currently much interest in the use of absorption enhancers such as bile salts, lecithins, surfactants, fatty acids and chelators. Two examples have been selected, both of which permit discussion of safety as well as efficiency. Illum *et al.* (1989) added 0.5% lysolecithin to a solution of insulin and demonstrated that a dose of 16.7 IU/kg depressed blood glucose levels in the same way as 1.33 IU/kg given subsutaneously (Fig. 15.10). A derivative of fusidic acid, sodium taurodihydrofusidate, was used by

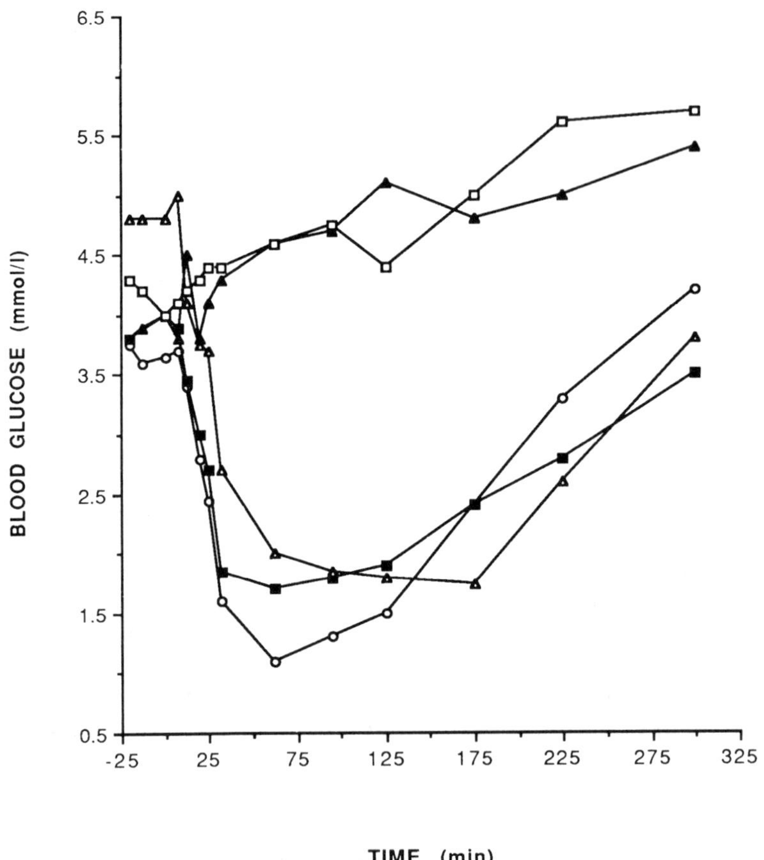

Fig. 15.10 — Blood glucose levels following intranasal (16.7 IU/kg) and subcutaneous (1.33 IU/kg) administration of sodium insulin (SHI) to rats. (□) i.n. blank phosphate buffer (control), (▲) i.n. SHI solution, (■) i.n. SHI solution containing 0.5% laureth-9, (△) i.n. SHI solution containing 0.5% L-α-lysophosphatidylcholine, (○) s.c. SHI solution.

Longeneckar *et al.* (1987) and the influence of concentration on the amount of insulin absorbed when aqueous solutions were instilled into the noses of sheep is shown in Fig. 15.11. These two materials are undoubtedly effective but the disruption of the

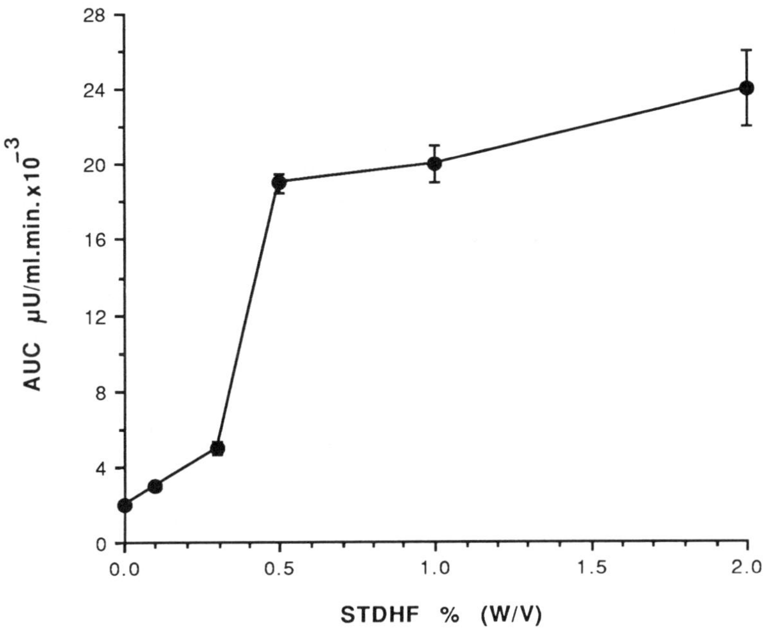

Fig. 15.11 — Effect of concentration of sodium taurodihydrofusidate on the nasal absorption of insulin in sheep.

membrane necessary to permit the passage of a molecule as large as insulin raises questions of both local and systemic toxicity. The examples quoted are claimed to minimize such effects. Illum *et al.* (1989) included in their studies polyoxyethylene-9-lauryl ether (Laureth-9), and effective and widely studied enhancer known to cause morphological damage to the mucosa. An efficiency similar to lysophosphatidylcholine was demonstrated (Fig. 15.10) but it was inferred that the latter caused no morphological damage. Similarly, although an effective enhancer, sodium taurodihydrofusidate was over 100 times less disruptive in an erythrocyte haemolysis test than Laureth-9.

Extensive studies to establish the long- and short-term safety of such enhancers are now necessary before they become available to the formulator. Modifications to the nasal membranes must be small and reversible and the ciliary mechanisms must be unaffected. Nevertheless, hopes are high that an enhancer, effective in low concentrations and of acceptable toxicity, will be found; this will greatly extend the utility of nasal presentation for peptides.

15.6 OTHER ROUTES FOR PEPTIDE DELIVERY

Since the realization of the full clinical potential of peptides depends so heavily on drug delivery, every conceivable route of administration has been explored. Non-parenteral delivery of insulin has often been the spur to innovation because the incidence of diabetes and the apalling nature of the sequelae justify research programmes in which the change of success is less. These alternatives are at a primitive stage, often complicated by the inclusion of adjuvants of uncertain toxicological status.

15.6.1 Pulmonary delivery

The clinical success of intranasal delivery naturally prompts an examination of the lower respiratory tract with its larger and more permeable surface. The alveoli are, however, guarded very effectively by the filtering effect of the conducting airways and an aerosol must be constituted of particles smaller than 2–3 μm for effective deposition. Constructing such as cloud with an adequate drug concentration has so far eluded experimentalists. Nor are the metabolic or immunological consequences of alveolar deposition, if achieved, clear. Nevertheless, Jones *et al.* (1988) demonstrated significant but unquantified absorption of insulin when administered to an endotracheally intubated rabbit from a conventional metered-dose inhaler. A rapid response, which contrasted with subcutaneous administration, was obtained, peak blood level occurring within 12 minutes of administration. The work of Kassem and Ganderton (1990), who showed that the respirable fraction of a cloud generated from a powder mixture could be greatly increased by modifying the carrier surface and conditions of fluidization, could be extended to give more efficient pulmonary delivery.

15.6.2 Buccal and rectal administration

Although the buccal mucosum is not naturally permeable to peptides, accessibility and patient acceptability have promoted some interest. Absorption enhancers are invariably necessary but these are easily co-administered and their effect can be confined to a specific area by means of a patch. A very broad study was carried out by Aungst and Rogers (1989) who evaluated 32 enhancers of the transport of insulin from the buccal cavity of surgically modified rats. Absorption efficiency ranged from zero to 30% when compared to intramuscular insulin, the most effective enhancers being Laureth-9 and certain steroidal detergents.

Rectal delivery, through less acceptable to the patient, has a long history of general application. Rectal membranes have low permeability and absorption enhancers will usually be necessary for useful absorption. A number of materials were evaluated by Aungst and Rogers (1988) and although efficiency of insulin absorption similar to that of buccal administration could be obtained, important differences, described in Fig. 15.12, were found. The intrinsic permeability of the rectum was higher than that of the buccal cavity but both became equally permeable in the presence of Laureth-9. Sodium salicylate, on the other hand, selectivity increased the permeability of the rectal mucosa, an effect also noted with EDTA. These epithelia, which differ in structure, respond differently to agents which may

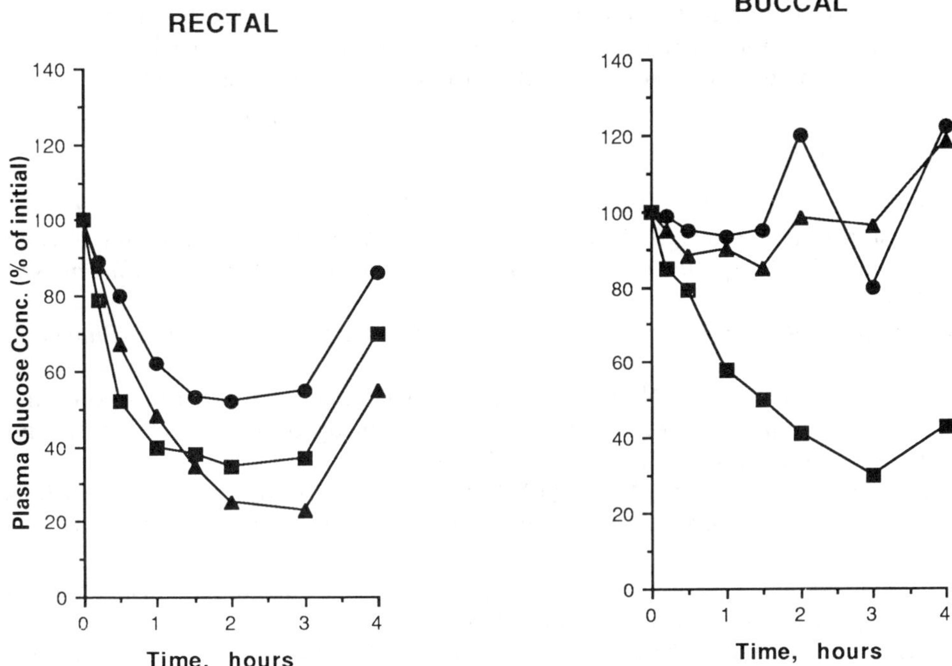

Fig. 15.12 — Plasma glucose concentrations in rats administered rectal or buccal insulin (10 IU/kg) with no adjuvant (●), or with Laureth-9 (■), or sodium salicylate (▲).

react with the membrane protein, lipid or ionic constituents. These effects are extremely complex and require elucidation before surfactants, chelators and salicylates can be used with confidence.

15.6.3 Percutaneous absorption of peptides

The skin presents a formidable barrier to the absorption of most drugs. It may, to a limited extent, be made permeable to peptides by the application of skin penetration enhancers such as dimethyl suphoxide. Transport may also be facilitated by iontophoresis. Kari (1986), for example, was able to demonstrate the control of blood glucose levels in diabetic rabbits by applying a current of 0.8 mA to an insulin reservoir 6.2 cm^2 in area, provided that the stratum corneum of the contact region was disrupted by scraping. It is doubtful whether these methods have any clinical feasibility.

15.7 CONCLUSIONS

Drug delivery has an important and perhaps dominant influence on the development of peptide medicines. In exploiting routes other than parenteral administration, one encounters barriers for which there is little or no intrinsic permeation of the peptide. Breaching these barriers raises important questions of toxicity and, until massive research is carried out on the mode of action of penetration enhancers and the reversibility of their effects, a major regulatory hurdle will be raised against their use. In the meantime, much can be done to refine parenteral dosage forms, both in terms of the time course of their action and the ease with which they are used.

REFERENCES

Anik, S. T., McRae, G., Nerenberg, C., Worden, A., Foreman, J., Hwang, J., Kushinsky, S., Jones, R. E. & Vickery, B. (1984) Nasal absorption of nafarelin acetate, the decapetide [D-Nal(2)^6] LHRH, in rhesus monkeys. I. *J. Pharm. Sci.* **73** 684–685.

Anonymous (1990) Roche repackages interferon to help home injection. *Pharm. J.* **244** 135.

Aungst, B. J. & Rogers, N. J. (1988) Site dependence of absorption-promoting actions of Laureth-9, Na salicylate, Na_2EDTA and aprotinin on rectal, nasal and buccal insulin delivery. *Pharm. Res.* **5** 305–308.

Aungst, B. J. & Rogers, N. J. (1989) Comparison of the effects of various transmucosal absorption promoters on buccal insulin delivery. *Int. J. Pharm.* **53** 227–235.

Brown, L., Siemer, L., Munoz, C. & Lauger, R. (1986) Controlled release of insulin from polymer matrices. *Diabetes* **35** 684–691.

Geigert, J., Panschar, B. M., Fong, S., Huston, H. N., Wong, D. E., Wong, D. Y., Taforo, C. & Pemberton, M. (1988) The long term stability of recombinant (Serline-17) human interferon β. *J. Interferon Res.* **8** 539–547.

Harris, A. S., Nilsson, I. M., Wagner, Z. G. & Alker, V. (1986) Intranasal administration of peptides: nasal deposition, biological response and absorption of desmopressin. *J. Pharm. Sci* **75** 1085–1088.

Heller, J., Penhale, D. W. H., Helwing, R. F. & Fritzinger, B. K. (1981) *Polym. Engng. Sci.* **21** 727–731.

Illum, L., Farraj, N. F., Critchley, H., Johansen, B. R. & Davis, S. S. (1989) Enhanced nasal absorption of insulin in rats using lysophosphatidylcholine. *Int. J. Pharm.* **57** 49–54.

Jones, A. L., Kellaway, I. W. & Taylor, G. (1988) Pulmonary absorption of aerosolised insulin in the rabbit. *J. Pharm. Pharmacol.* **40** 92P.

Kari, B. (1986) Control of blood glucose levels in alloxan-diabetic rabbits by iontophoresis of insulin. *Diabetes* **35** 217–221.

Kassem, N. & Ganderton, D. (1990) The influence of carrier surface on the characteristics of inspirable powder aerosols. *J. Pharm. Pharmacol.* **42** Supplement, 11p.

Lee, V. H. & Yamamoto, A. (1990) Penetration and enzymic barriers to peptide and protein absorption. *Advanced drug delivery reviews* **4** 171–207.

Leong, K. W., Brott, B. C. & Langer, R. (1985) Bioerodible polyanhydrides as a drug-carrier matrices. I: Characterization, degradation and release characteristics. *J. Biomed. Mater. Res.* **19** 941–955.

Longeneckar, J. P., Moses, A. C., Flier, J. S., Silver, M. C. & Dubovi, E. J. (1987) Effects of sodium taurodihydrofusidate on nasal absorption of insulin in sheep. *J. Pharm. Sci.* **76** 351–355.

Sanders, L. M., Kent, J. S., McRae, G. I., Vickery, R. H., Rice, T. R. & Lewis, D. H. (1984) Controlled release of a Luteinising hormone-releasing hormone analogue from poly (*d*,*l*-lactide-*co*-glycolide) microspheres. *J. Pharm. Sci.* **73** 1294–1297.

Su, K. S. E., Campanale, K. M., Mendelsohn, L. G., Kerchner, G. A. & Gries, C. L. (1985) Nasal delivery of polypeptides I: Nasal absorption of enkephalins in rats. *J. Pharm. Sci.* **76** 394–398.

16

The development of 'Zoladex' — a case history

F. J. T. Fildes, F. G. Hutchinson, B. J. A Furr
ICI Pharmaceuticals, Mereside, Alderley Park, Macclesfield, Cheshire, SK10 4TG

16.1 INTRODUCTION

The isolation and characterization of the peptide, luteinizing hormone-releasing hormone (LHRH), in 1971 allowed researchers to investigate the complex nature of the feedback control of pituitary gonadotrophin secretion. The hormone initially found use in fertility control, and even now pulsatile LHRH is the preferred method of ovulation induction in women suffering from hypogonadotrophic hypogonadism. However, the naturally occurring peptide has a very short half-life, and so synthetic analogues have been sought, which would be more resistant to enzymatic degradation, and which might also, perhaps, allow new avenues of administration.

ICI Pharmaceuticals has been at the forefront of this research, and after extensive chemical, biological and toxicological tests, has identified an analogue which is ideal in terms of its potency and safety. This analogue, christened 'Zoladex'† (ICI 118630, goserelin; D-Ser (Bu^t),^{6}, $Azgly^{10}$-LHRH) incorporates structural modifications at the LHRH residues 6 and 10 (see Fig. 16.1). These modifications

Glu–His–Trp–Ser–Tyr–Gly–Leu–Arg–Pro–Gly–NH_2
LH–RH

Glu–His–Trp–Ser–Tyr–D–Ser(Bu^t)–Leu–Arg–Pro–Azgly–NH_2
Zoladex ICI 118,630

Fig. 16.1 — Primary structures of Zoladex and LHRH.

† Zoladex is a trademark, the property of Imperial Chemical Industries plc.

provide increased resistance to enzymatic degradation, and make Zoladex 100 times more potent than LHRH.

The endocrinological studies on Zoladex and other potent LHRH agonists have shown that continued use induces pituitary desensitization following an initial gonadotrophin stimulation. Whereas a single bolus injection of Zoladex will therefore stimulate the pituitary–gonadal axis causing an *increase* in serum sex hormone concentrations, its chronic administration results in a down-regulation of the LHRH receptors, and so leads to a *suppression* of gonadotrophin secretion, and ultimately to a suppression of gonadal function. Zoladex thus causes selective chemical castration, and may be beneficial in the treatment of sex-hormone-responsive tumours, such as prostate cancer in men, and breast cancer in premenopausal women. It may also have some utility in the treatment of non-malignant conditions in women, including endometriosis and uterine fibroids.

Whatever the indication for which Zoladex is used, there still remains the major problem of delivery of this agent to the body; Zoladex is a polypeptide and in order to realize its full therapeutic and commercial potential, the pharmaceutical scientist has to identify and manufacture practical and effective formulations. At present, the full potential of polypeptide drugs is not realized, either because of our inability to direct them selectively to their site of action, or because of the difficulties of achieving temporal control of their systemic or tissue concentrations in relation to biological need. The very high potency of LHRH analogues such as Zoladex, however, renders them eminently suitable for incorporation into, and release from, sustained-delivery dosage forms.

These macromolecular agents are usually ineffective when administered orally, as they are rapidly degraded by proteolytic enzymes in the gastrointestinal tract. Moreover, even if they are stable to enzymatic digestion, their molecular weights are generally too high to allow their absorption through the intestinal wall. There would be obvious benefits to accrue, therefore, if formulations could be developed which would give an acceptable bioavailability of the drug, together with the convenience of oral administration. This technical target continues to provide impetus for expanding research programmes centred around oral peptide delivery. However, in the short term the inherent physicochemical and biological properties of peptides and proteins will inevitably prejudice this route as a general solution, and such research programmes can be expected to meet with only limited success. The same can also be said of other routes of administration, including intranasal (Anik *et al.* 1984, Petri *et al.* 1984), buccal (Anders *et al.* 1983), intravaginal (Okada *et al.* 1982, 1983a,b 1984), rectal (Yoshikawa *et al.* 1985) and percutaneous (Uda and Yamada 1984). All of these alternative routes are associated with a low and variable bioavailability. Consequently polypeptides and proteins are normally administered parenterally (either by subcutaneous, intramuscular or intravenous injection), but even by this route major problems are encountered. These macromolecular agents have very short elimination half-lives and so frequent injections are required to produce an effective therapy.

For polypeptide hormones, where the pharmacology of the agent is compatible with sustained release, the most appropriate dosage form is one that is capable of releasing drug continuously at a controlled rate over a period of weeks or even months. Until recently the major thrust for development of sustained delivery systems for macromolecular drugs was directed towards the use of hydrolytically and enzymatically stable polymers. Much of this work concentrated on the use of synthetic high molecular weight materials as the rate-controlling barrier or matrix, and the various carriers considered are summarized in Table 16.1. The carrier

Table 16.1 — Non-erodible carriers for sustained release of polypeptides

Hydrogels
Cross-linked polyacrylamide (Davis 1972, Davis 1974)
Cross-linked polyvinyl alcohol (Langer & Folkman 1976)
Cross-linked poly(hydroxyethyl methacrylate) (Langer & Folkman 1976)
Hydrophobic polymers
Ethylene/vinyl acetate copolymer (Langer 1984, Cohen *et al.* 1984, Hsu and Langer, 1985)
Silicone elastomers (Lotz and Syllwasschy 1979, Hsieh *et al.* 1985)
Microporous polypropylene (Kruisbrink & Boer 1984)
Cross-linked (meth) acrylates (Yoshida *et al.* 1985)

systems based on these types of polymers have the major disadvantage that they require surgical retrieval of the expired delivery system. Since it is preferable that the carrier system ultimately disappears 'naturally' from the site of administration, biodegradable polymers are desirable.

The biodegradable polymers that are currently being evaluated as carriers for the sustained release of low molecular weight drugs are presented in Table 16.2.

Table 16.2 — Biodegradable polymers used in drug delivery

Polylactic acid (Polylactide)
Polyglycolic acid (Polyglycolide)
Poly (lactic acid-*co*-glycolic acid)
Poly (ε-caprolactone)
Poly (hydroxybutyric acid)
Poly ortho-esters
Poly acetals
Poly dihydropyrans
Poly cyanoacrylates
Synthetic polypeptides
Cross-linked polypeptides

Experience over many years with homo- and co-polymers of lactic and glycolic acids has shown that these materials are inert and biocompatible, and that they are degraded to toxicologically acceptable products (Wise *et al.* 1979). In the initial design of (biodegradable) parenteral sustained delivery systems, therefore, these polymers are invariably selected, particularly when release over many weeks is required. It is this approach that has been adopted by ICI, and ICI researchers were the first to identify and characterize the mechanisms of transport which allow the movement of polypeptide drugs from biodegradable polyester formulations (Hutchinson 1982). Such formulations may be developed as solid injectable depots or as injectable suspensions of microparticles.

The utility of these types of formulations is well-illustrated by the LHRH analogue Zoladex. Solid injectable depot formulations of this analogue have been developed, and have been used in the treatment of hormone-responsive prostate and mammary tumours in animals and man. (The same type of formulation has been demonstrated to be applicable to other polypeptide drugs.) Zoladex depot formulations are approximately 1 cm long by 1 mm diameter, and release 3.6 mg of drug over 28 days.

16.2 RATIONALE OF ZOLADEX FORMULATION

16.2.1 Problems with peptide formulation

The successful development of sustained-release biodegradable delivery systems for peptide drugs, such as Zoladex, requires the recognition and resolution of a number of major problems:

(1) The mechanism most commonly used to achieve sustained release, namely controlled diffusion through a matrix or membrane, may not be appropriate for a high molecular weight polypeptide.

Design of a sustained-release dosage form must take into account the properties of both the rate-controlling polymer and the drug. In order for the drug to diffuse through the polymer it must have some solubility in the high molecular weight carrier. This is often the case for low molecular weight drugs, but not so for polypeptides. It is well-established that, in the absence of specific chemical interactions, polypeptides will either be insoluble in, or incompatible with, structurally disimilar polymers like polyesters, because of adverse entropic and enthalpic factors (Bohn 1975). The low or negligible solubility of the macromolecule in a polymer, such as polyester, will prevent transport of the agent through the polymer phase.

With regard to the properties of the drug, the most important of these are its size, shape and solubility (Baker and Lonsdale 1974). There is an approximate log–log correlation between molecular weight (M) and diffusion coefficient (D) where $\log D = a - b \log M$ (where a and b are arbitrary constants) such that D decreases as molecular weight increases. For polypeptides, M is large and the diffusion coefficient becomes vanishingly small because the diffusant cannot be accommodated by the free volume of polymer arising from its rotational and translational segmental mobility. Thus, for reasons of incompatibility and molecular size, polymers such as polyesters are not likely to allow partition-dependent diffusion of polypeptides through the polyester phase.

(2) Polypeptides are biologically labile and can be readily degraded by tissue enzymes. They must, therefore, be effectively protected at the depot site if active drug is to be released continuously. The difficulty of achieving this is emphasized by the fact that synthetic polypeptides have actually been used as biodegradable carriers for drugs such as steroids and narcotic antagonists (Mitra *et al.* 1979, Sidman *et al.* 1983).
(3) The excipients used to achieve sustained release of macromolecular agents might provoke an adjuvant-induced immunological response, which may be related to the nature of the excipient, the delivery rate, or profile of release. There is some evidence that sustained release of large proteins may be an effective means of raising antibodies (Langer 1981).
(4) Long-lasting depots might become encapsulated by fibrous tissue, thus inhibiting further release of drug. This is certainly the case for non-degradable silicone elastomer implants (Anderson *et al.* 1981).

These imposing problems for the sustained delivery of Zoladex have been resolved by the design of biodegradable delivery systems based on polyesters such as poly(*d*, *l*-lactide) and poly(*d*, *l*-lactide-*co*-glycolide), to give formulations which allow release of the polypeptide over an extended period of time. These studies have encompassed polymer synthesis to give polyesters of a defined structure and molecular weight, characterization of the degradation processes leading to erosion and ultimate disappearance of the biodegradable carrier, morphological studies on the degrading polymer, and drug release studies and biological evaluation of defined dosage forms.

16.2.2 Biodegradable polyesters derived from lactic and glycolic acids

These simple biodegradable homo- and co-polymers are prepared at elevated temperature by the ring-opening polymerization of dry, freshly prepared acid dimers, *d*, *l*-lactide and glycolide, by using organo-tin compounds as catalysts. Control of molecular weight is achieved by using a chain transfer agent such as *d*, *l*-lactic acid. Polymers of variable composition are therefore obtained, having intrinsic viscosities ranging from <0.1 to >1 (Fig. 16.2). The polymers can be further

Polyesters

$H(O\underset{\displaystyle CH_3}{\underset{\mid}{CH}}CO)_N OH$	Polylactic acid/Polylactide
$H(O\,CH_2\,CO)_M OH$	Polyglycolic acid/Polyglycolide
$H\left[(O\underset{\displaystyle CH_3}{\underset{\mid}{CH}}CO)_n (O\,CH_2\,CO)_m-\right]_p OH$	Poly (lactide–*co*–glycolide)

Fig. 16.2 — Polymers and co-polymers of lactic and glycolic acids.

characterized by size exclusion chromatography (using polystyrene standards) to define their number average molecular weight (M_n), weight average molecular weight (M_w), and polydispersity ($P=M_w/M_n$).

Number average molecular weight is defined as

$$M_n = \frac{\sum n_i M_i}{\sum n_i}$$

where n_i is the number of molecules of polymer having molecular weight M_i, and weight average molecular weight is defined as

$$M_w = \frac{\sum W_i M_i}{\sum W_i} = \frac{\sum n_i M_i^2}{\sum n_i M_i}$$

where W_i is the weight of the polymer molecules having molecular weight M_i.

Additionally, the polymers can be characterized by ^{13}C nuclear magnetic resonance spectroscopy. This technique permits definition of the distribution of co-monomers and polymer structure (that is, the average values for *n* and *m* of co-polymers shown in Fig. 16.2).

As polypeptides have high molecular weights and are water-soluble, it is unlikely that they will be released from these biodegradable polyesters by classical partition-dependent diffusion through the polyester. Consequently, degradation of the poly(*d*, *l*-lactide) or poly(*d*, *l*-lactide-*co*-glycolide) will be a critical factor in determining transport of the high molecular weight polypeptide from the dosage form. The degradation of these polymers in the absence of drug has been characterized in terms of molecular weight and distribution, weight loss, water uptake and morphology of the hydrated and degraded polymer.

Degradation of the polymers *in vitro* in buffer at pH 7.4 results in progressive changes in molecular weight and molecular weight distribution. Under these conditions degradation is not enzyme-mediated and must occur by simple hydrolytic cleavage of ester groups. The profile of weight loss and change in molecular weight, for polymers having the most probable distribution (*P* approximately 2), are consistent with this. High molecular weight polymers degrade to lower molecular weight species (as measured by changes in viscosity), but retain their water insolubility (Fig. 16.3). Only after an extended period of degradation does any weight loss occur. In contrast, very low molecular weight polymers can degrade with weight loss immediately. Similar results are obtained with high-lactide-containing polymers, except that the time scale of events is more extended for these more hydrolytically

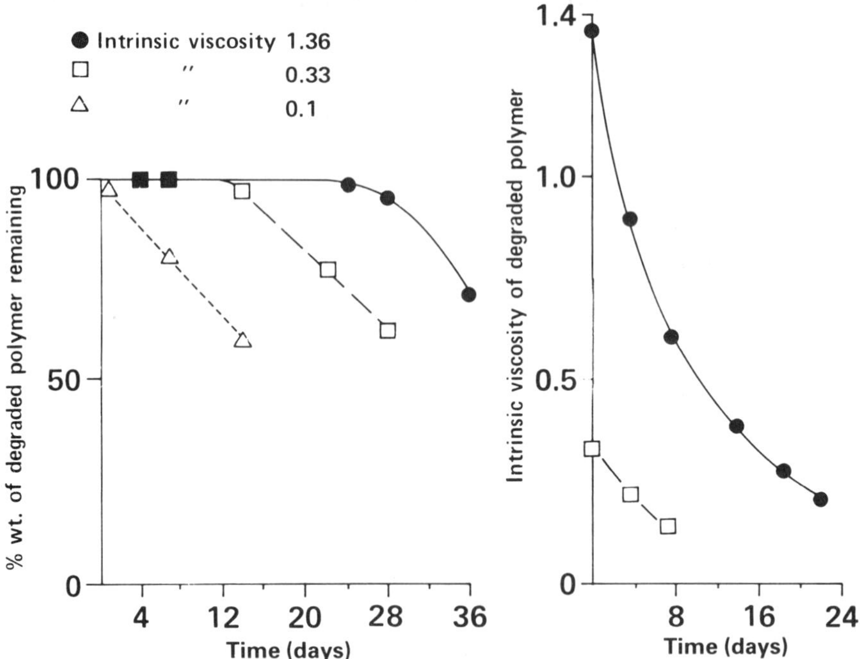

Fig. 16.3 — *In vitro* degradation of 50/50 molar (*d, l*-lactide-*co*-glycolide) at 37°C in buffer at pH 7.4.

stable polymers. These results are consistent with bulk hydrolysis in the *in vitro* condition, and this correlates broadly with the polymer degradation *in vivo*, suggesting that even in subcutaneous tissue, enzyme-mediated degradation is significantly less important than simple hydrolysis. In this event, polylactides could effectively protect polypeptides at the depot site from the influence of degradation enzymes.

For such degradation experiments, an essentially linear relationship between the logarithm of the number average molecular weight and time holds for high molecular weight polymers. A discontinuity arises at extended times of degradation (Fig. 16.4).

Pitt and Schindler (1980) have observed similar behaviour with poly(*d, l*-lactide), but did not recognize that the discontinuity occurs because of water-uptake by the degrading polymer. For an amorphous polyester, the water-uptake appears to be governed by the intrinsic hydrophilicity of the repeat units and end-group effects. With these polyesters the end groups are alkoxylic and carboxylic groups and the proportion of these increases as the polymer molecular weight falls. That is, as degradation proceeds the essentially hydrophobic polymer becomes more hydrophilic. The profile of water uptake at 37°C in buffer at pH 7.4, for polymers which have been dried rigorously, has been studied as a function of time. For these systems the water-uptake is determined by two events. The first of these is simple diffusional ingress and, in the absence of degradation, this occurs to a level that is characteristic of the equilibrium swelling of this kind of polyester. However, these polymers are

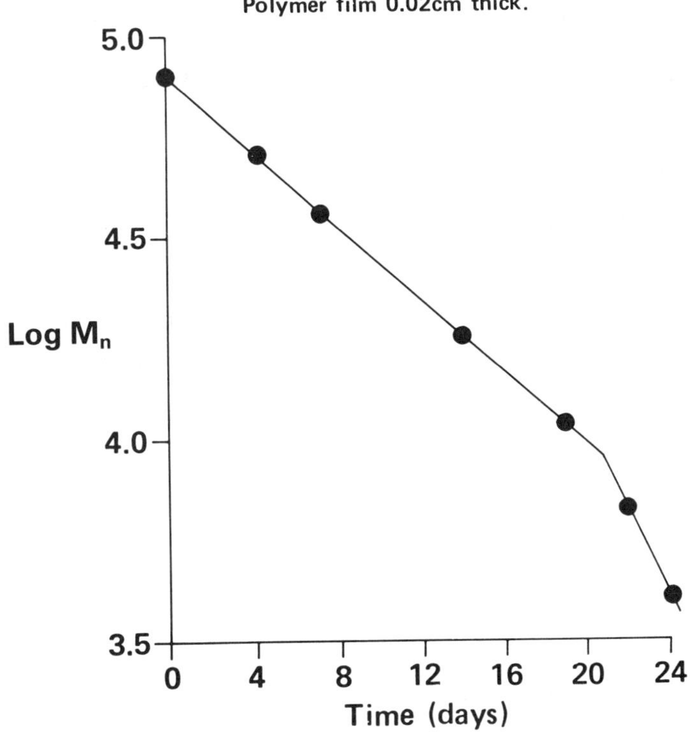

Fig. 16.4 — *In vitro* degradation of poly(*d*, *l*-lactide-*co*-glycolide) at 37°C in buffer at pH 7.4 and dependence of number average molecular weight on time of degradation.

hydrolytically unstable and following, or even during, this initial diffusion phase, the polymer can degrade and so take up more water. For high molecular weight polymers, having a normal distribution ($P=2$), these two phases of water-uptake are separated by an interval during which water-uptake remains virtually constant. In contrast, low molecular weight polymers have essentially a continuous water-uptake.

It can be shown empirically that the water-uptake for a thin polymer film having a molecular weight M_n and a polydispersity P, in the absence of significant hydrolytic degradation, is described approximately by the hyperbolic function:

$$[H_2O] = a + \frac{b}{PM_n}$$

where a and b are constants related to polymer composition.

If the initial diffusional ingress into thin films is assumed to be instantaneous, then approximate expressions can be derived for the degradation-induced change of molecular weight and water-uptake, using a similar (but modified) model proposed by Pitt and Schindler (1980).

Degradation of poly(lactide-*co*-glycolide) proceeds by hydrolytic scission of ester groups, generating polymers containing one terminal carboxyl group per chain. Defining degradation as the appearance of $-CO_2H$ and applying the normal kinetic equation governing ester hydrolysis, we have:

$$\frac{d[CO_2H]}{dt} = K[H_2O][\text{ester}][CO_2H] \tag{1}$$

where $[CO_2H] \propto 1/M_n^t$ and M_n^t is the number average molecular weight at time t.

For all practical purposes the concentration of ester can be considered a constant and since

$$[H_2O] = a + \frac{b}{PM_n}$$

equation (1) reduces to

$$\frac{d[1/M_n^t]}{dt} = k\left(a + \frac{b}{PM_n^t}\right)\frac{1}{M_n^t} \tag{2}$$

where $k = K[\text{ester}]$

At $t=0$, $M_n^t = M_n^0$ is the initial number average molecular weight and the solution to equation 2 is obtained as:

$$M_n^t = M_n^0\, e^{-akt} + \frac{b(e^{-akt} - 1)}{aP} \tag{3}$$

with

$$[H_2O]_t = a\left[1 + \frac{b}{aPM_n^0\, s^{-akt} + b(e^{-akt} - 1)}\right] \tag{4}$$

It should be noted that these expressions are derived assuming that the initial polymer has a normal distribution (i.e. $P=2$) and that the hydrolysis of the chains is essentially random. The equations also show clearly that degradation of the polymer depends not only on molecular weight and composition but also on molecular weight distribution.

It can be seen from Fig. 16.5 that the derived equation for water-uptake correlates broadly with experimentally determined events. Thus, hydrolytic degradation is characterized by reduction in molecular weight, enhanced water-uptake and ultimately weight loss of polymer. All these events occur at a temperature which is below or near to the glass transition temperature of the polyester. This in turn implies that morphological changes are likely to occur as the polymer undergoes hydrolysis. This is confirmed by development of porosity within the degrading polyester (Fig. 16.6).

These studies demonstrate that the degradation of poly(*d*, *l*-lactide) and poly(*d*, *l*-lactide-*co*-glycolide) is dependent on molecular weight, polydispersity, geometry, polymer composition and polymer structure, and ultimately leads to enhanced water-uptake and the generation of porosity. Thus, water-soluble polypeptides such as Zoladex may be released from these biodegradable polyesters since enhanced water uptake and the generation of porosity should facilitate transport of polypeptide from the dosage form. This is likely to involve diffusion through aqueous pores generated in the drug/polymer matrix. In this event, the release of polypeptide will occur by a different mechanism from the processes thought to occur during release of steroids, narcotic antagonists and antimalarials from poly *d*, *l*-lactide) and poly(*d*, *l*-lactide-*co*-glycolide) (Wise *et al.* 1979). Whereas these low molecular weight drugs will diffuse, by a simple partition-dependent process, through intact polymer membranes in diffusion cell experiments, these same polymer membranes are totally impermeable to polypeptides.

16.3 RELEASE STUDIES WITH ZOLADEX

Chronic administration of LHRH analogues, such as Zoladex, has been shown to cause a reversible chemical castration which leads to the regression of hormone-responsive animal and human mammary and prostate tumours (Furr and Hutchinson 1985). Because of low oral potency, the drugs have usually been administered parenterally, once or more times daily. Clearly, a biodegradable formulation, based on a subdermal depot or injectable suspension of poly(lactide-*co*-glycolide), capable of delivering the drug over a period of 28 days or longer, would be more acceptable. In the particular case of Zoladex (molecular weight 1269) solid depot formulations were selected, because these were thought likely to afford a clearer understanding of the physicochemical parameters controlling drug transport from the dosage form.

The continuous release of the polypeptide *in vivo* can be measured qualitatively by the biological effect elicited in regularly cycling adult female rats. Normally, these rats have an oestrous cycle of four days and the occurrence of oestrus is indicated by the presence of cornified cells in vaginal smears. In rats given subdermal depots of Zoladex, the release of drug at an effective rate will cause a fall in circulating oestrogens, which in turn leads to a suppression of oestrus and an absence of cornified cells in vaginal smears. Rats therefore show an extended period of dioestrus.

With respect to polymer composition, our studies have shown that, for amorphous homo- and co-polymers of approximately the same high molecular weight, increasing lactide content results in slower degradation. This is reflected in the biological effect elicited in rats, using subdermal depots containing small amounts of drug in high molecular weight carriers (Fig. 16.7). When administered to female rats

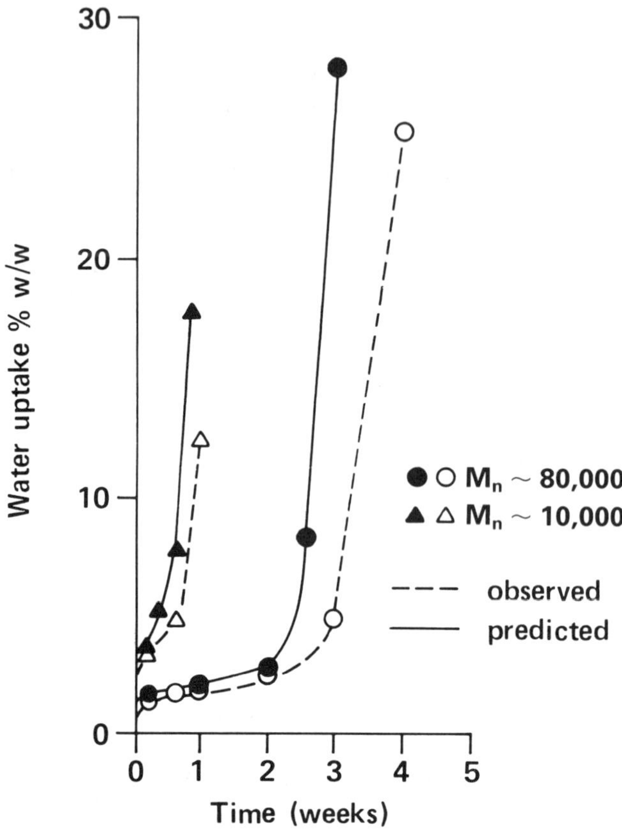

Fig. 16.5 — *In vitro* degradation of poly(*d, l*-lactide-*co*-glycolide) at 37°C in buffer at pH 7.4. Effect of degradation on water-uptake by polymer.

some drug is released initially (as judged by the biological response) but this immediate release soon ceases as the biological effect is not maintained. There then follows a period during which the drug is either not released at all or else is released at an ineffective rate. At some later time point release recommences and continues until the depot has been fully depleted of drug. For polymers of similar molecular weight and distribution it can be seen that the interval between the two phases of release is shortest for the most rapidly degradable polymer.

On the basis of degradation studies, transport of drug from these depots is likely to be governed by various properties of the rate-controlling polyester. These properties include polymer composition, molecular weight and distribution as well as the level of drug incorporation, morphology of the drug/polymer mixture, degradation characteristics of the polymer and geometry.

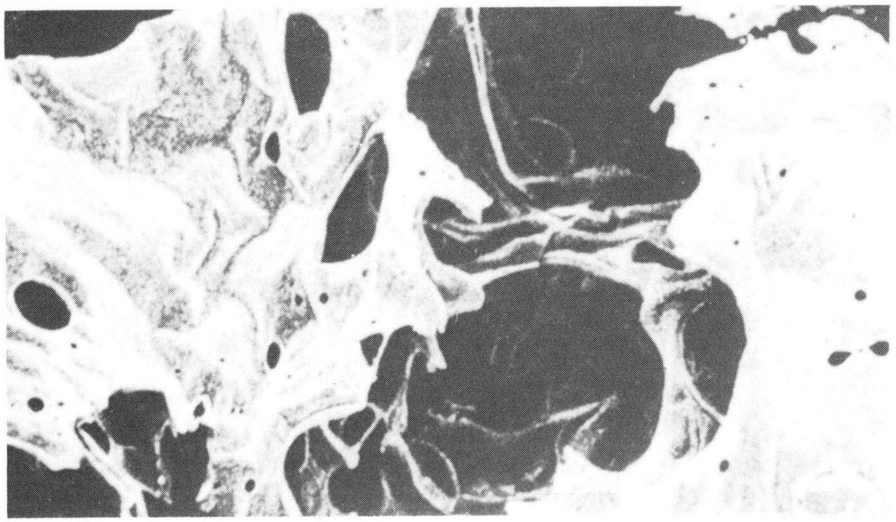

Fig. 16.6 — Electron photomicrograph of surface of degraded polymer. Polymer film (0.02 cm thick) incubated in pH 7.4 buffer at 37°C for 14 days.

The initial molecular weight of the biodegradable polymer is also an important parameter controlling the degradation of the polyester. Thus for a given level of drug incorporation the interval between the two phases of release is reduced as the molecular weight of the polymer decreases. It is apparent from Fig. 16.8 that the net result of reducing the molecular weight is to bring forward the degradation-induced release. At low molecular weights there is virtual overlap of the two phases of release.

A further important component of drug release is the level of the drug incorporation throughout the biodegradable polyester. For a given molecular weight, increased drug loading results in an extended initial phase of release, presumably due to the increased continuity, or contiguity, of drug domains, with the surface of the dosage form. Extending the time of this initial release can ensure overlap with the degradation-induced phase of release. At incorporations of above 15% continuous release can be obtained using high molecular weight polymers (Fig. 16.9).

It can be shown that release of polypeptide from these biodegradable polyesters occurs by diffusion through aqueous pores generated in the dosage form (Fig. 16.10). These aqueous channels, which facilitate drug release, are generated by two distinct and separate mechanisms. The first involves leaching of drug from polypeptide domains at or near the surface of the delivery system, and this is essentially a dissolution/diffusion-controlled event. However, drug within the body of the depot, existing in isolated domains not continuous or contiguous to the surface, cannot be released until the second mechanism becomes operative. This second mechanism involves degradation of the polyester and is associated with the generation of microporosity in, and enhanced water-uptake by, the degrading polymer.

The parameters that control the initial phase of release include the drug loading, morphology and geometry, whereas the second phase is intrinsically related to the

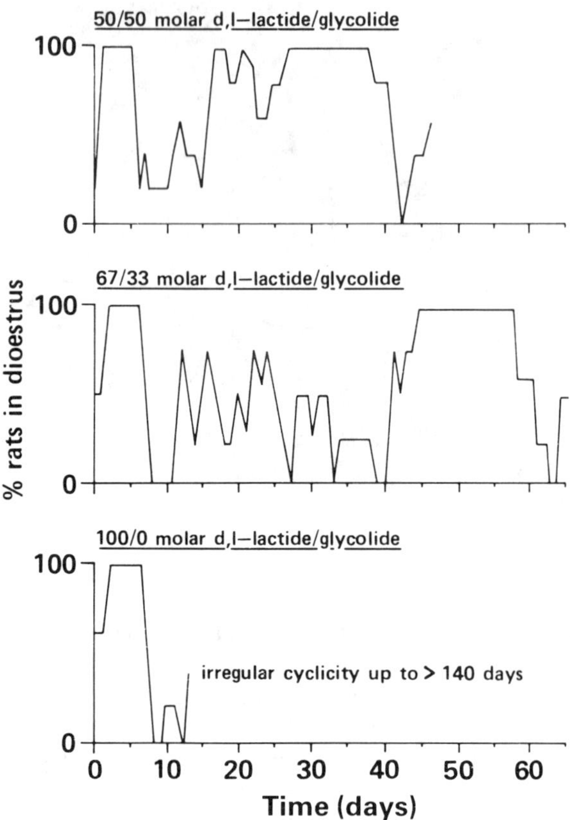

Fig. 16.7 — Regularly cycling female rats treated with 100 μg Zoladex using subcutaneous depots containing 3% wt/wt drug in high molecular weight polymer.

degradation properties of the polyester. When these two phases of release do not overlap, then discontinuous release is observed (Fig. 16.11(a)). However, by controlling the properties of the polymer, the initial phase of release can be made to overlap with the second phase and depots can be defined which give continuous release over 28 days both *in vitro* and *in vivo* (Fig. 16.11(b)). These depots have been used to induce a castration-like effect in rats and thereby to inhibit the growth of mammary and prostate tumours (Furr and Hutchinson 1985).

Although the primary objective in this work was to produce a formulation of the drug which was more convenient to adminster, and which would secure improved compliance, it also appears to have yielded depot formulations which give an improved drug efficacy. This was demonstrated in rats which were given a single subcutaneous bolus dose of 50 μg Zoladex after either pretreatment with daily subcutaneous injections of saline or 50 μg Zoladex for six weeks or, alternatively, at the start of the experiment and at four weeks, a single subcutaneous depot calculated to release a dose of Zoladex equivalent to 50 μg daily. Rats were killed at various times after the final treatment and serum LH was determined by radioimmunoassay in blood collected from the dorsal aorta. The results are shown in Fig. 16.12. As

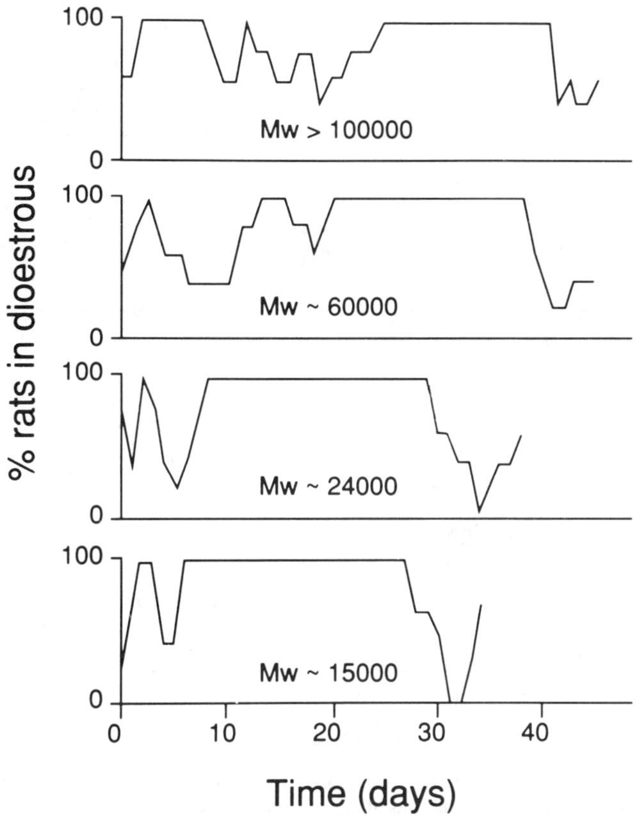

Fig. 16.8 — Regularly cycling fremale rats treated with 300 μg Zoladex using subcutaneous depots containing 10% wt/wt drug in 50/50 molar poly(*d*, *l*-lactide-*co*-glycolide) of different molecular weights.

expected, the bolus dose of Zoladex elicited a massive secretion of LH in the rats pretreated with saline. In spite of six weeks' pretreatment with 50 μg Zoladex daily, the bolus dose of the drug still caused a substantial release of LH, although there was a clear degree of desensitization of the pituitary gland. In contrast, there was negligible LH secretion in response to the bolus injection in rats pretreated with the Zoladex depot, indicating an improved level of pituitary desensitization with this formulation.

Studies in the pigtailed monkey have also shown that the subcutaneous (s.c.) administration of a depot containing Zoladex is more effective at suppressing serum testosterone than daily s.c. injections of the drug (Furr and Hutchinson 1985).

16.4 USE OF ZOLADEX DEPOTS IN ANIMAL MODELS FOR HORMONE-RESPONSIVE PROSTATE AND MAMMARY TUMOURS

The efficacy of this depot formulation of Zoladex was tested in two sex hormone-responsive tumour models. The first was the dimethyl-benzanthracene (DMBA)-

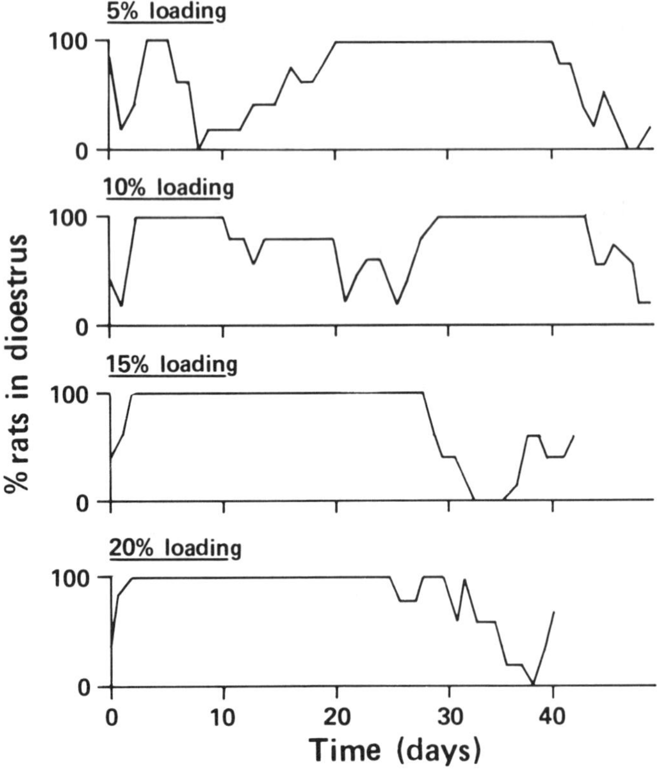

Fig. 16.9 — Regularly cycling female rats treated with 300 μg Zoladex using subcutaneous depots containing various drug loadings in high molecular weight 50/50 molar poly(*d, l*-lactide-*co*-glycolide).

induced rat mammary carcinoma (Huggins *et al.* 1959), which is known to be dependent on both oestrogen and prolactin (Jordan 1982). A single s.c. depot containing 500 μg Zoladex caused an inhibition of oestrogen secretion, the disappearance of cornified cells from vaginal smears and regression of DMBA-induced mammary tumours (Fig. 16.13). Half of the tumours present at the start of the experiment were not palpable at 28 days but all save one of them reappeared between 40 and 60 days as the single Zoladex depot became exhausted. In contrast, the DMBA-induced mammary tumours increased in size by more than 50% in control animals given a placebo depot.

If single depots of the drug were given s.c. at weeks, 0, 4, and 8 of the study, the regression was more impressive, and no tumour was palpable by week 11 (Fig. 16.14). Again, by week 16 regrowth of the tumours occurred because treatment stopped at week 8. By week 20 the tumours had re-attained pretreatment size. When given at 30, 58 and 86 days after administration of the carcinogen, single depots containing 500 μg Zoladex delayed the appearance of tumours for a period of around 100 days, i.e. for the expected duration of action of such a treatment regimen (Fig. 16.15). When given at 28-day intervals, starting on day 30 after administration of the

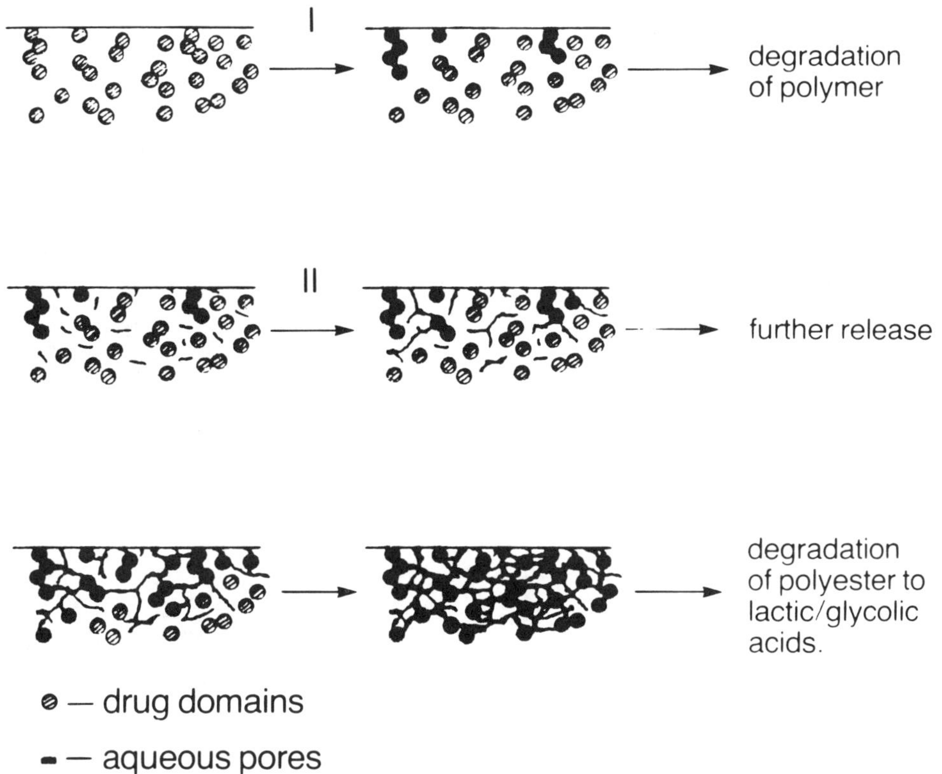

Fig. 16.10 — Mechanisms of release of polypeptides from formulation based on poly(*d*, *l*-lactide) and poly(*d*, *l*-lactide-*co*-glycolide).

carcinogen, single s.c. depots of Zoladex caused a more profound inhibition of tumour appearance and only 9 out of 21 rats had mammary tumours at the end of the study on day 450. Those tumours which were found did not regress following ovariectomy and so were classified as non-hormone-responsive. It is concluded from this study, which approximates the clinical adjuvant therapy setting, that Zoladex may be of benefit as a treatment for primary breast cancer post-mastectomy in premenopausal women.

The second tumour model used to check drug efficacy was the Dunning R3327H rat transplantable prostate adenocarcinoma. This tumour is androgen-responsive and has been used extensively as a model for the human disease (Smolev *et al*. 1977). Single s.c. depots containing 1 mg Zoladex given every 28 days to rats bearing Dunning R3327H prostate tumours implanted on each flank caused a marked inhibition of tumour growth indistinguishable from that in surgically castrated rats (Fig. 16.16). Twenty-one days after the eighth depot was given, the rats were killed and the weights of the sex organs recorded and serum hormone concentrations measured by radioimmunoassay (Table 16.3). Testes weights were about 10% those

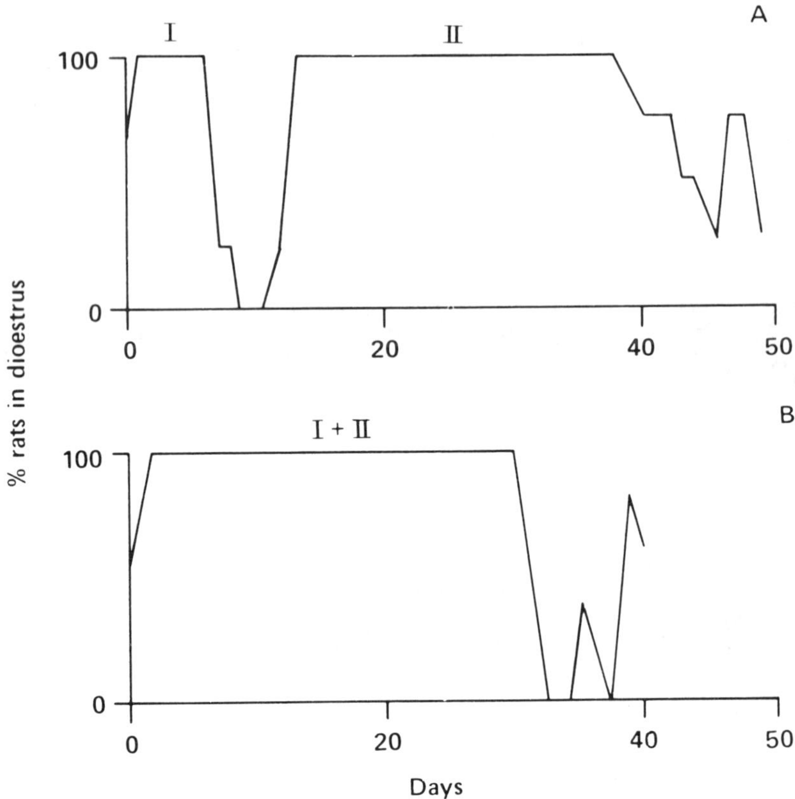

Fig. 16.11 — Effect of subdermal depots containing 300 μg Zoladex administered to regularly cycling adult female rats.

of control rats of a similar age and weight and showed atrophic histological changes. Ventral prostate gland and seminal vesicle weights were identical to those in the surgically castrated group and, histologically, were also completely atrophic.

In the group given a Zoladex depot, the serum LH and testosterone levels were undetectable, and serum FSH was decreased by 60–70% (Table 16.3). As with surgically castrated animals, the Zoladex-treated rats also showed a doubling of serum prolactin levels, probably as a consequence of androgen withdrawal. This is in contrast with the effect of Zoladex in female rats, where there is a significant reduction in the serum prolactin following oestrogen withdrawal (Furr & Nicholson 1982).

The promising results achieved in animal studies have now been substantiated in clinical trials in men suffering from prostatic carcinoma (Ahmed *et al.* 1983, Allen *et al.* 1983, Walker *et al.* 1983, 1984, Robinson *et al.* 1985, Donelly 1984, Furr & Milsted 1988) and in premenopausal women with breast cancer (Williams *et al.* 1986).

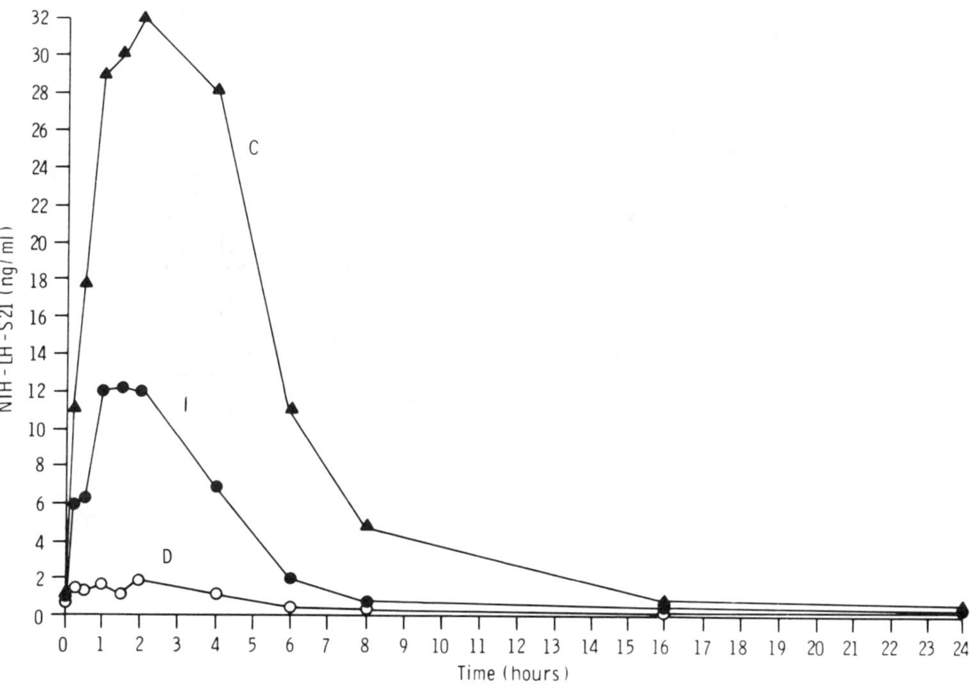

Fig. 16.12 — Serum LH release in response to a bolus of 50 μg Zoladex in rats pretreated with saline (C) or 50 μg Zoladex (I) daily for six weeks. The response after treatment at the start of the experiment and at four weeks with single subcutaneous depots, calculated to release 50 μg Zoladex daily, is also shown (D). Serum LH was measured by radioimmunoassay and the values given are the means of five samples at each time point.

Table 16.3 — Sex organ weights and serum hormone concentrations in Zoladex-treated, and surgically castrated rats bearing Dunning R3327H prostate tumours. Control values for rats of a similar age and weight are shown for comparison

	Zoladex	Castrated	Control
Testes wt (mg)	366.5 ± 10.2	—	3500
Ventral prostate wt (mg)	21.3 ± 1.1	19.9 ± 0.7	250
Seminal vesicle wt (mg)	54.3 ± 1.0	53.8 ± 0.8	350
Serum LH (ng/ml)	<0.2	12.9 ± 1.0	1.5
Serum FSH (ng/ml)	174 ± 6	1413 ± 24	400
Serum prolactin (ng/ml)	63.2 ± 4.5	60.9 ± 7.5	30
Serum testosterone (ng/ml)	<0.25	<0.26	3

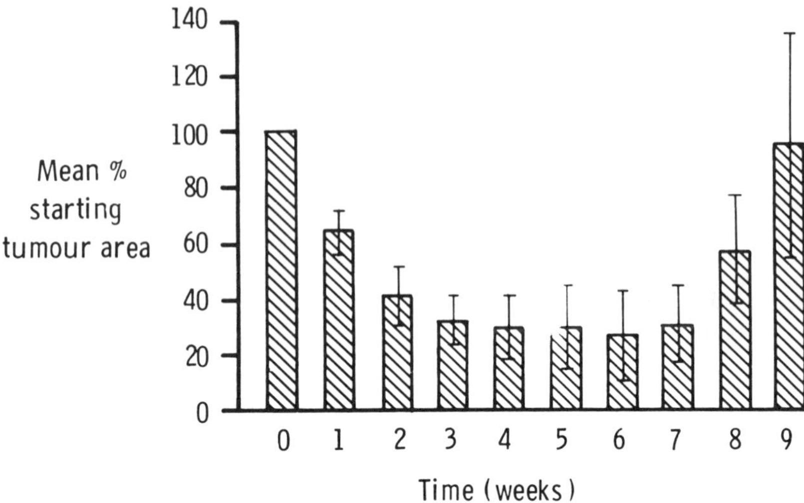

Fig. 16.13 — Effect of a single subcutaneous depot containing 500 μg Zoladex given at time zero on growth of DMBA-induced rat mammary tumours. The values shown are means ± SEM for ten rats.

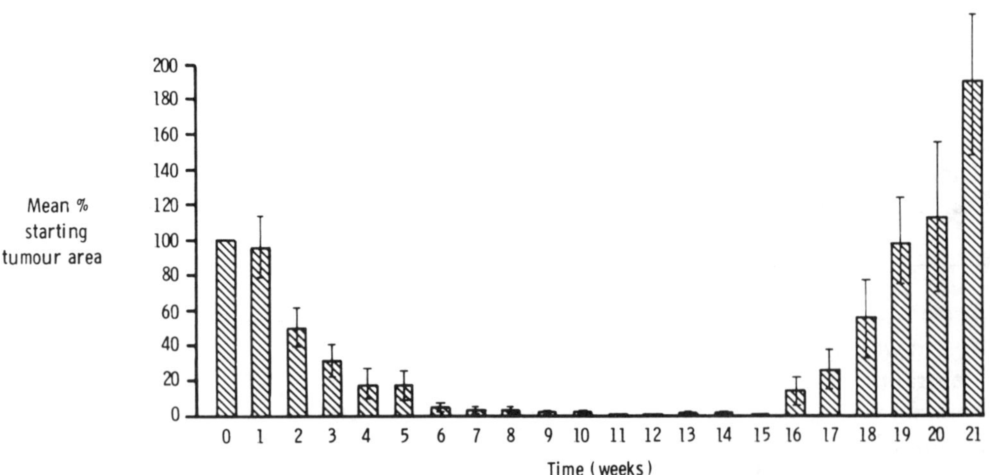

Fig. 16.14 — Effect of single subcutaneous depots containing 500 μg Zoladex given at 0, 4 and 8 weeks on growth of DMBA-induced rat mammary tumours. The values are the means ± SEM for ten rats.

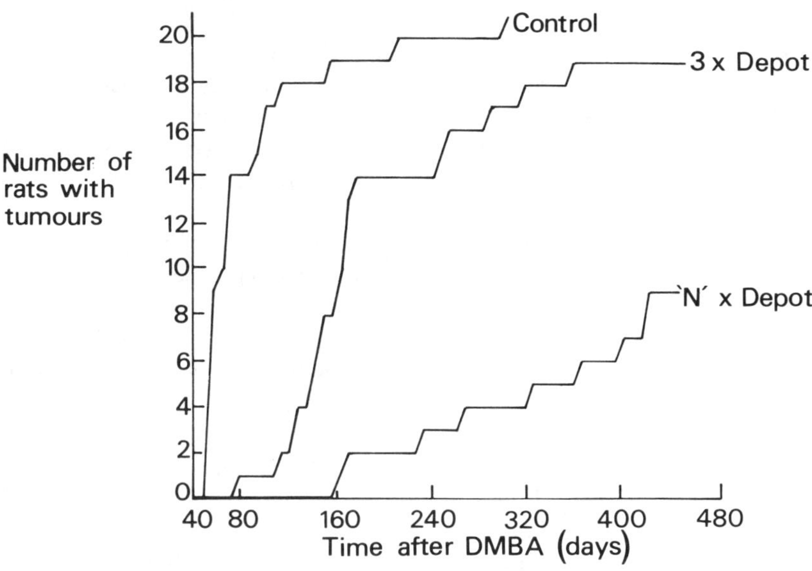

Fig. 16.15 — Effect of depots containing 500 μg Zoladex on the appearance of mammary tumours induced in rats by DMBA. The number of rats with tumours in groups given placebo depots (Control), single depots at days 30, 58 and 86 (3 × Depot) or single depots every 28 days starting at day 30 (N × Depot) are shown. Each group comprised 21 rats.

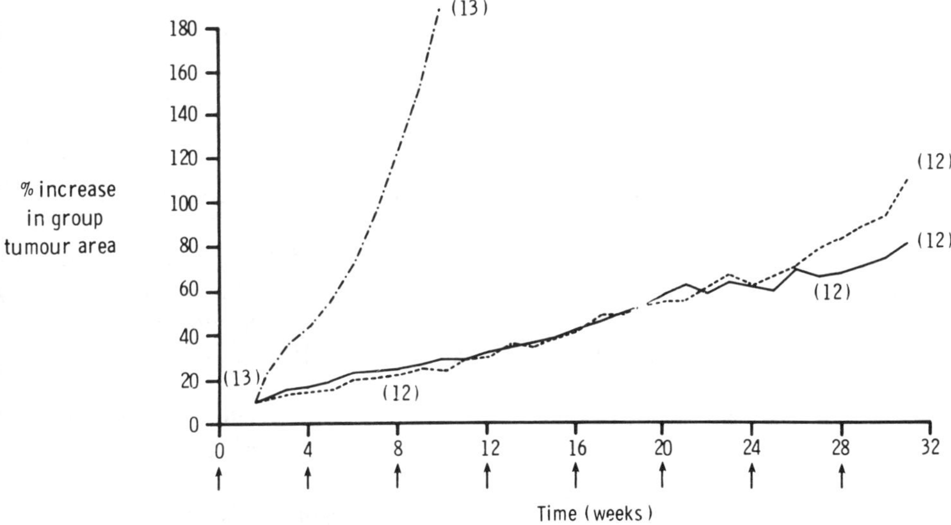

Fig. 16.16 — Effect of single subcutaneous depots containing 1 mg Zoladex on the growth of Dunning R3327H transplantable rat prostate tumours. The depots were given every 4 weeks as shown by the arrows. The control group (— · — ·) had 13 animals and the Zoladex-treated (- - - -) and surgically castrated(——) groups 12 animals each, as shown by the figures in parentheses.

REFERENCES

Ahmed, S. R., Brooman, P. J. C., Shalet, S. M., Howell, A., Blacklock, N. J. & Rickards, P. (1983) Treatment of advanced prostate cancer with LHRH analogue ICI 118,630: clinical response and hormonal mechanisms. *Lancet* **ii** 415–419.

Allen, J. M., O'Shea, J. P., Mashiter, K., Williams, G. & Bloom, S. R. (1983) Advanced carcinoma of the prostate: treatment with a GnRH agonist. *Br. Med. J.* **286** 1607–1609.

Anders, R., Merkle, H. P., Schurr, W. & Ziegler, R. (1983) Buccal absorption of protirelin: an effective way to stimulate thyrotropin and prolactin. *J. Pharm. Sci.* **72** 1481–1483.

Anderson, J. M., Niven, H., Pellagalli, J., Olanoff, L. S. & Jones, R. D. (1981) The role of the fibrous capsule in the function of implanted drug–polymer sustained release systems. *J. Biomed. Mater. Res.* **15** 889–902.

Anik, S. T., Sanders, L. M., Chaplin, M. D., Kushinsky, S. & Nerenberg, C. (1984) Delivery systems of LHRH and its analogs, in *LHRH and its Analogs: Contraceptive and Therapeutic Applications* (Eds B. H. Vickery, J. J. Nestor Jr. & E. S. E. Hafez), pp. 421–435, MTP Press Limited, Boston, USA.

Baker, R. W. & Lonsdale, H. R. (1974) Controlled release: mechanisms and rates, in *Controlled release of Biologically Active Agents*, Volume 47 in Advances in Experimental Medicine and Biology (Eds A. C. Tanquary & R. E. Lacey), pp. 15–71, Plenum Press, New York.

Bohn, L. (1975) Compatible polymers, in *Polymer Handbook* 2nd Edition (Eds. J. Brandrup and E. H. Immergut), III 211, John Wiley and Sons, New York.

Cohen, J., Siegel, R. A. & Langer, R. (1984) Sintering technique for the preparation of polymer matrices for the controlled release of macromolecules. *J. Pharm. Sci.* **73** 1034–1037.

Davis, B. K. (1972) Control of diabetes with polyacrylamide implants containing insulin, *Experientia,* **28,** 348.

Davis, B. K. (1974) Diffusion in polymer gel implants, *Proc. Nat. Acad. Sci., USA* **71** 3120.

Donnelly, R. J. (1984) Continuous subcutaneous administration of 'Zoladex' (ICI 118,630 — an LHRH analogue) to patients with advanced prostatic cancer. *J. Steroid Biochem.* **20** A21, 1375.

Furr, B. J. A. & Hutchinson, F. G. (1985) Biodegradable sustained release formulation of the LHRH analogue 'Zoladex' for the treatment of hormone-responsive tumours in *EORTC Genitourinary Group Monograph 2, Part A: Therapeutic Principles in Metastatic Prostatic Cancer* (Eds F. H. Schroder & B. Richards), pp. 143–153, A. R. Liss, Inc.

Furr, B. J. A. & Milsted, R. A. V. (1988) LHRH analogues in cancer treatment. In *Endocrine Management of Cancer 2*. Contemporary Therapy. (Ed B. A. Stoll), pp. 16–29, S. Karger, Basle.

Furr, B. J. A. & Nicholson, R. I. (1982) Use of analogues of LHRH for treatment of cancer. *J. Reprod. Fertil.* **64** 529–539.

Hsieh, D. S. T., Chiang, C. C. & Desai, D. S. (1985) Controlled release of macromolecules from silicone elastomer. *Pharm. Tech.* **9** 39–49.

Hsu, T. T. & Langer, R. (1985) Polymer for the controlled release of macromolecules: effect of molecular weight of ethylene–vinyl acetate copolymer. *J. Biomed. Mater. Res.* **19** 445–460.

Huggins, C., Briziarelli, G. & Sutton, H. (1959) Induction of mammary carcinoma in the rat and influence of hormones on the tumours. *J. Exp. Med.* **104** 25.

Hutchinson, F. G. (1982) Continuous-release pharmaceutical compositions, European Patent 58481B.

Jordan, V. C. (1982) Laboratory models of hormone dependent cancer. *Clin. Oncol.* **1** 21–40.

Kruisbrink, J. & Boer, G. J. (1984) Controlled long-term release of small peptide hormones using a new microporous polypropylene polymer: its application to vasopressin in the Brattleboro rat and perinatal use. *J. Pharm. Sci.* **73** 1713–1718.

Langer, R. (1981) Polymers for the sustained release of macromolecules: their use in a single step method of immunisation. *Meth. Enzymol.* **73** 57–75.

Langer, R. (1984) Controlled drug release systems. *Biophys. J.* **45** A26.

Langer, R. & Folkman, J. (1976) Polymers for the sustained release of proteins and other macromolecules. *Nature* **263** 797.

Lotz, W. & Syllwasschy, B. (1979) Release of oligopeptides from silicone rubber implants in rats over periods exceeding 10 days. *J. Pharm. Pharmacol.* **13** 649–650.

Mitra, S., Van Dress, M., Anderson, J. M., Peterson, R. V., Gregonis, D. & Feijen, J. (1979) Pro-drug controlled release from polyglutamic acid. *Polym. Prep. A.C.S., Div. Polym. Chem.* **20** 32–35.

Okada, H., Yamazaki, I., Ogawa, Y., Hirai, S., Yashiki, T. & Mima, H. (1982) Vaginal absorption of a potent luteinizing hormone-releasing hormone analogue (Leuprolide) in rats I: absorption by various routes and absorption enhancements. *J. Pharm. Sci.* **71** 1367–1371.

Okada, H., Yamazaki, I., S., Yashiki, T. & Mima, H. (1983a) Vaginal absorption of a potent luteinizing hormone-releasing hormone analogue (Leuprolide) in rats II: mechanism of absorption enhancement with organic acids. *J. Pharm. Sci.* **72** 75–78.

Okada, H., Yashiki, T. & Mima, H. (1983b) Vaginal absorption of a potent luteinizing hormone-releasing hormone (Leuprolide) in rats III: effect of estrous cycle on vaginal absorption of hydrophilic model compounds. *J. Pharm. Sci.* **72** 173–176.

Okada, H., Yamazaki, I., Yashiki, T., Shimamoto, T. & Mima, H. (1984) Vaginal absorption of a potent luteinizing hormone-releasing hormone analogue (Leuprolide) in rats IV: evaluation of the vaginal absorption and gonadotrophin responses by radioimmunoassay. *J. Pharm. Sci.,* **73** 298–302.

Petri, W., Seidel, R. & Sandow, J. (1984) Pharmaceutical approach to long-term therapy with peptides. *Int. Cong. Ser.-Excerpta Medica* **656** 63–76.

Pitt, C. G. & Schindler, A. (1980) The design of controlled drug delivery systems based on biodegradable polymers, in *Progress in Contraceptive Delivery Systems* (Eds. E. S. E. Hafez & W. A. A. van Os), Vol. 1, pp. 17–46, MTP Press, Lancaster, UK.

Robinson, M. R. G., Denis, L., Mahler, C., Walker, K., Stitch, R. & Lunglmayr, G. (1985) An LHRH analogue ('Zoladex') in the management of carcinoma of the prostate: a preliminary report comparing daily subcutaneous injections with monthly depot injections. *Eur. J. Surg. Oncol.* **11** 159–165.

Sidman, K. R., Steber, W. D., Schwope, A. D. & Schnaper, G. R. (1983) Controlled release of macromolecules and pharmaceuticals from synthetic polypeptides based on glutamic acid. *Biopolymers* **22** 547–556.

Smolev, J. K., Heston, W. D. W., Scott, W. W. & Coffey, D. S. (1977) Characterisation of the Dunning R3327H prostatic adenocarcinoma: an appropriate model for prostatic cancer. *Cancer Treat. Rev.* **61** 273–297.

Uda, Y. and Yamada, M. (1984) Percutaneous pharmaceutical compositions for external use', European Patent Application 127426.

Walker, K. J., Nicholson, R. I., Turkes, A. O., Griffiths, K., Robinson, M., Crispin, Z. & Dris, S. (1983) Therapeutic potential of the LHRH agonist, ICI 118,630, in the treatment of advanced prostatic carcinoma. *Lancet* **ii** 413–414.

Williams, M. R., Walker, K. J., Turkes, A., Blamey, R. W. & Nicholson, R. I. (1986) The use of an LHRH agonist (ICI 118,630; 'Zoladex') in advanced premenopausal breast cancer. *Brit. J. Cancer* **53** 629–636.

Wise, D. L., Fellman, T. D., Sanderson, J. E. & Wentworth, R. L. (1979) Lactic/glycolic acid polymers in *Drug Carriers in Biology and Medicine*, (Ed. G. Gregoriadis), pp. 237–270, Academic Press, London.

Yoshida, M., Asano, M., Kaetsu, I., Nakai, K., Yamanaka, H., Shida, K. & Shiraishi, A. (1985) Slow release composite and process for producing the same. European Patent Application 130409.

Yishikawa, H., Satoh, Y., Naruse, N., Takada, K. & Muranishi, S. (1985) Comparison of disappearance from blood and lymphatic delivery of human fibroblast interferon in rat by different administration routes. *J. Pharmacobio-Dyn.* **8** 206.

17

Site-specific proteins

E. Tomlinson
Somatix Corporation, One Kendall Square, Cambridge, Massachusetts, USA

17.1 INTRODUCTION

Selective drug delivery and targeting seeks to improve on the risk/benefit ratio associated with all drugs, by effecting the optimal arrival of drug at its site of action, in a manner that is appropriate for the disease and the drug, and which leads to a significant reduction in the potential for side-effects. Approaches being adopted to achieve these effects include the control of drug dispersion by incorporating the drug into a (macro)molecular carrier (or by altering the drug's own (macro)molecular structure), and the control of the input of the drug into the body to ensure an appropriate *format* or *pattern* of dosing. Through the application of recent developments in molecular genetics there is now an unprecedented opportunity to explore the pathogenesis of disease as well as to design new types of drugs that are able to cure rather than simply treat symptoms. By the end of the millenium, clinicians will undoubtedly employ an array of drug administration systems, most of which will favour an improvement in dosing regimen, compliance and a reduction in side-effects. However, by then, dramatic new benefits may be accruing through the use of *site-specific delivery* systems. These opportunities, and new ways for achieving a selective dispersion of both conventional drugs and therapeutic proteins are highlighted in the following sections, and views are given on the probable therapeutic (delivery) systems of the future.

17.2 SELECTIVE DRUG DELIVERY AND TARGETING

Numerous examples of disease can be evidenced for which improved drug delivery to a poorly accessible region would lead to an improvement in drug use; as for example with many intracellular infections, diseases of the central nervous system, diseases of the immune systems, cancerous states, some cardiovascular-related diseases, haematopoietic diseases and arthritic disease. In addition, many of the intricate dosing regimens and the high doses of drug which are often applied are frequently chosen because of a poor perfusion of the site of action (e.g. with rheumatoid joints), coupled with an inappropriate pharmacodisposition of the drug. Since the latter may

may lead to untoward metabolism, all of these effects often combine to produce deleterious effects (Tomlinson 1987).

In addition, many enzyme and hormone drugs are able to be produced using recombinant DNA technology and other methods. Biotechnologists are able to make such complex structures as enzymes, hormones and physiological receptors, whose important features may be then identified. Proteins can act as drugs *per se*, or as templates in the production of agonist and/or antagonist drugs. Protein drugs often have poor chemical stability and/or are cleared rapidly from the general circulation. Also, many natural proteins are produced on demand in the body to act locally within a very short distance of their site of production. Many are unable either to withstand a trip throughout the blood stream (as would occur if they were to be administered as drugs), or to leave the blood compartment and to pass through the various membranes and barriers of the body to reach their sites of action. To make a copy of a biologically active natural material and then administer it is no guarantee of success. These new agents may act as part of a strictly timed cascade of events, and hence both the *timing and staging of delivery* are key for their correct use, as is the need (in most instances, at least) to be able to achieve a local action without the risk of interaction with non-target cell populations (Tomlinson, 1989, 1990).

17.3 SITE-SPECIFIC DELIVERY OF THERAPEUTIC PROTEINS

The physicochemistry, chemical lability and site of action of polypeptides and proteins intended for therapeutic use often indicates the use of a carrier or a change in protein structure. Some of the attempts at controlling the biological dispersion and reactivity of proteins are now described.

17.3.1 Deletion mutants

Site-directed mutagenesis can be used to create novel proteins which may or may not resemble endogenous materials. These approaches are being used not only to improve on the stability and intrinsic specificity of such endogenous proteins, but also (and increasingly), to achieve a selective and often prolonged delivery of the polypeptide/protein to its site of action. For example, recent work has pointed to the altered pharmacokinetic and thrombolytic properties of deletion mutations of human tissue-type plasminogen activator (tPA) in rabbits. Wild-type tPA is characterized by a rapid clearance by the liver, with an alpha distribution phase half-life of a few minutes. Using a series of deletion mutants (which included removal of fibronectin-like, epidermal growth factor type and glycosylation-site regions), researchers have demonstrated that regions within tPA responsible for its liver clearance, its fibrin affinity, and its fibrin specificity are not localized in the same structures. It thus appears possible to alter specific functions of tPA related to its poor pharmacokinetics without decreasing its efficacy. This approach to producing site-specific (bio)polymer drugs is likely to gain much attention in the future.

17.3.2 Hybrid protein delivery systems

Knowledge of the structure, position and function of many of the operational receptors in the body in controlling the extracellular and intracellular disposition of proteins is leading to the design of *hybrid* proteins which have the combined or re-ordered features of one or more proteins, in order to have both effector functions as well as protection and recognition properties. Intracellular recognition signal structures have been used for designing synthetic peptides able to mimic (secretory) events. Site-specific hybrid proteins may be produced by either synthetically linking protein fragments, or using ligated gene-fusion processes. Table 17.1 gives examples of hybrid protein delivery systems which have been produced and/or suggested for therapeutic use.

Table 17.1 — Hybrid protein delivery systems

Fragments of:	
Recognition portion	Effector portion
Gene fusion products	
Interleukin 2	Diphtheria toxin
Growth factor	Toxin (e.g. ricin A)
Cell-specific polypeptide	
(α/β MSH; substance P)	Restructured diphtheria toxin
Antitumour Fab immunoglobulin	Fragment A of diphtheria toxin
CD_4	*Pseudomonas* exotoxin
	γ-interferon and β-tumour necrosis factor
Chemical linkage of fragments/proteins	
Human placental lactogen hormone	Diphtheria toxin A chain
β Chain of human chorionic gonadotrophin hormone	Ricin A chain
Insulin	Diphtheria A
Epidermal growth factor	Ricin A
IgG(2a) fragments	Gelonin
	Diphtheria toxin
	Pseudomonas toxin
HIV-specific Ab	Ricin A
Antifibrin Ab	Tissue Plasminogen Activator
Anti-T cell antibody	Ricin A chain
Anti-endothelia IgG	Glucose oxidase
Anti-epithelia Ig/fragments	Ricin and other toxins
Anti-sigM-IgM	Saporin-6
Anti-transferrin receptor Ab	Ricin A chain

The gene-fusion strategy is exemplified by the proposal of targeting either bacterial or plant toxins to specific cells using hybrids created by combining toxin and growth factor genes. Such an approach relies on the deletion of the toxin gene sequence encoding the toxin's normal cell-binding site, allowing the protein to display the cell specificity of the growth factor. The toxin is then targeted to cells which interact with the growth factor (e.g. cancer cells). Included in this category of novel proteins are the recently described immunoadhesins (Capon *et al.* 1989). These are molecules containing the glycoprotein 120 (gp120)-binding domain of the receptor (CD_4) for human immunodeficiency virus combined with portions of antibodies. They have been shown to block HIV-1 infection of T-cells and monocytes by intercepting the virus. These novel hybrid delivery systems have a long plasma half-life, and retain the important properties of both parent molecules; namely, they bind gp120 and block infection of T-cells and monocytes by HIV-1. It is claimed that the attainment of a high steady-state level of immunoadhesin makes it likely that effective concentrations of the hybrid will be attained in lymph and lymphatic organs, where HIV may be most active (Capon *et al.* 1989).

The principle of targeting to a surface marker of a cell in a particular compartment has been used in the development of a treatment of graft versus host disease, which can occur after organ transplantation. The cytotoxic T-cells of the immune system that come into play at times of rejection do so with their surface receptor for interleukin-2 (IL-2) exposed. Using biotechnology, that portion of IL-2 which interacts with its receptor on the T-cell can be linked to a toxic material (e.g. a portion of diphtheria toxin) to create a site-specific hybrid fusion protein, which can enter and then kill T-cells just as they prepare to reject the graft (Fig. 17.1). Similar approaches are being considered in the treatment of HIV infection by routing drugs into affected cells with carriers that use the same pathway as the virus (Chaudhary *et al.* 1988).

Frequent attention has been given to using monoclonal antibodies as targeted drug carriers, because of the remarkable selectivity that these agents for intra- and extracellular structures. In particular, many workers have considered the use of monoclonals for the targeting of toxic materials to human tumours. Clinical trials of toxic materials (e.g. radioisotopes) linked to antibodies date back to the 1950s. Such systems are used in tumour imaging, but their therapeutic potential has not been fully realized. The major reasons for this include antigen specificity and tumour cell heterogeneity, as well as an inability of the macromolecular system to diffuse or to percolate into the tumour mass. Apart from oncofoetal antigens, no complete tumour-specific antigens have been identified in human neoplasms — the antigens are also found in normal tissues, albeit in trace amounts. In addition, there may not be sufficient of the antigen on the target cell to allow delivery of the necessary quantity of therapeutic agent. A further restriction is that tumour cells often show heterogeneity in expression of the 'tumour antigen', which could result in only a proportion of cells being targeted and the deleterious development of other tumour cell populations. A recent approach has been to link the antibody to an enzyme, rather than a drug, so that the enzyme is localized at the tumour. A prodrug which is cleaved specifically by this enzyme is then administered, with putative selective release of active drug at the tumour cell surface (Bagshawe *et al.* 1988) (Fig. 17.2). Whether this can lead to an enhanced concentration of drug within the tumour mass in humans is still to be resolved.

Concern about the immunogenicity of such systems (which are generally murine in origin) has led to the development of 'humanized' antibodies which are fashioned

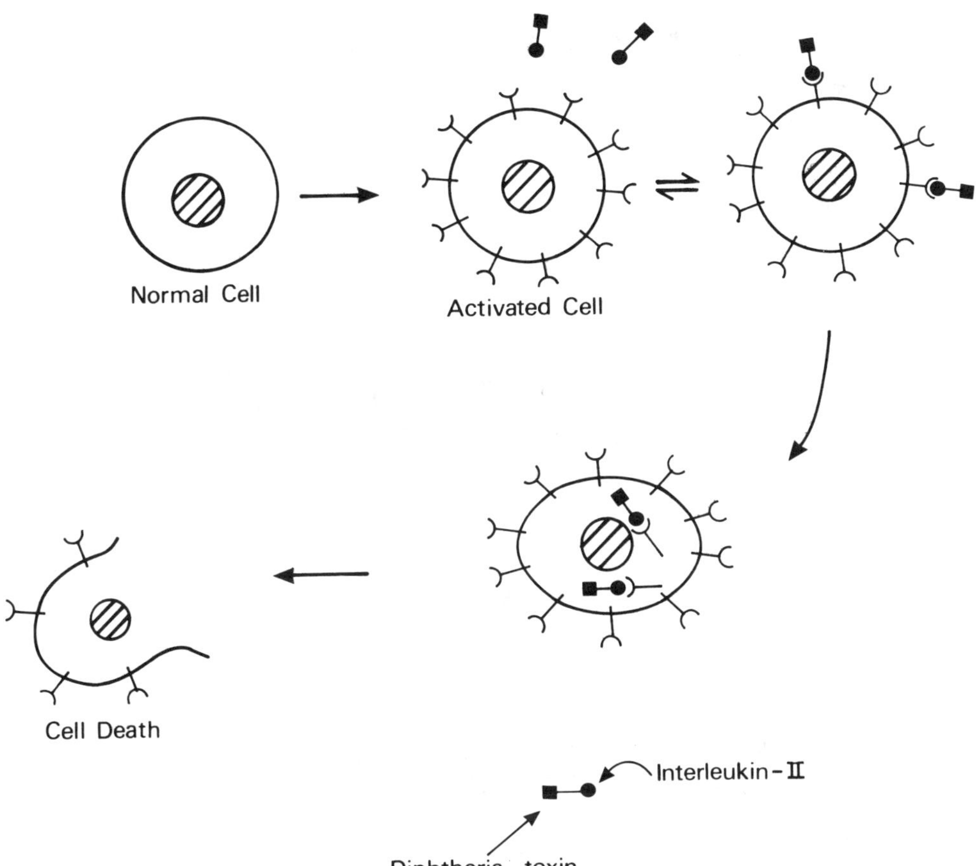

Fig. 17.1 — Scheme showing the interaction of an interleukin-2/toxin hybrid protein delivery system with activated T-cells, as proposed for treatment of graft-versus-host rejection disease.

by recombinant DNA technology to be human-like rather than murine-like (Fig. 17.3) (Jones *et al.* 1986, Larrick 1989). However, there are still serious doubts about the repeated dosing of patients with even this type of antibody, although for some life-threatening circumstances this approach may become possible to a limited extent. Recently, clinical trials have taken place with 'humanized' antibody conjugates for use in various haematological and splenic cancers; extremely low (perhaps even femtomole) dose levels have been proposed.

In an attempt to decrease the molecular size of the immunoglobulin, $F(ab)_2$ and Fab fragments (Fig. 17.3) have been investigated. For diagnostic imaging of highly vascularized regions, more rapid clearance of Fab molecules permits more rapid visualization. Fab molecules also penetrate tumours more effectively, thereby providing a more even distribution (Skerra and Pluckthun 1988). Further developments along these lines can be expected, for instance, it is possible to engineer single chain Fv molecules (Fig. 17.3) (Bird 1988) and even isolated domains, such as V_H, may find application (Ward *et al.* 1989).

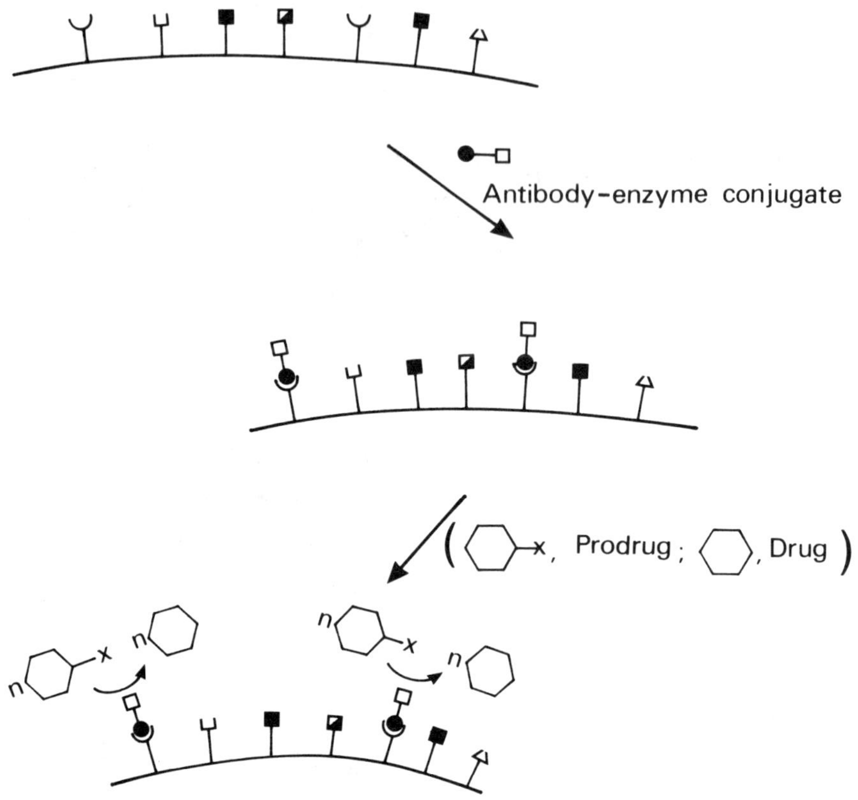

Fig. 17.2 — Scheme displaying the enzyme–antibody/prodrug concept. A tumour cell is considered with five different proteins located on the outer surface of the plasma membrane. The tumour-specific antigen binds the anitbody–enzyme conjugate, which when exposed to the prodrug generates a high local concentration of chemotherapeutic agent.

17.3.3 Ligated gene fusion hybrid delivery systems

Gene-fusion techniques may be used to produce distinct therapeutic proteins which combine the varied properties of parent proteins. This is exemplified by recent stategies which have been proposed for the targeting of bacterial and/or plant toxins to specific cells using hybrids created by ligating toxin and growth factor genes. Such an approach relies on the deletion of the toxin gene sequence encoding the cell-binding site, which allows the hybrid-fusion protein to display the cell specificity of the growth factor. Recently, a hybrid protein between interferon-γ and tumour necrosis factor-β (TNF-β) has been shown to have a greatly increased antiproliferative activity *in vitro*, compared with either interferon-γ or TNF-β alone, whilst still retaining their antiviral activity and cytotoxic effects. Hybrid protein delivery systems may involve not just adduction of a protein (fragment) to a recognition moiety, but also to a re-ordering of the structures of the effector portions of therapeutic proteins in order to enhance their pharmacological action.

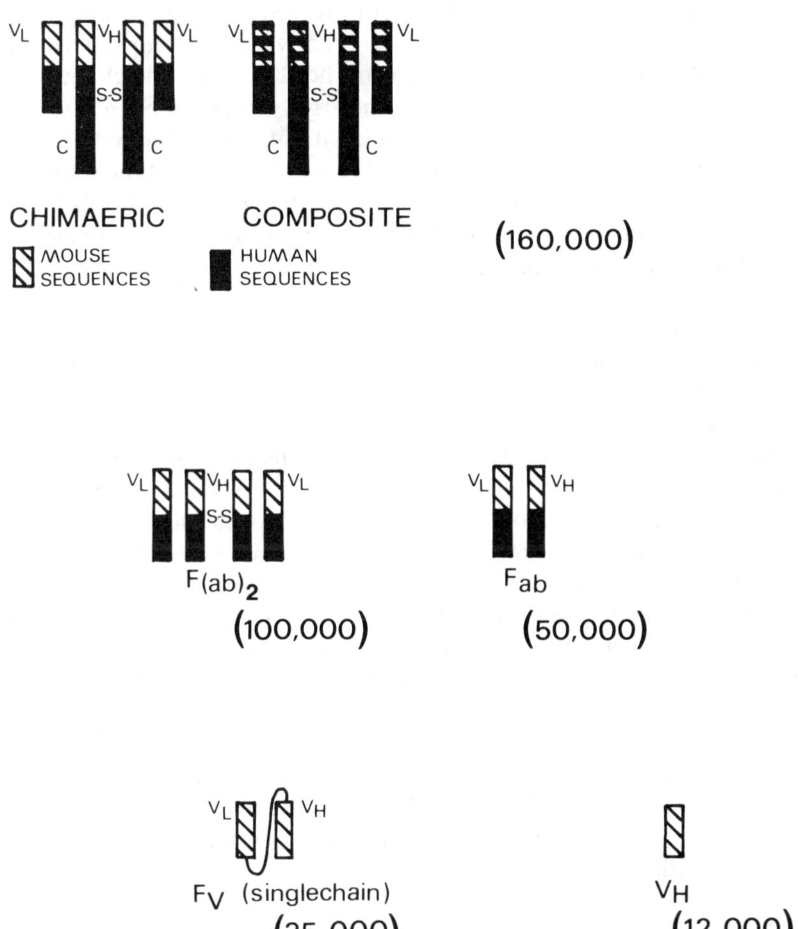

Fig. 17.3 — Schematic diagram showing construction of chimeric and composite human monoclonal antibodies. Approximate molecular weights given in brackets. Chimeric monoclonal antibodies are constructed by grafting the rodent heavy- and light-chain variable regions onto human constant regions. Composite monoclonal antibodies are constructed by grafting the rodent hypervariable regions onto the human framework regions.

Enzymatic cleavage of antibodies yields $F(ab)_2$ units, which on mild reduction yield Fab molecules. The two peptide chains of the Fab molecule are held together by hydrophobic interactions. The Fv fragment is the smallest structure which contains the entire antigen binding site. As isolated, Fv molecules are unstable but they can be engineered as single-chain polypeptides which are more stable. The variable domain of the heavy chain V_H can be used as an antigen binding moiety.

17.3.4 Synthetically linked hybrid conjugates

Large changes in the biological disposition of proteins have been reported upon their chemical linkage to other protein (fragments). For example, the toxin gelonin, which has a circulation half-life in mice estimated at 3.5 min, when conjugated to immunoglobulin (fragments) has a terminal phase blood half-life in the order of days, with only a slight variation in this time as the conjugated immunoglobulin (fragment) is changed. The immunotoxin field provides many relationships between protein structure and deposition. Point mutations in the B polypeptide chain of diphtheria toxin that block non-specific binding to non-target cells have been produced. Upon linking this entity to an anti-T-cell monoclonal antibody, it may be demonstrated that, because of a change in the non-target tissue distribution of the toxin, it becomes orders of magnitude less toxic than the native toxin to non-target cells (*in vitro*). The availability of monoclonal antibodies has led many workers to consider these for hybrid protein-mediated targeting of toxic materials to human tumours.

17.3.5 Protein (re)glycosylation

As described above, the size, surface character and chemical reactivity of proteins largely control their dispersion in the body. In addition, it has been demonstrated that many endogenous glycoproteins (i.e. serum glycoproteins, lysosomal enzymes and perhaps also sulphated pituitary glycoproteins such as chorionic gonadotrophin), interact through their specific carbohydrate residues complexing with (oligosaccharide-specific) recognition systems on the plasma surfaces of target cells. Glycosylation patterns are thus signals used by the body to regulate the dispersion of its own glycoproteins, as implicated for both enzyme and hormone disposition as well as immune surveillance and coagulation. Three biological properties of glycoproteins may be adjusted upon altering their surface distribution of carbohydrates, namely, circulation blood half-life (and potentially duration of action), immunogenicity, and ability to access (cellular) sites of action. Expressed glycoproteins may be engineered in order to affect their selectivity for target cells. Modifications of proteins are usually carried out via enzymatic synthesis. Post-expression protein remodelling has been proposed for resurfacing some proteins, inclucing factor VIII, erythropoietin, colony-stimulating factors, β- and γ-interferons, interleukins I–III, and even antibodies.

17.3.6 Hydrophilic polymer protectants

Polymer chemistry approaches may be adopted for altering the pharmacodisposition of polypeptides and protein drugs. These serve to increase their apparent size and/or reduce their (untoward) interactions with blood and tissue components. Immunosurveillance is mediated through physicochemical interaction between a (therapeutic) protein and components of the immune system. Frequently, *opsonization* by fibrinogen, fibronectin and other blood components serves as a prelude to the recognition and removal of the formed complex by cells; antigen–antibody interaction and Fc-mediated removal also occurs. Opsonized materials are taken into cells by engulfment after adherence to, and vesiculation of, phagocytosing cell membranes. Both opsonization and adherence can be diminished if the attractive forces between the interacting therapeutic protein, blood macromolecule and, for example, a cell-surface macromolecule are diminished. For colloids, a high potential energy barrier can be formed by creating a *sterically stabilized* surface upon introducing a hydrated (i.e. hydrophilic) polymer at the surface of the colloid. Hence surface modifications to proteins can be made to improve their *tolerance* within the vasculature — owing largely to the formation of a surface which makes it energetically unfavourable for other macromolecules to approach. Table 17.2 gives examples

Table 17.2 — Hydrophilic polymer protectants for therapeutic proteins

Protein	Proposed use of conjugate
Polyethylene glycols	
Islet-activating protein	Insulinogenic activity
Superoxide dismutase	Kidney transplantation, burns, reperfusion damage
L-Asparaginase	Malignant haematological disorders
Adenosine deaminase	Adenosine deaminase deficiency
Urokinase	Coagulant, fibrinolytic
Proteins	Radioprotection
Interleukin-2	Cancer
Uricase	Altered antigenicity
tPA	Thrombolysis
Bilirubin oxidase	Bilirubinaemia
Octylphenoxy polyethoxy ethanols	
Interleukin-2	
Polyoxyethylene sorbitans	
Interleukin-2	
Dextran	
Urokinase	Fibrinolytic
Purine nucleosides	Inhibitors of adenosine deaminase
α-1,4-Glucosidase	
*neo*glycoproteins	Enzyme replacement therapy
Carboxypeptidase G_2	Enzyme replacement therapy
β-Galactosidase	Enzyme replacement therapy
L-Asparaginase	Cancer (lower antigen reactivity and increased circulatory persistence)
Albumin	
Asparaginase	Cancer
Poly-D-alanyl peptides	
Asparaginase	Cancer
N-(2-*hydroxypropyl*)*methacrylamide* (*HPMA*)	
Antibodies	Cancer-seeking agents

of hydrophilic (bio)polymeric *protectants* that have been described for conjugation to therapeutic proteins. Synthetic and biological materials have been used or suggested, and include polyethylene glycols, poloxamers, poloxamines, albumin, immunoglobulin G, carboxymethycellulose, natural xanthans and sorbitans. Conjugations of proteins with hydrophilic polymers have often been reported as being successful in altering their *potency* as well as for reducing their imunogenicity and increasing their duration of action. This technology has been used for stabilizing therapeutic proteins by forming protein conjugates with hydrophilic polyethylene glycol chains, as well as for extending the blood half-lives of a number of peptidergic mediators such as interleukin-2 (IL-2), and enzymes, including catalase, asparaginase and urokinase, whilst still maintaining their reactive functionalities. (It is useful to recall here that simply to prolong blood levels of polypeptides/proteins is not necessarily desirable, and this may be the case with IL-2, since a non-physiological pattern of exposure to non-target cells may occur.) Significant changes to biological dispersion can be made, as demonstrated by the use of pegoylated asparaginase in rabbits; the pegoglated protein increases the circulating half-life over the native enzyme, from about 20 hours to between 125 and 160 hours; with a greatly reduced clearance from the blood dropping from about 100 ml kg^{-1} day^{-1} to about 3 ml kg^{-1} day^{-1}, and with an almost 40-fold increase in the plasma area under the curve. The pegoylated-asparaginase conjugate is well-tolerated in patients that have already been treated with native asparaginase and have neutralizing anti-asparaginase antibodies.

A decrease in immunogenicity of therapeutic proteins may result from either a reduction in their aggregation, or simply by a masking of any antigenic determinants. Conjugation of proteins with hydrophilic polymers can also increase their chemical and physicochemical stability, and is known to enhance resistance to proteolysis and heat denaturation. It has been found that oxidized dextrans conjugated to asparaginase give the enzyme increased circulatory half-times and a lowered antigen reactivity. The dextrans bound to asparaginase protect the enzyme from inactivation by proteases and enzyme-specific antibodies, as well as reducing the antigen reactivity of the enzyme *in vivo*. This increased stability may be due to the modification of the lysine residues present in the antigenic sites of the enzyme or at trypsin binding sites, or to the soluble dextran sterically hindering the interaction between the enzyme and these destructive elements. *In vivo*, the antigen reactivity of the conjugate generally falls with an increase in the size of the dextran. A marked variability exists in the relative abilities of the various protectants to affect the action, and/or the immunogenicity and antigenicity of therapeutic proteins. Avoidance of uptake by macrophages and the reticular endothelial system should enable therapeutic proteins to remain for considerably longer periods within the circulation. This can be considered as useful for increasing the statistical probability for a competing process to occur (e.g. extravasation or interaction with an intravascular target cell), and/or for the development of a long-term circulating depot of active protein.

17.4 GENE DELIVERY

The ability to transfect genes into cells and to cause their expression is leading to new forms of human gene therapy, in which functionally active genes are inserted into the (somatic) cells of a person requiring the expression of a given protein.

In order to ensure that a translated gene gives a constant (and lifelong) supply of its product, then it must generally be delivered into cells of the body that proliferate during the entire life of the recipient. Numerous target cells have been considered and studied, including pluripotent stem cells in bone marrow, endothelia and the hepatocytes of the liver. Possible candidates are given in Table 17.3.

Table 17.4 — Biochemical lesions and disease targets for gene therapy

Abnormal gene	Disease/tissue
β-Globin	Sickle cell/blood β-Thalassemia/blood
α-Globin	α-Thalassemia/blood
Pyruvate kinase	Pyruvate kinase deficiency/blood
Glucocerebrosidase	Gaucher's disease/blood; bone
Adenosine deaminase	Immunodeficiency/blood; liver; spleen
Hypoxanthine-guanine phosphoribosyl transferase	Lesch–Nyhan Syndrome/blood; nervous system
Purine nucleoside phosphorylase	Immunodeficiency/blood
Factor VIII	Haemophilia/blood
Factor XI	Christmas disease/blood
Phenylalanine hydroxylase	Phenylketonuria/liver; nervous system
LDL receptor	Familial hypercholesterolaemia/liver

17.4.1 Gene correction and gene augmentation

Two approaches are being actively considered, either gene correction, involving specific correction of a mutant gene sequence, or gene augmentation, where a correctly functioning gene is inserted into non-specific sites in a genome, whilst leaving the malfunctioning gene alone or uncorrected. The former approach, which is essentially *in situ* repair of the genome, may be carried out via a process called homologous recombination, in which foreign DNA, introduced into target cells *ex vivo*, combines with specific sites in the genome, owing to the presence of specific enzyme and other intracellular pieces of enabling machinery (Friedmann 1989).

17.4.2 Site-specific gene delivery

For gene augmentation, genes can be grossly delivered into cells either by chemical methods or physical methods such as direct microinjection into the nucleus. The use of viral vectors bears most current promise for delivering genes to alter mutant target genes (Wilson *et al.* 1989). The retroviral gene delivery system consists of an RNA copy of the replacement gene, packaged into a viral component of a pathogenic virus. This system has many advantages, but primarily the potential for infecting most if not all target cells. Because of their size and content, murine and avian retroviruses have now become the vehicles of choice to deliver foreign genes into mammalian cells. It is important to appreciate that (disabled) retroviral systems are capable of infecting a broad class of cells, and that target cell replication and DNA synthesis must be occurring in those cells.

Other viral vectors are receiving increasingly significant attention, including those derived from several classes of non-integrating viruses such as herpes virus. Non-cytopathic herpes-based vectors could enable large amounts of foreign sequences and even whole genes to be transferred (Friedmann 1989).

Recently a novel delivery method has been proposed using skin fibroblasts which have been infected with a retroviral vector delivery system containing (human Factor IX) cDNA (St Louis and Verma, 1988). After the gene was delivered *ex vivo*, these cells were implanted under the epidermis in experimental mice and were then found able to synthesize (and to secrete into the body) the gene product. Although retroviral vectors will target to the genome of the cells, other approaches have been developed using fusion lipid vesicles for delivery of mRNA into the cytoplasm of cells. This approach of using genetically altered cells to deliver systemic proteins has been mooted for a number of proteins including tPA and factor VIII. Apart from skin fibroblast cells, endothelial cells, stem cells and hepatocytes are currently recieving attention as target cells for gene delivery and subsequent expression.

During 1987 it was proposed that genetically altered lymphocytes be placed into humans (Anderson *et al.* 1987). After receiving recent approval from the USA regulatory bodies, this first gene addition experiment has resulted in a patient with advanced melanoma receiving an infusion of his own white blood cells containing the foreign gene. This is a landmark in the development of gene therapy approaches, since, although not a therapy in itself (it is being proposed to track the cytotoxic progression of tumour infiltration lymphocytes (TIL) in the body), it is the first time that genetically altered cells have been returned to the body. It is believed that this study will pave the way for further studies of adoptive immunotherapy in cancer, with the ultimate hope that delivery techniques will be developed to deliver genes which can express products that will cure such debilitating diseases as cystic fibrosis, muscular dystrophy, and sickle cell anaemia. Current views are that, although great technical problems remain, particularly with amplifying the expression of the gene and delivering it selectively into cells, it is probably achievable by the end of this century.

Ex vivo approaches coupled with cell implantation, are attractive technically, but others have suggested that direct *in vivo*. targeting of genetic material could become a reality. Given the difficulty in effecting simple control of the dispersion of low molecular weight drugs (as described in our previous articles), the present author views this possibility as being remote indeed.

These are early days for modern approaches to gene therapy, and clearly both *in vitro* and *in vivo* site-specific delivery are key components for its successful clinical introduction. Many years, if not decades, of further technical and clinical development must pass before these new approaches come into general clinical practice. The issue of gene therapy is a complex one, evoking memories of the debates on ethical, legal and moral issues which took place two decades ago when the first heart transplantation took place.

17.5 SUMMARY

Increasing attention is being given to controlling the dispersion of proteins intended for therapeutic use. This contribution has examined largely biotechnological and chemical means for achieving this. Approaches therefore include site-directed mutagenesis, proteolysis, ligated gene fusion, protein aggregation and/or conjugation with (other) biologically active effector functions.

REFERENCES

Anderson, W. F., Blaese, R. M., Nienhuis, A. W. & O'Reilly, R. J. (1987) Human gene therapy: preclinical data document. (Document submitted to the Human Gene Therapy Subcommittee, of the US NIH Recombinant DNA Advisory Committee, 24.4.87, and available from the Office of Recombinant DNA Activities of NIH.)

Bagshawe, K. D., Springer, C. J. & Searle, F. (1988) A cytotoxic agent can be generated selectively at cancer sites, *Br. J. Cancer* **58** 700–703.

Bird, R. E. (1988) Single-chain antigen-binding proteins. *Science* **242** 423–426.

Capon, D. J., Chamow, S. M., Mordenti, J., Marsters, S. A., Gregory, T., Mitsuya, H., Byrn, R. A., Lucas, C., Wurm, F. M., Groupman, J. E., Broder, S. & Smith, D. H. (1989) Designing CD4 immunoadhesins for AIDS therapy. *Nature* **337** 525–531.

Chaudhray, V. K., Mizukami, T., Fuerst, T. R., Fitzgerald, D. J., Moss, B., Pastau, R. & Berger, E. A. (1988) Selective killing of HIV-infected cells by recombinant human CD4 *Pseudomonas* exotoxin hybrid protein. *Nature* **335** 369–372.

Friedmann, T. (1989) Progress trowards human gene therapy, *Science* **244** 1275–1281.

Jones, P. T., Dear, P. H., Foote, J., Neuberger, M. S. & Winter, G. (1986) Replacing the complementarity-determining regions in a human antibody with those from a mouse. *Nature* **321** 522–525.

Larrick, J. W. (1989) Potential of monoclonal antibodies as pharmacological agents. *Pharmacol. Rev.* **41** 539–557.

St Louis, D. & Verma, I. M. (1988) An alternative approach to somatic cell gene therapy. *Proc. Natl. Acad. Sci. USA* **85** 3510–3514.

Skerra, A. & Pluckthun, A. (1988) Assembly of a functional immunoglobulin Fv fragment in *Escherichia coli*. *Science* **240** 1038–1041.

Tomlinson, E. (1987) Theory and practice of site-specific drug delivery. *Advanced Drug Delivery Reviews*, **1** 87–198.

Tomlinson, E. (1989) Considerations in the physiological delivery of therapeutic proteins. In: Prescott, L. F. & Nimmo, W. S. (eds) *Novel Drug Delivery and its Therapeutic Applications*. John Wiley and Sons Limited, Chichester, pp. 245–262.

Tomlinson, E. (1990) Selective delivery and targeting of therapeutic proteins. In: Harris, T. J. R. & Hentschel, C. C. G. (eds) *Protein Production, by Biotechnology*. Elsevier Applied Science, Amsterdam, pp. 207–225.

Ward, E. S., Güssow, D., Griffiths, A. D., Jones, P. T. & Winter, G. (1989) Binding activities of a repertoire of single immunoglobulin variable domains secreted from *Escherichia coli*. *Nature* **341** 544–546.

Wilson, J. M., Birinyi, L. K., Salaomon, R. N., Libby, R., Callow, A. D. & Mulligan, R. C. (1989). Implantation of vascular grafts lined with genetically modified cells. *Science* **244** 1344–1346.

Appendix 1
Reduction of diketopiperazine formation in Fmoc-peptide synthesis

L. E. Cammish
Applications Specialist, Waters Chromatography Division, The Boulevard, Blackmoor Lane, Watford, WD1 8YW.

A1.1 Introduction

Sequential addition of amino acid residues during solid phase peptide synthesis where proline is the carboxy-terminal residue attached to the solid support can result in a special case of deletion in which there is simultaneous loss of two residues, namely the amino acid attached to the solid support via the benzyl ester linkage together with the subsequent residue (Pedrose *et al.* 1986).

After addition of the second amino acid onto the carboxy-terminal proline residue and subsequent removal of the *N*-α-amino Fmoc-protecting group, a cyclization reaction can occur. The free amino group at the dipeptide stage can attack the ester bond to the peptide support creating a stable six-membered diketopiperazine (DKP) ring which is eliminated from the polymer therefore resulting in a dramatic decrease in the overall yield of the desired peptide (Fig. A1.1). Ring formation of this description is particularly pronounced where proline is the carboxy-terminal amino acid since the cyclic proline side-chain lies in the plane of the DKP molecule and does not hinder ring closure. MilliGen/Biosearch have devised a novel method to solve this problem which rests not only in the ability to selectively programme flowrates on the MilliGen/Biosearch 9050 PepSynthesiser (Fig. A1.2) but also on the continuous flow nature of the instrument. By increasing the flowrates during the Fmoc-deprotection and the subsequent DMF wash, the lifetime of the free amino group at the dipeptide stage can be reduced. This, in effect, greatly decreases the degree of DKP formation, therefore significantly increasing the overall yield of the peptide. Practical examples to illustrate the success of this methodology as well as a description of its use are given below.

A1.2 DIKETOPIPERAZINE FORMATION

Initial studies were carried out on the MilliGen/Biosearch 9050 PepSynthesiser by alteration of the synthesis conditions for assembly of the Gly–Gly–Pro tripeptide.

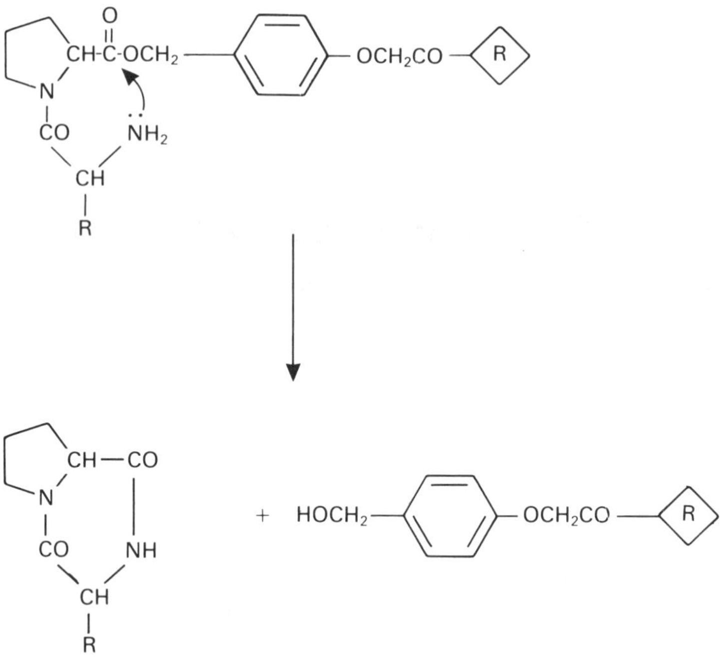

Fig. A1.1 — DKP formation with a carboxy-terminal proline residue.

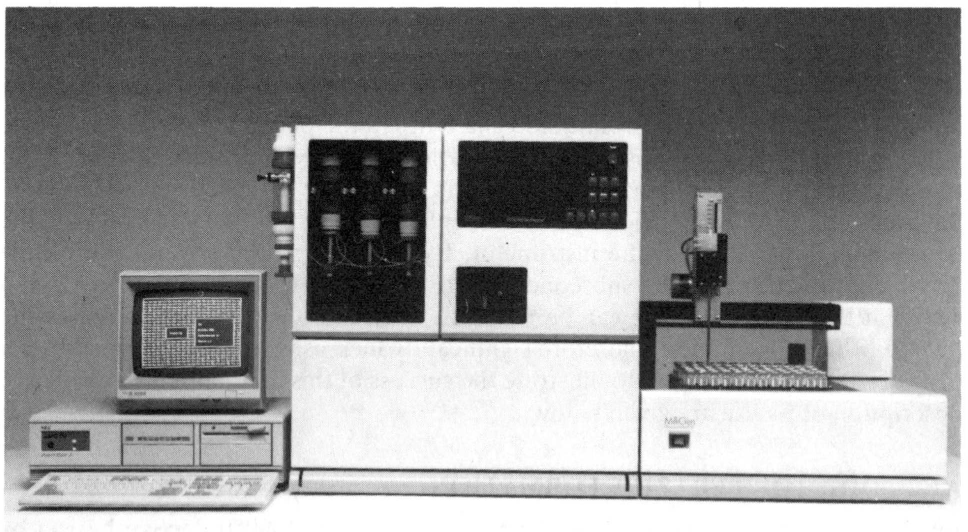

Fig. A1.2.

This sequence was chosen since a carboxy-terminal proline residue combined with unhindered glycine residues would present a case of maximum DKP formation for test purposes.

The synthesis of the tripeptide was performed three times using different synthesis conditions for control and optimization purposes. In each synthesis, the starting resin was 2.0g of Fmoc-*L*-Pro-PepSyn-KA support with a 0.09 mmol/g loading. After each synthesis, quantitative amino acid analysis was performed using a Waters Pico Tag amino acid analysis system to determine the final loading of proline on the PepSyn KA support and therefore define the degree of DKP formation.

A1.3 NEW METHODOLOGY

1. Method 1

The standard chemistry for the MilliGen/Biosearch 9050 PepSynthesiser was employed using Fmoc-amino acid pentafluorophenyl esters on a 0.2-mmol-scale synthesis. Standard coupling cycles were used incorporating 30-min acylations at a re-circulating flowrate of 5 ml/min. Fmoc deprotection using 20% piperidine/DMF was carried out for 7 min at a flowrate of 5 ml/min and the subsequent DMF wash to remove all traces of piperidine prior to the next coupling step was performed for 12 min at 5 ml/min (Table A1.1).

Table A1.1 — The standard continuous flow protocol and the protocol for minimizing DKP formation

Function	Flowrate (ml/min)	Duration (minutes)
Standard protocol		
DMF wash	5	0.5
Dissolve AA	—	—
Base wash	5	7.0
DMF wash	5	12.0
Inject AA	5	—
Modified protocol		
DMF wash	5	0.5
Dissolve AA	—	—
DMF wash	5	3.0
Base wash	15	1.5
DMF wash	15	3.0
Inject AA	5	—

2. Method 2

Pentafluorophenyl esters were again used for this synthesis. Deprotection of the carboxy-terminal proline, together with subsequent washes, and acylation of the glycine on to the proline residue were performed using the standard conditions. Subsequent steps were reprogrammed such that deprotection of the second residue on the resin was shortened to 90 s and the flowrate increased to 15 ml/min with the following DMF wash shortened to 3 min at 15 ml/min. The programme was also modified to ensure that the next amino acid derivative was introduced onto the Fmoc-deprotected resin in the column directly after completion of the DMF wash therefore keeping the lifetime of the free amine at the dipeptide stage to a minimum and reducing the possibility for DKP formation (see Table A1.1).

The acylation step was carried out for the standard 30-min duration with subsequent steps also performed using standard protocols.

3. Method 3

The use of Fmoc-amino acids with BOP/HOBt as coupling reagents has been reported to result in a very rapid and highly efficient method of activation in peptide synthesis (Hudson *et al.* 1988). The tripeptide was synthesized for a third time by applying this chemistry using equimolar quantities of BOP, HOBt and Fmoc-Gly-OH dissolved in 2 equivalents of *N*-methylmorpholine in DMF. The synthesis conditions were, otherwise, exactly as in Method 2.

A1.4 RESULTS

Quantitative amino acid analysis of the tripeptide product from Method 1 revealed that only 0.04 mmol/g proline was present (Table A1.2). Comparing this to the

Table A1.2 — Assessment of DKP formation from syntheses of Gly–Gly–Pro on the MilliGen/Biosearch 9050 Pepsynthesiser

	Method 1 (PFP esters standard	Method 2 (PFP esters modified)	Method 3 (BOP/HOBt modified)
Loading of starting Pro-resin (mmoles/g)	0.090	0.090	0.090
Final loading obtained after synthesis of Gly–Gly–Pro†	0.040	0.078	0.083
Estimated percentage of DKP formation	56%	13%	8%
Fmoc deprotection peak height for Gly-1 (mm)	62	130	139

†Substitution values obtained from amino analysis data taking an average of the proline and glycine figures.

original proline loading of 0.09 mmol/g confirms that the overall yield of the product had been reduced by over one half owing to cyclization at the dipeptide stage to form DKP, which was eliminated from the resin. In Method 2 amino acid analysis gave a final proline loading of 0.078 mmol/g, indicating that the alteration in flowrate and duration of the deprotection and wash steps had resulted in a substantial reduction in DKP formation and therefore a significant improvement in overall yield of the desired product. Method 3 gave results which were even more favourable. The final proline loading in this case was 0.083 mmol/g, indicative of a further reduction in DKP formation and improvement in overall yield.

During all three syntheses, continuous UV monitoring of the Fmoc-deprotection peaks showed peak heights, and peak areas, which were in general agreement with the amino acid analysis results.

A1.5 CONCLUSIONS

Flowrate programming in Fmoc continuous-flow peptide synthesis, involving increasing flowrates and decreasing the duration of the deprotection and wash steps, is clearly a very effective way of minimizing DKP formation and increasing peptide yield in the synthesis of peptides where there is a carboxy-terminal proline.

The programmable features of the MilliGen/Biosearch 9050 PepSynthesiser are essential to the implementation of this strategy.

REFERENCES

Hudson, D. (1988) Methodological implications of simultaneous solid phase peptide synthesis. 1. Comparison of different coupling procedures. *J. Org. Chem.*, **53** 617–624.

Pedroso, E., Grandas, A., de las Heras, X., Eritja, R. & Giralt, E. (1986) Diketopiperazine formation in solid phase peptide synthesis using *p*-alkoxybenzyl ester resins and Fmoc amino acids. *Tetrahedron Letters*, **27** 743–746.

Appendix 2
Capillary electrophoresis

Tom Large
Technical Support Manager, Applied Biosystems Ltd. Birchwood Science Park, Warrington, Cheshire WA3 7PB

A2.1 INTRODUCTION

High performance capilliary (free solution or capillary zonal) electrophoresis represents the first of a novel range of electrophoretic techniques in which molecular separations are achieved using applied voltages in narrow capillaries. The other related techniques now available include gel-filled capillary electrophoresis, micellar electrokinetic chromatography, and capillary electrokinetic chromatography.

Within the scope provided by these various techniques it is possible to analyse both large and small molecules, ranging from pharmaceutical preparations to recombinant proteins. Moreover, all of the techniques employ equipment that is mechanically simpler than that used in 'traditional' electrophoretic work, and most can be performed without the use of organic solvents (which provides worthwhile savings on waste disposal and running costs).

A2.2 OVERVIEW

Free solution capillary electrophoresis

Free solution capillary electrophoresis, also called capillary zonal electrophoresis (CZE) involves the separation of solutes on the basis of their charge-to-mass ratio. The separations are carried out in a fused silica capillary which is strengthened by a polyamide coating. The dimensions of the capillary can be varied, but are usually in the range 25–100 μm (internal diameter, i.d.) by 50–100 cm (length). Since the detector cell is provided by the capillary itself, the path length and sensitivity of the system are directly proportional to the capillary internal diameter. The choice of internal diameter is therefore a balance of sensitivity and current produced under the applied voltage (which is proportional to the ionic strength of the buffer which is employed). If too wide a capillary is used to overcome sensitivity problems, then the separation suffers through the current-related heat production (which increases as a function of the square of the capillary radius). This can limit the ionic strength of the buffer that can be used, which in turn reduces the efficiency of the separation. (Excessive heat production will also demand the use of cooling systems which would otherwise be unnecessary.)

The polarity of the instrument is selected so as to achieve migration towards the appropriate electrode (commonly the cathode) situated at the detection end of the capillary. Under the applied voltage, a process termed electroendosmosis (Eo) then takes place, which creates a flow of buffer towards the cathode. At high pHs, the rate of this buffer flow can exceed the rate of analyte migration. In this situation, a negatively charged peptide which, in the absence of Eo, would be expected to travel towards the anode, will actually move towards the cathode. (Under these circumstances the system polarity should be kept negative at the detection end of the capillary.) If two negatively charged species are present, then they will still be separated by virtue of the fact that the species with the lower charge-to-mass ratio will be swept along by Eo at a greater speed.

The separations in CZE are achieved, due to the differing velocities of solutes arising from variances in their charge-to-mass ratio. This varies according to the pH of the buffer in the system and is akin to the selectivity changes in high-performance liquid chromatography (HPLC). Changing the applied voltage within the ranges available on most CZE instruments is similar to changing the flowrate down an HPLC column; it is primarily the time scale of the separation that is effected, although some changes in efficiency are also observed.

The benefit provided by CZE is that it gives the same high resolving power as obtained with conventional electrophoresis, but without the limitations of gels which restrict the voltages applied owing to heating effects. The higher applied voltages allow for greater field strengths and give concomitant increases in resolution and peak efficiency. In practical terms, this permits separations with 200 000–500 000 plates per metre, which is some five times greater than allowed by HPLC.

The process of Eo originates from the deprotonation of the silanol groups in the walls of the fused silica capillaries. This generates O^- groups on the inner surface of the capillary, with the result that cations from the buffer associate at the wall, forming a double layer down the length of the capillary (Fig. A2.1). When a voltage is then applied, these cations migrate towards the cathode and, since they are located at the wall of the capillary, their migration pulls the buffer as a plug of liquid towards the cathode. In contrast with the conventional pumped systems (as used in liquid chromatography (LC) for example) the buffer flow in CZE is a 'true' plug flow, because the driving force is provided at the capillary wall, where the resistance to flow due to friction is at its greatest. (The frictional losses between the outer charged layer and the buffer core are minimal.)

The Eo flow rate is proportional to the pH and field strength (voltage and capillary length) and is at its greatest at alkaline pHs where the deprotonation of the wall is greatest. The most significant additional factor affecting Eo is temperature. This is because:

(1) increasing the temperature decreases the viscosity of the solution and therefore increases the flow of liquid driven by Eo.

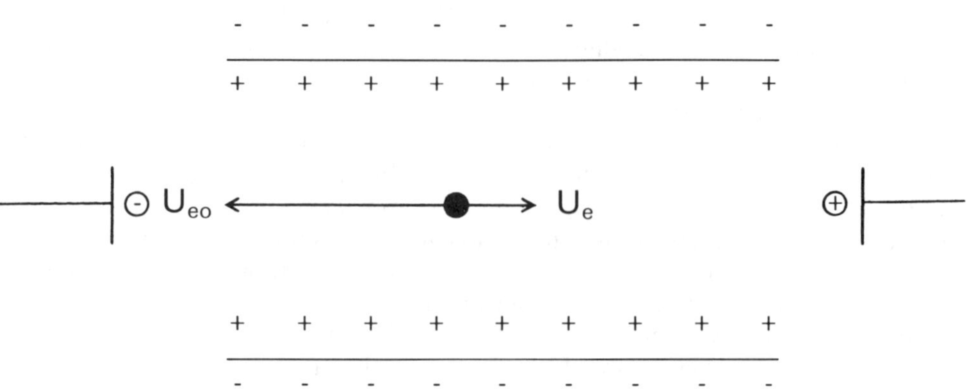

Fig. A2.1 — Electroendosmosis: the bulk flow of liquid due to the effect of the electric field on the positively charged double-layer adjacent to the negatively charged capillary wall.

(2) the pHs of buffers vary with temperature (some buffers being more susceptible than others).

Thus, by programming the temperature during a run, it is possible to control and vary the pH and selectivity. This can be used to sharpen and accelerate peaks at the latter part of a peptide map for example.

Gel-filled capillary electrophoresis

Gel-filled capillary electrophoresis (GFCE) has recently become commercially available. The technique offers certain advantages: in addition to allowing rapid molecular weight determinations, the system provided will also offer a new range of separation applications in which compounds such as chiral additives are bound to the matrix.

Micellar electrokinetic capillary chromatography (MECC)

If there is no native charge on the solutes, and none can be created, then separations may be accomplished by partitioning the neutral species with surfactant micelles added to the buffer system. The most commonly used surfactant is sodium dodecyl sulphate (SDS), which is an anionic surfactant with a critical micellar concentration (CMC) of 8 mM/l. The SDS micelles have a negatively charged outer shell and a hydrophobic core (see Fig. A2.2).

In the presence of an electric field the SDS micelles migrate towards the anode. As previously explained, however, Eo will cause a flow of buffer towards the cathode. The migration of the SDS micelles is thus opposed to the direction of the Eo flow. At pHs below about 6.5, the rate of micelle migration exceeds the opposing force of Eo, resulting in the micelles moving towards the anode. As the pH of the buffer is raised, this rate of movement reduces, until at approximately pH 6.5, the electrophoretic migration and the Eo flow become equal and cancel out, so that the micelles remain stationary in the capillary (Fig. A2.2).

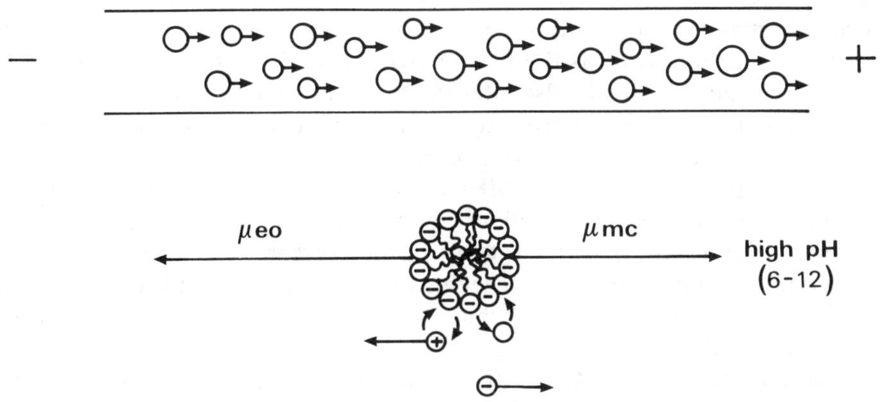

Fig. A2.2 — Schematic illustration of the separation mechanism involved in micellar electrokinetic capillary chromatography (MECC). The lower panel shows a stylized cross-sectional view of an SDS micelle, with the dodecyl chains forming a hydrophobic cure, and the anionic head groups forming a polar shell. Positively charged analytes may associate with the surface of the micelles; neutral analytes will partition in to the micelle core; and negatively charged analytes will remain in the buffer. Over the pH range 6–12, there will be a net movement of the micelles towards the cathode, because the electrophoretic mobility (μmc, upper panel) exceeds and opposed the mobility due to electroendosmosis (μeo).

Under these conditions any analyte introduced into the capillary that can partition into the micelles will be detected later than an analyte having no affinity for the micelles (the detection point being at the cathode end of the capillary, downstream of the Eo flow). This is analogous to HPLC separations, and provides a means of separating neutral species according to their partition rates into micelles: the greater the capacity of the micelles for the analyte, the slower the rate of the analyte migration, and the later it is observed at the detector. The addition of an organic solvent such as methanol will alter this partitioning and vary the degree and time of separation.

This technique has a significant advantage over HPLC as anyone who has injected a compound down an HPLC column never to see it again will understand. The method does not utilise a stationary phase. As noted above, the micelles in the CZE system are stationary at around pH 6.5; if the pH is raised above 6.5, the micelles are forced towards the cathode by Eo, the rate of their movement increasing as the pH is increased. This means that in the event of analyte being totally included within the micelles (i.e. with a micelle capacity approaching infinity), then the analyte will still be detected (and observable) as the micelle it is contained in passes the detection point. (The time taken for this passage will be pH dependent.)

MECC therefore provides the potential for 100% analysis with a characteristic 'capacity corridor'. The extent of this corridor can be found by introducing into the capillary a zero capacity marker (e.g. water, which causes a negative peak at 200 nm) and an infinite capacity marker (e.g. Sudan III, which resides totally within the micelles). These markers provide the maximum and minimum run times for the system and these data can be found even before the first sample is attempted.

When an organic solvent is added to the system to alter the micelle partition rates, the effect is to make the extra-micellar environment more hydrophobic, thereby shifting the equilibrium towards the free analyte, away from the micelle-included analyte. The organic solvent will also alter the potential at the capillary wall, and so the micelle capacity will be further reduced because of the diminution in Eo flow rate which causes an extension of the capacity corridor.

The other key parameter affecting MECC is temperature. The evidence obtained so far suggests that raising the capillary temperature increases the rate of mass transfer and accelerates the partition rate giving much greater efficiencies. Although further experimentation is required, it seems likely that this effect is caused by the reduction in viscosity of the system, which leads to a greater permeability of the micelles to the analytes.

The major advantage of MECC (compared with CZE for example) is that positively charged, negatively charged and neutral species can all be separated in the same run (see Fig. A2.3).

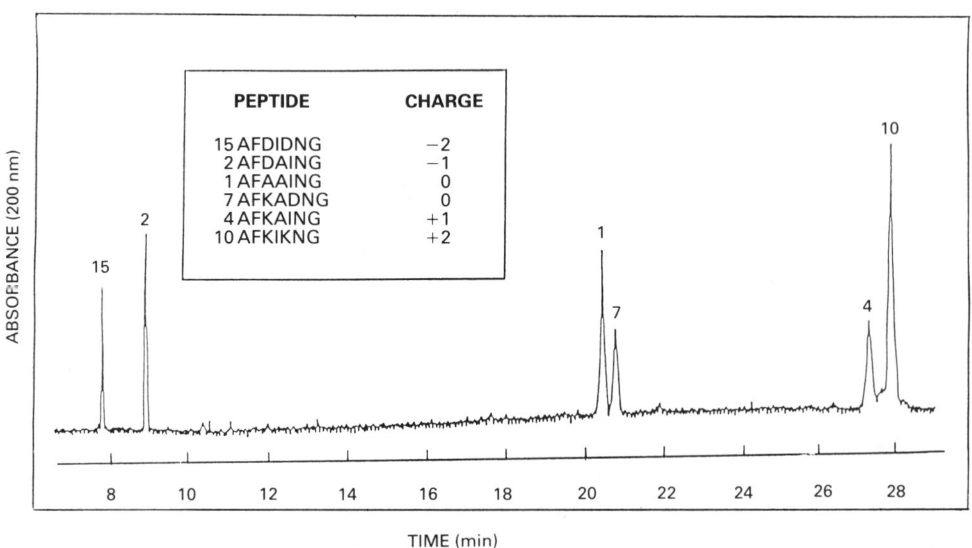

Fig. A2.3 — A typical MECC trace showing the resolution of a mixture of six peptides with differing charges.

A typical micellar separation would be performed using a pH-6.5 buffer containing 50-nM phosphate/25-mM borate/50-mM SDS, with a temperature of 60°C and an applied voltage of 20 kV. If extra resolution is required the pH may be increased to 7.5.

Capillary electrokinetic chromatography

Research is under way to develop a technique which utilizes the driving force provided by Eo (with buffer included in the mobile phase) to pump solvent systems over capillaries filled with LC packing materials. This will allow separations to be achieved by conventional partitioning techniques, and is termed capillary electrokinetic chromatotraphy (CEC) (Knox and Grant, 1987, Knox, 1988).

A2.3 INSTRUMENT DESIGN CONSIDERATIONS

Sample introduction

The dimensions of the capillaries used in these systems are such that very small volumes of liquid are involved, e.g. a capillary measuring 50 μm (ID) by 72 cm (length) has a volume of approximately 1.2 μl. For practical purposes, therefore, the maximum recommended volume of sample introduced into the system (to minimize any loss in separation efficiency) is about 100 nl. However, the dynamics of the system are such that much smaller volumes than this give optimal results, and so injection volumes of 4–10 nL are the norm. Such small volumes demand very careful handling, and require a precise sample introduction technique. There are essentially two modes of sample introduction:

1. *Electrokinetic injection*

Here, the capillary and the electrode are placed in the sample and a loading voltage of around 5 kV is applied. This causes the components to pass into the capillary by electrophoresis. The capillary and the electrode are then placed into the buffer and the analytical voltage is applied.

This is a very simple method which does not require any complex (and therefore expensive) electronics or valving systems, but does suffer from several disadvantages:

(1) The components are not introduced into the capillary in the same relative concentrations as found in the sample, i.e. if a sample contains 50% A and 10% B, it can appear that equal amounts of the two compounds are present upon subsequent analysis, owing to the non-linearity of the injection technique. This discrepancy will arise if the two compounds have differing mobilities, that is different charge-to-mass ratios. In the example above, therefore, if compound B has a greater mobility than compound A, then more of this compound will enter the capillary during the injection process. As a result, the composition of the sample remaining in the sample vial will be altered, and duplicate runs from the same sample vial will be made invalid.

(2) The sample matrix, unless itself charged, wil not be introduced into capillary, and will not therefore be included during the analysis.

(3) neutral species will not be introduced into the capillary (unless Eo is present under the loading conditions).

2. *Vacuum (pressure driven) and gravity (syphoning/hydrostatic) introduction*

All of these methods physically draw sample and matrix into the capillary. Here, the key issue is the reproducibility of the results from run to run. If the integrated areas (as in LC) are not less than 2%, then it may be questionable as to the benefit of either the injection method or the operation of the system utilizing it.

Detection

Here again performance must be considered just as with an HPLC detector. This is probably one of the most technically critical areas of capillary electrophoresis, and most of the early work was carried out using fluorescence detectors. While these are not ideal as general detectors, owing to the limited numbers of compounds that have native or derived fluorescence, they do give an optical 'null' detection system, recording far less of the systematic noise than UV systems, and giving a higher sensitivity (subject to fluorescence activity) with simpler optics. Fluorescence detection therefore was the initial choice.

The development of high sensitivity monochromatic detectors means that variability with low detection levels are achievable. It is normal to expect a level of sensitivity that is within at least an order of magnitude of an LC method. This has been achieved by designing the instrument around the technique (and therefore the UV detector around the capillary) in order to maximize performance. One of the advantages of having a pathlength of 50 μm ID is that the contribution of absorbance from the buffer/solvent system is minimal. This allows for the use of the 190–200-nm region for routine detection, with the opportunity to go lower wavelengths if the monochromator is purged with nitrogen (Fig. A2.4).

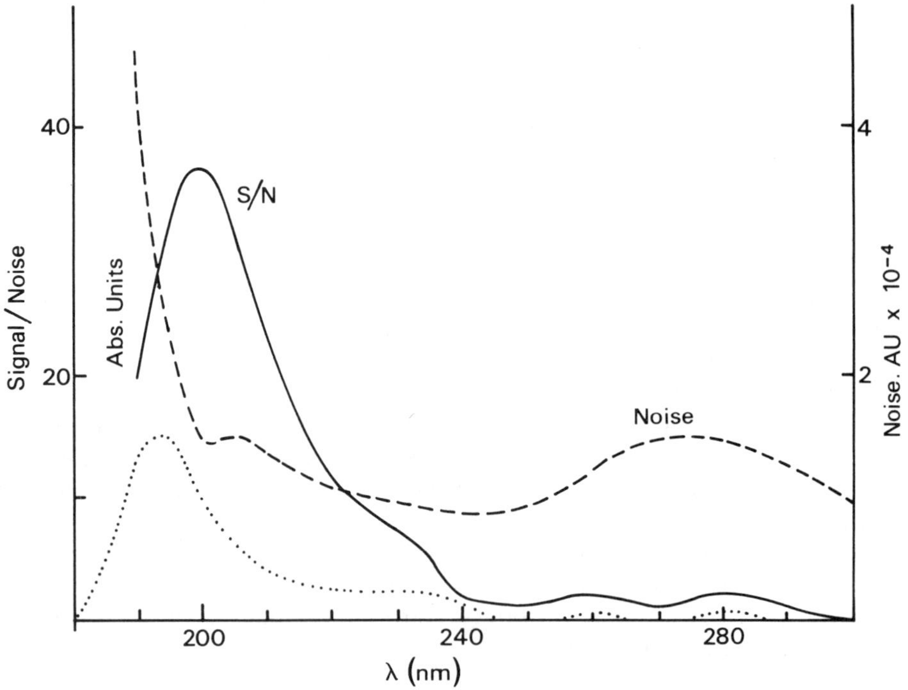

Fig. A2.4 — Detection sensitivity as a function of instrument noise versus component absorption. The intrinsic detector noise and typical protein sample absorption curves (shown as dashed and dotted lines respectively) are integrated to give the observed signal/noise (S/N) curve. The S/N curve peaks at around 200 nm and so this wavelength provides the optimum detection point.

As a measure of sensitivity a normal working concentration of sample would be 100–200 μg/ml when using a 5 nl injection volume on a 0.015 AUFS detector setting. This should allow for detection of minor components at the 0.5% level in addition to the primary components. Detector linearity is important to achieve this and should be in the region of three orders of magnitude.

Data capture is through analogue outputs identical to those of HPLC detectors, and in terms of integration, requirements are similar to the criteria for 'fast LC', or capillary gas chromatography. Consequently most of the commercially available integrators and personal computer packages should be adequate.

The power supply, while being the driving force behind the technique, is akin to the pump in an HPLC system. It should prove to be far more robust and reliable than its HPLC counterpart. Here the main constaint must be that of safety and it is this that influences specification. While the voltages normally employed (up to 30 000 V) at the current present (typically up to 300 mA) are not particularly dangerous, neither are they desirable. There is an inherent distrust of using instruments producing this sort of voltage. The safety devices used at minimum, should include interlocks which act as contact breakers in the event of access to the electrode/capillary compartment, and a second line of protection utilizing captive bolts, which prevent entry into the machine during a run may be desirable.

A2.4 CONCLUSIONS

Electro-separation (ES) techniques will offer analysts in a broad range of applications areas an opportunity to gain new insights into separation problems (to date unresolvable), to gain added confidence of purity by comparing the separations by LC and ES, and the opportunity to use a lower costing (and more 'environmentally friendly') technique in terms of consumables and solvent disposal.

REFERENCES

Grossman, P. D., Colburn, J. C., Lauer, H. H., Nielsen, R. G., Riggin, R. M., Sittampalam, G. S. & Rickard, E. C. (1989) Application of free-solution capillary electrophoresis to the analytical scale preparation of proteins and peptides. *Anal. Chem.* **61** 1186–1194.

Hjerten, S. (1985) High-performance electrophoresis — elimination of electroendosmosis and solute adsorption. *J. Chromat.* **347** 191–198.

Jorgensen, J. N. & Lukacs, K. D. (1983) Capillary zone electrophoresis. *Science* **222** 266–272.

Knox, J. H. (1988) Thermal effects and band spreading in capillary electroseparation. *Chromatogrophia* **26** 329–337.

Knox, J. H. & Grant, I. H. (1987) Miniturisation in pressure and electroendosmotically driven liquid chromatography: Some theoretical considerations. *Chromatographia* **24** 135–143.

Terabe, S., Yashima, T., Tanaka, N. & Araki, M. (1988) Separation of oxygen isotopic benzoic acids by capillary zone electrophoresis based on isotope effects on the dissociation of the carboxyl group. *Anal. Chem.* **60** 1673–1677.

Index